Data-Driven Techniques in Speech Synthesis

Edited by

R.I. Damper
University of Southampton

KLUWER ACADEMIC PUBLISHERS
BOSTON / DORDRECHT / LONDON

A C.I.P. Catalogue record for this book is available from the Library of Congress.

ISBN 978-1-4419-4733-8

Published by Kluwer Academic Publishers,
P.O. Box 17, 3300 AA Dordrecht, The Netherlands.

Sold and distributed in North, Central and South America
by Kluwer Academic Publishers,
101 Philip Drive, Norwell, MA 02061, U.S.A.

In all other countries, sold and distributed
by Kluwer Academic Publishers,
P.O. Box 322, 3300 AH Dordrecht, The Netherlands.

Printed on acid-free paper

This book is dedicated to the memories of Frank Fallside (1931–1993) and John Holmes (1929–1999), two pioneers of the science and technology of speech synthesis.

Contents

Preface

Modern speech synthesis began in the 1950s with the development of electronic formant synthesisers, such as PAT (Parametric Artificial Talker) designed by Walter Lawrence in the UK and OVE designed by Gunnar Fant in Sweden. Many others followed and, with the widespread introduction of fast digital computers, became implemented as computer programs. The best known of these was designed by Dennis Klatt in the US.

In the 1960s, John Holmes demonstrated that such synthesisers, provided they were driven by appropriate, hand-crafted, control signals (formant frequencies and bandwidths, voicing information, etc.), were capable of generating speech signals which were perceptually indistinguishable from natural speech in all but the most stringent listening conditions. The priority in speech synthesis research then switched to the automatic generation of the control signals. The first rule-based system which produced intelligible speech from a sequence of phonetic symbols was described in 1964 in a paper by John Holmes together with John Shearme and Ignatius Mattingly based partly on the work on speech perception carried out at the Haskins Laboratories in New Haven, Connecticut.

A complete text-to-speech (TTS) system now required automatic translation from text (a sequence of letter) to an equivalent sequence of phonetic symbols. This proved to be a more difficult problem than first thought, at least for a language like English. The memory limitations of computers at that time precluded the use of a large dictionary, so context-dependent rule systems were developed. The first such rule systems were published in the early 1970s.

With the re-emergence of neural networks in the 1980s, data-driven techniques began to be developed. Sejnowski and Rosenberg's NETtalk was not only the first attempt to apply neural networks to speech synthesis but also one of the largest neural networks at that time. It inspired others to apply machine learning techniques to speech synthesis. The research since that time is the subject matter of this book.

In the first chapter, Bob Damper presents a useful introduction to TTS systems. He describes early work on developing rule systems and some of the problems encountered. Data-driven speech synthesis is exemplified by NETtalk and its derivative NETspeak. Other parts of TTS systems, namely prosodic assignment and the techniques available for generating the speech signal from a symbolic description, are also discussed.

The next four chapters are devoted to techniques for converting text into a phonetic transcript. Ghulum Bakiri and Thomas Dietterich discuss a number of extensions of NETtalk which lead to more accurate letter-to-sound conversion. An alternative technique based on analogy is introduced by Kirk Sullivan. If a word is encountered that is not in the pronouncing dictionary, similarly spelled words are searched for and the unknown pronunciation is deduced by analogy.

In the next chapter, Helen Meng describes a framework for integrating a variety of sources of linguistic knowledge. This leads to a system which can be used for pronunciation-to-spelling as well as spelling-to-pronunciation generation. In Chapter 5, Robert Luk and Bob Damper show that letter-to-phoneme conversion can be effectively performed by stochastic transducers.

One of the more popular units for synthesising speech is the diphone, which begins in the middle of one phoneme and ends in the middle of the next, thereby preserving the important transitional information. Sabine Deligne, Françcois Yvon and Frédéric Bimbot develop and evaluate a system based on a generalisation of the diphone – the multiphone. Walter Daelemans and Antal van den Bosch discuss TREETALK, their system based memory and decision tree learning. Andrew Cohen suggests that there are advantages of TTS conversion via non-symbolic intermediate (phonetic) representations.

The next group of chapters move away from the text-to-phoneme stage of a TTS system. Alan Black, Kurt Dusterhoff and Paul Taylor describe their Tilt model of intonation and how it was trained and tested with a corpus of recorded broadcast speech. The next chapter is rather different but potentially extremely useful. John Coleman and Andrew Slater discuss the Klatt synthesiser and the ways in which it can be driven to generate natural sounding speech. In Chapter 11, Julia Hirschberg gives a detailed account of experiments on accent and phrasing assignment. Finally Andrew Cohen describes the remaining part of his system from the intermediate phonetic representation to the speech signal, and some preliminary experiments in producing actual output.

It is sometimes remarked that speech synthesis, being a one-to-many process (there are many acceptable ways of pronouncing a word) is easier than speech recognition which is a many-to-one process. However, as the length of the utterance to be synthesised increases, from word to sentence to paragraph, it is becoming increasingly obvious that for a given speaker and intended meaning the number of acceptable pronunciations is severely limited. The variety of data-driven techniques described in this book demonstrate that this problem is being actively addressed and give confidence that acceptable solutions will soon become available for the multitude of applications for which high quality speech synthesis is required.

William A. Ainsworth
MacKay Institute of Communication and Neuroscience
University of Keele
Staffordshire ST5 5BG, UK

June, 2001

List of Contributors

Ghulum Bakiri
Department of Computer Science
University of Bahrain
Isa Town
Bahrain
Email: microcen@batelco.com.bh

Frédéric Bimbot
IRISA/INRIA Rennes
Campus Universitaire de Beaulieu
Avenue du Général Leclerc
35042 Rennes cedex
France
Email: frederic.bimbot@irisa.fr

Alan W. Black
Language Technologies Institute
Carnegie Mellon University
5000 Forbes Avenue
Pittsburgh
PA 15213
USA
Email: awb@cs.cmu.edu

Andrew D. Cohen
39 Avonmore Road
London W14 8RT
United Kingdom
Email: arcanalpha@yahoo.com

John S. Coleman
Phonetics Laboratory
University of Oxford
41 Wellington Square
Oxford OX1 2JF
United Kingdom
Email: john.coleman@phonetics.oxford.ac.uk

Walter Daelemans
Center for Dutch Language and Speech
University of Antwerp
Universiteitsplein 1
B-2610 Wilrijk
Belgium
Email: walter.daelemans@uia.ua.ac.be

Robert I. Damper
Image, Speech and Intelligent Systems (ISIS) Research Group
Department of Electronics and Computer Science
University of Southampton
Southampton SO17 1BJ
United Kingdom
Email: rid@ecs.soton.ac.uk

Sabine Deligne
Human Language Technologies
IBM TJ Watson Research Center
PO Box 218
Yorktown Heights
NY 10598
USA
Email: sabine@watson.ibm.com

Thomas G. Dietterich
Department of Computer Science
Oregon State University
Corvallis
Oregon 97331
USA
Email: tgd@cs.orst.edu

Kurt E. Dusterhoff
Phonetic Systems UK Ltd
Millbank
Pullar Close
Stoke Road
Bishops Cleeve
Gloucestershire GL52 8RW
United Kingdom
Email: kdusterhoff@phoneticsystems.com

Julia Hirschberg
Human-Computer Interface Research
AT&T–Research
180 Park Avenue
PO Box 971
Florham Park
NJ 07932-0971
USA
Email: julia@research.att.com

Robert W. P. Luk
Department of Computing
Hong Kong Polytechnic University
Hung Hom
Kowloon
Hong Kong
Email: csrluk@comp.polyu.edu.hk

Helen Meng
Human-Computer Communications Laboratory
Department of Systems Engineering and Engineering Management
The Chinese University of Hong Kong
Shatin, NT
Hong Kong
Email: hmmeng@se.cuhk.edu.hk

Andrew Slater
Phonetics Laboratory
University of Oxford
41 Wellington Square
Oxford OX1 2JF
United Kingdom
Email: andrew.slater@phon.ox.ac.uk

Kirk P. H. Sullivan
Department of Philosophy and Linguistics
Umeå University
SE-901 87 Umeå
Sweden
Email: kirk@ling.umu.se

Paul A. Taylor
Centre for Speech Technology Research
University of Edinburgh
80 South Bridge
Edinburgh EH1 1HN
United Kingdom
Email: pault@cstr.ed.ac.uk

Antal van den Bosch
ILK/Computational Linguistics
Room B332
Tilburg University
PO Box 90153
NL-5000 LE Tilburg
The Netherlands
Email: antalb@kub.nl

François Yvon
Départment Informatique
Ecole Nationale Supérieure des Télécommunications (ENST)
46 rue Barrault
75634 Paris cedex 13
France
Email: yvon@inf.enst.fr

Chapter 1

LEARNING ABOUT SPEECH FROM DATA: BEYOND NETTALK

Robert I. Damper
Department of Electronics and Computer Science, University of Southampton

Abstract Speech synthesis is an emerging technology with a wide range of potential applications. In most such applications, the message to be spoken will be in the form of text input, so the main focus of development is text-to-speech (TTS) synthesis. Strongly influenced by the academic traditions of generative linguistics, early work on TTS systems took it as axiomatic that a knowledge-based approach was essential to successful implementation. Presumed theoretical constraints on the learnability of their native language by humans were applied by extension to machine learners to conclude the futility of trying to make useful 'blank slate' inferences about speech and language simply from exposure. This situation has changed dramatically in recent years with the easy availability of computers to act as machine learners and large databases to act as training resources. Many positive achievements in machine learning have comprehensively proven its usefulness in a range of natural language processing tasks, despite the negative assumptions of earlier times. Thus, contemporary speech synthesis relies heavily on *data-driven* techniques.

This chapter introduces and motivates the topic of data-driven speech synthesis, and outlines the concepts that will be encountered in the rest of the book. The main problems that any TTS system must solve are: automatic generation of pronunciation, prosodic adjustment, and synthesis of the final output speech. The first of these problems has been quite well-studied and it is here that machine-learning techniques have been most obviously applied. Indeed, the problem of text-phoneme conversion (the 'NETtalk' problem) has become something of a benchmark in machine learning and, hence, we will have most to say on this topic. As the utility of data-driven methods becomes ever more widely accepted, however, attention is starting to turn to the use of these techniques in other areas of synthesis, most particularly modelling and generation of prosody, and the generation of the output speech itself.

R. I. Damper (ed.), Data-Driven Techniques in Speech Synthesis, 1-25.

1. Introduction

For many years, scientists and technologists have dreamed of building machines able to converse with their creators, by endowing them with some measure of 'intelligence' or 'understanding' together with input/output capabilities of speech recognition and speech synthesis. This book is concerned with the latter technology – output of artificially-generated speech. Even if we never manage to build a machine with the native intelligence to converse sensibly, there remains an important market for the more restricted capability of producing speech artificially. Important example applications include audio ('speaking') information services, reading machines for blind people, and aids for first or second language learners. In the vast majority of such applications, the most convenient input to the speech synthesis system will be message text, represented in some appropriate format, although occasionally some other input (e.g. high-level representation of concepts, cf. Young and Fallside 1979) might be appropriate. Hence, a main focus of this book is text-to-speech (TTS) synthesis. At this stage of the subject's development, most work in TTS synthesis has been done in connection with languages having an alphabetic writing system (see Gelb 1952 and Sampson 1985 for useful background information on writing systems). Indeed, as the remainder of this book will make clear, more has been done on English text-to-speech synthesis than any other language, reflecting its cultural and commercial importance (but see Sproat 1998 for a thorough review of multilingual synthesis). As we shall see, it is noteworthy that the vagaries of the English spelling system and its relation to the sound system mean that the problems of text-to-speech conversion are especially severe for this language.

Although the technology has made steady progress over the last twenty five or so years, enabled by the revolutionary advances in digital computer technology, automatic synthesis of speech is still constrained by the quality (naturalness and intelligibility) of the final output. This limits the acceptability to everyday consumers who constitute the mass market for TTS systems. Hence, the continued development of speech synthesis from its current state is primarily a search for improved quality output.

In this author's view, two widespread misconceptions have unfortunately conspired to hold back the development of speech synthesis. First, synthesis and recognition have often been seen as direct and inverse forms of essentially the same problem. According to Kirsch (1996, Preface), "we call two problems *inverse* to each other if the formulation of each of them requires full or partial knowledge of the other" although "it is obviously arbitrary which of the two problems we call the direct and which we call the inverse problem". Typically, however, the direct problem will be easier and have been studied earlier in more detail, while the inverse problem is ill-posed and/or inferential in nature, having no unique solution or known method of unique solution. According to the first misconception, then, the 'direct' synthesis problem is relatively easy, whereas only the 'inverse' recognition problem is truly challenging and worthy of the attentions of the best researchers. This led to a view of speech synthesis as an effectively solved problem, requiring only minor improvements before practical exploitation. The remainder of this book will attest to the mistaken nature of this view. We will see that speech synthesis abounds with its own inverse problems, especially when we attempt to learn solutions (i.e. extract speech knowledge) from example data.

The second misconception, strongly influenced by the generative linguistics tradition, centres on the view of speech and language as governed by innate, tacit knowledge of rules (Chomsky and Halle 1968; Pinker 1999). According to this view, whose main theoretical underpinning was the work of Gold (1967), starting from scratch with a 'blank slate', little or nothing of value can be learned about natural language (including speech) from example data, because the learning problem is intractable. Rather, essential knowledge in the form of rules must be supplied *a priori*. When building speech technology devices, it is the task of the designer to implant this rule-based knowledge, so the job becomes one for the expert linguist. This view has come under increasing attack in recent years, as a consequence of the availability of relatively large text and speech databases (Young and Bloothooft 1997), and the involvement and achievements in speech technology of workers from areas such as function approximation, statistical inference and machine learning, who have not accepted the foundational assumptions and arguments of Chomskyian linguistics. To quote Elman et al. (1996, p. 385): "Connectionist simulations of language learning ... have already shown that the impossible is in principle possible". Consequently, a debate or tension has developed between 'knowledge' and 'statistics' in the design and implementation of speech technology (cf. Fant 1989). The philosophy underpinning this book is that the latter approach – which we will characterise as *data-driven* or *corpus-based* – is worthy of serious consideration. Yet widespread acceptance of the knowledge-based paradigm certainly held back the development of speech science and technology over many years.

So the theme developed here is that the production of high-quality synthetic speech is a challenging intellectual problem with important technological applications, and that much can be gained by adopting data-driven methods in which we attempt to learn solutions to our problems from representative datasets. This is not to denigrate the importance and role of prior knowledge, which can greatly ease the problems of learning from data. For instance, Luk and Damper (1996, p. 134) found, in relation to work on so-called stochastic transduction (Luk and Damper 1991, 1994), "that the provision of quite limited information about the alternating vowel/consonant structure of words aids the inference process significantly". As Sproat, Möbius, Maeda, and Tzoukermann (1998, p. 77) write: "It is unreasonable to expect that good results will be obtained from a system trained with no *[human]* guidance ... ". The challenge is to ensure that the prior knowledge, or guidance, is both robust and integrated appropriately into the learning process (cf. Gildea and Jurafsky 1996).

Given the above background, this first chapter aims to introduce the topic of data-driven speech synthesis, and to outline key concepts which will be encountered in the rest of the book. We start (Section 2) by outlining a typical architecture for a TTS system. For good engineering reasons, modular architectures are popular so it is logical to consider modules in turn. Although it can be somewhat arbitrary to decide where one module ends and another starts, any complete system must at a minimum include modules for pronunciation, prosody, and the synthesis of final output speech. Accordingly, Section 3 next details the problem of automatic pronunciation generation. It is here that machine-learning techniques have probably had most impact on speech synthesis, so we will have most to say on this topic. Central to our discussion will be the seminal work of Sejnowski and Rosenberg (1987) on NETtalk, which has also impacted profoundly on the fields of machine learning, cognitive science and the psychology of reading – probably more so than on speech synthesis, in fact. Subsequently, Section 4 deals with the problem of data-driven prosody which has received rather less attention than pronunciation. Section 5 then presents a brief overview of concatenative synthesis before briefly concluding (Section 6).

2. Architecture of a TTS System

The tasks which need to be accomplished in any TTS system are largely decomposable into separate modules. There is considerable practical advantage in separately developing, implementing and evaluating modules. Hence, current systems are almost exclusively modular in their architecture. Typical modules are text preprocessing and normalisation, morphological analysis, automatic pronunciation, syntactic analysis, prosody generation and final synthesis. Different system designers and architects will treat the boundaries between these modules differently – for instance, text normalisation may be integrated with pronunciation generation – but the pronunciation, prosody and synthesis modules are virtually indispensable. For this reason, attention will focus on these three aspects of synthesis in this book.

While there is widespread agreement on the advantages of modularity, there is less of a consensus on how the modules should be fitted together into an overall system architecture. Figure 1.1 shows two possibilities: (a) a mainly serial or pipelined architecture, in which the output of one module becomes the input to the next module in sequence; and (b) a multilevel organisation,

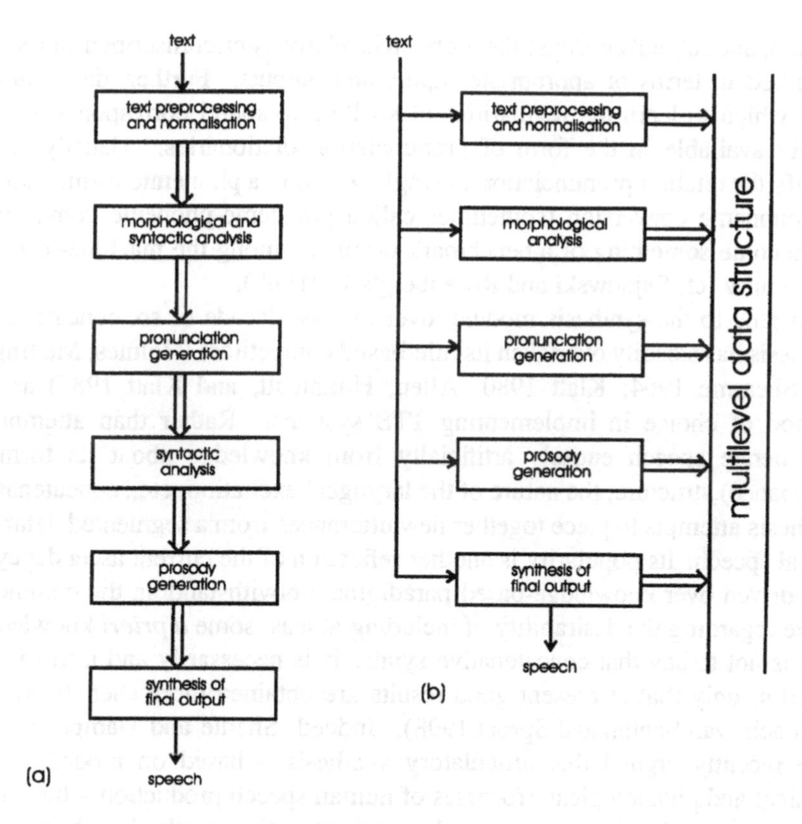

Figure 1.1. Two possible modular architectures for a text-to-speech system: (a) Serial, sequential or pipelined; (b) Multilevel.

in which the modules communicate via a common multilevel data structure. Although the sequential architecture is popular in practice – e.g. the Bell Labs TTS system (Sproat and Olive 1997; van Santen and Sproat 1998) – it is often considered somewhat old-fashioned. A major issue is how to achieve the necessary separation of function such that each module depends only on those that precede it in the cascade. Thus, Dutoit (1997, Fig. 3.2, p. 59) refers to "an old . . . strategy" and points out (pp. 61–2) the many potential advantages of the multilevel approach including, for example, its greater flexibility in that it can actually implement the totally serial architecture if need be.

We have identified the pronunciation, prosody and synthesis modules as vital parts of any TTS system. In principle, the implementation of each of these can be attacked by attempting to learn the relevant solution from example text and/or speech data. To date, however, the three problems have received different degrees of attention from researchers working in the data-driven paradigm.

Automatic pronunciation generation from text has been well studied (see Damper 1995 for a fairly recent review and van den Bosch 1997 for an update), undoubtedly because the problem is relatively circumscribed and easily described in terms of appropriate inputs and outputs. Further, data sources from which to learn the regularities of spelling-to-sound correspondence are widely available in the form of pronunciation dictionaries. Usually, these specify the citation pronunciation of single words in a phonemic form. Indeed, text-phoneme conversion (sometimes called grapheme-phoneme conversion) has become something of a benchmark problem among the machine-learning community (cf. Sejnowski and Rosenberg's NETtalk).

Turning to the synthesis module, over the last decade or so, concatenative synthesis has steadily overtaken its rule-based competitors (Holmes, Mattingly, and Shearme 1964; Klatt 1980; Allen, Hunnicutt, and Klatt 1987) as the method of choice in implementing TTS systems. Rather than attempting to generate speech entirely artificially from knowledge about its formant (resonance) structure, the nature of the laryngeal excitation, etc., concatenative synthesis attempts to piece together new utterances from a segmented database of real speech. Its popularity is another reflection of the current ascendancy of data-driven over knowledge-based paradigms, notwithstanding the comments above regarding the desirability of including at least some *a priori* knowledge. This is not to say that concatenative synthesis is necessarily and intrinsically superior, only that at present good results are obtained more cheaply by this approach (van Santen and Sproat 1998). Indeed, Shadle and Damper (2001) have recently argued that articulatory synthesis – based on modelling the physical and physiological processes of human speech production – has much greater potential in the long term than concatenative synthesis. At present, however, this potential is far from being achieved with any current articulatory synthesis system. According, very little is said about the topic in this book.

Finally among these three problems, prosody generation has received the least attention, almost certainly because it is the hardest. For instance, Sproat, van Santen, and Olive (1998, p. 249) describe "the task of proper prominence and phrase boundary assignment from paragraph-length text" as "formidable" and argue that "performing this task at high levels of accuracy will not be possible without major advances in several areas of computational linguistics and artificial intelligence ...". Indeed, many of the problems "may ... well have no general solution". It will be no surprise, then, to find that there is considerably less material in this book on prosody generation that on automatic pronunciation. Data-driven prosody generation is definitely a topic for the future.

3. Automatic Pronunciation Generation

Figure 1.2 shows a typical schematic for the pronunciation module of a TTS system. The first stage is text normalisation, whereby abbreviations and

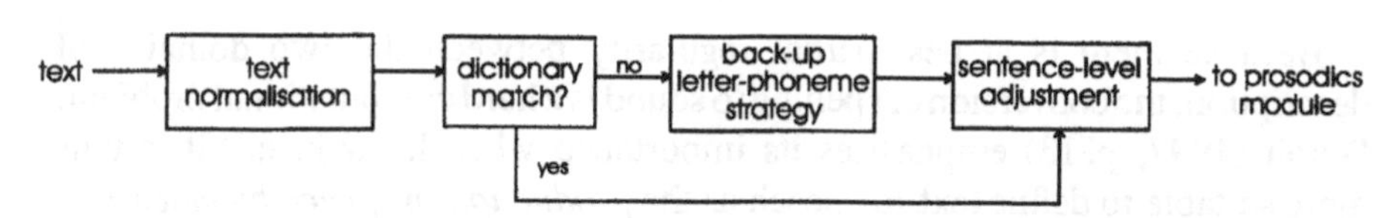

Figure 1.2. Typical arrangement of components of the pronunciation module.

typographical conventions (such as *Dr.* for *Doctor*, '&' for *and* and so on) are expanded into 'normal' text. Also, a decision must be made whether to sound abbreviations letter-by-letter (e.g. USA) or pronounce them as words (e.g. BAFTA). This apparently simple, straightforward stage can actually be deceptively complex, and its success is crucial to the overall quality of output from the final system.

For many important languages like English and French, there is only partial regularity of correspondence between the spelling and sound systems. Consequently, the pronunciation module of a TTS system aims to convert input text into some specification which is much closer to the actual sounds to be uttered. Most usually, although there are exceptions, this specification will be in terms of phonemes: abstract (or 'logical') sound units, whose role is to denote sound differences that are important in distinguishing between words. Although phonemes are strictly abstract, they are nonetheless extremely useful in specifying a reference or canonical pronunciation of words – sometimes called the *baseform* (Lucassen and Mercer 1984). Hence, we will often specify pronunciations in terms of phonemes as if the latter were physical, acoustic units, rather than abstract. Some authors (e.g. Black, Lenzo, and Pagel 1998) prefer the term *phone* in this situation, but this usage is slightly unusual and so we do not adopt it here. In this book, except where there is good reason to do otherwise, we will use the IPA notational system of the International Phonetic Association (1999) to specify pronunciations.

In essence, we are treating the canonical form as an adequate representation from which, after further processing, acoustic output can be synthesised. Part of this further processing will be sentence-level adjustment, illustrated in Fig. 1.2 as the final stage of the pronunciation module. Typical sentence level phenomena are elision in English whereby one or more sounds are omitted at the junctures between words in connected speech (e.g. *great terror*) and liaison in French whereby sounds which are absent in the citation form are introduced at junctures (e.g. *les Anglais*). As a further example, this component would also have to determine whether the English word *lead* should be pronounced /lid/ or /lɛd/, and the definite article *the* pronounced /ðə/ or /ði/. Again, the subdivision shown in the figure is a little arbitrary; we could as well consider the sentence-level adjustment to be part of the prosodics module.

Because there is at best partial regularity between the two domains of description, the conversion of spelling to sound is a hard computational problem. Dutoit (1997, p. 13) emphasises its importance when he writes: "It is thus more suitable to define text-to-speech as the *production of speech by machines, by way of the automatic phonetization of the sentences to utter*". For most words encountered in the input of a TTS system, a canonical pronunciation is easily obtained by dictionary look-up (Fig. 1.2), not forgetting that some words have variable pronunciations (cf. the example of *the* above). However, it is not possible simply to list all the words of a language and their pronunciations (Damper, Marchand, Adamson, and Gustafson 1999). This is because language is highly generative – new words are being created all the time. If an input word is absent from the system dictionary, but is a morphological derivative of a dictionary word, well-established techniques exist to infer a pronunciation (Allen, Hunnicutt, and Klatt 1987). Our concern here is the situation in which the input word is 'novel' – such that dictionary look-up (possibly in conjunction with morphological analysis) fails to produce an output. In this commonly-encountered situation, a default, back-up strategy must be employed.

3.1. The Nature of the Problem

Let us consider the problem as it arises in English TTS synthesis. In 'normal' writing (orthography), we use the 26 letters of the roman alphabet (plus a few extra symbols, such as the apostrophe) to spell the words of English, yet something like 45–60 phonemes are required to specify the sounds of English, at an appropriate level of abstraction/detail (i.e. so that we can make necessary logical distinctions between different words, and give a canonical pronunciation for each). It follows that the relation between letters and phonemes cannot be simple, one-to-one. In fact, English is notorious for the lack of regularity in its spelling-to-sound correspondence (Venezky 1965; Carney 1994). For instance, the letter *c* is pronounced /s/ in *cider* but /k/ in *cat*. On the other hand, the /k/ sound of *kitten* is written with a letter *k*.

A further complication is that there is no strict correspondence between the number of letters and the number of phonemes. Miller (1981, p. 49) puts it as follows:

> "... the pronunciation of part of a word is not generally part of its pronunciation: if x and y are segments of a word, then Pronunciation(x) + Pronunciation(y) $\neq$ Pronunciation($x + y$)."

Thus, the letter combination *ough* is pronounced /ʌf/ in *enough*, but /ɔ/ in *thought* and as dipthong /aʊ/ in *plough*. The digraph *ph* is pronounced as the single phoneme /f/ in *phase* but is pronounced as two phonemes (/p/ and /h/) in *uphill*. In the common case that a letter grouping corresponds to a single phoneme – as in the example of *phase* – we often refer to this letter grouping as a *grapheme* (see Henderson 1985 for comprehensive discussion of this term). Because this happens frequently, there are usually fewer phonemes than letters in a word but there are exceptions, e.g. (*six*, /sɪks/).

English has noncontiguous markings (Venezky 1970) as, for instance, when the letter *e* is added to (*mad*, /mad/) to yield (*made*, /meɪd/) ... also spelled *maid*! The final *e* is not sounded; rather it indicates that the vowel is lengthened or dipthongised. Such markings can be quite complex, or long-range, as when the letter *y* is added to *photograph* or *telegraph* to yield *photography* or *telegraphy* respectively: The final letter affects, or marks, the initial vowel. As a parting comment, English contains many proper nouns (place names, surnames) which display idiosyncratic pronunciations, and loan words from other languages which conform to a different set of regularities. These further complicate the problem.

3.2. Rules: The Traditional Solution

Given these problems, how is it possible to perform automatic translation of text to phonemes at all? It is generally believed that the problem is largely soluble provided sufficient context is available. That is, referring back to the quote from Miller (1981), the pronunciations of segments x and y are constrained by context such that Pronunciation(x) plus Pronunciation(y) does become equal to Pronunciation($x + y$). Thus, the traditional, rule-based approach has used a set of *context-dependent* rewrite rules, each of the form popularised by Chomsky and Halle (1968, p. 14):

$$A[B]C \rightarrow D \tag{1}$$

where B is the target letter substring, A and C are the left- and right-context respectively of B, and D is the phoneme substring representing the pronunciation which B receives. The B substring is variable in length: It can be a single letter, one or more graphemes (each corresponding to a single phoneme substring D), a complete word, etc.

Because of the complexities of English spelling-to-sound correspondence outlined above, more than one rule generally applies at each stage of translation. The potential conflicts which arise are resolved by maintaining the rules in sublists, grouped by (initial) letter and with each sublist ordered by specificity. Typically, the most specific rule is at the top and most general at the bottom. In the scheme of Elovitz, Johnson, McHugh, and Shore (1976), for instance, translation is a one-pass, left-to-right process. For the particular target letter (i.e. the initial letter of the substring B), the appropriate sublist is searched from top-to-bottom until a match is found. This rule is then fired (i.e. the corresponding D substring is right-concatenated to the evolving output string), the linear search terminated, and the next untranslated letter taken as the target. The last rule in each sublist is a context-independent default for the target letter, which is fired in the case that no other, more specific rule applies.

Employing this traditional back-up strategy, the set of phonological rules is supplied by a linguist (e.g. Ainsworth 1973; McIlroy 1973; Elovitz et al. (1976); Hunnicutt 1976; Divay and Vitale 1997). However, the task of manually writing such a set of rules, deciding the rule order so as to resolve conflicts appropriately, maintaining the rules as mispronunciations are discovered etc., is very considerable and requires an expert depth of knowledge of the specific language.

To anyone with a background in a numerate branch of science, it is a remarkable fact that traditional linguists' rules in the form of (1) eschew any kind of numerical measure, such as rule probabilities, frequency counts, etc., which could improve robustness and control overgeneration without recourse to elaborate rule-ordering schemes and similar devices (Damper 1991). This is a reflection of the long-standing suspicion of, or hostility to, probability and statistics in mainstream linguistics. Some feel for this hostility can be gleaned from the following (Chomsky 1969, p. 57):

> "But it must be recognized that the notion 'probability of a sentence' is an entirely useless one, under any interpretation of this term."

While it is undoubtedly true that strict interpretations of the concept of probability pose difficulties (as many texts on the philosophy of science make clear), this is surely very different from the concept being "entirely useless". For instance, a speech recognition system will routinely compute the probability of competing sentence hypotheses relative to underlying acoustic-phonetic and language models, and use the result to select among the hypotheses. Chomsky attempts to show the futility of this by arguing:

> "... we can of course raise the conditional probability of any sentence as high as we like, say to unity, relative to 'situations' specified on *ad hoc*, invented grounds."

But, quite obviously, no practically-minded speech technologist would use "invented grounds". Rather, the probability computation is relative to a model of training data, with efforts made to ensure that these data are representative. Many of the contributions in this book will attest convincingly to the usefulness of the concepts of probability and statistics in the study of speech and language – especially when we are aiming to produce a working technological artifact, like a speech synthesiser.

3.3. Letter-Phoneme Alignment

We refer to the mapping between *B* and *D* in (1) as a *correspondence*. Given a set of correspondences, we can *align* text with its pronunciation. For example, take the word (*make*, /meɪk/). A possible alignment, requiring the addition of a null phoneme '–' to make the number of letters and phonemes equal, is:

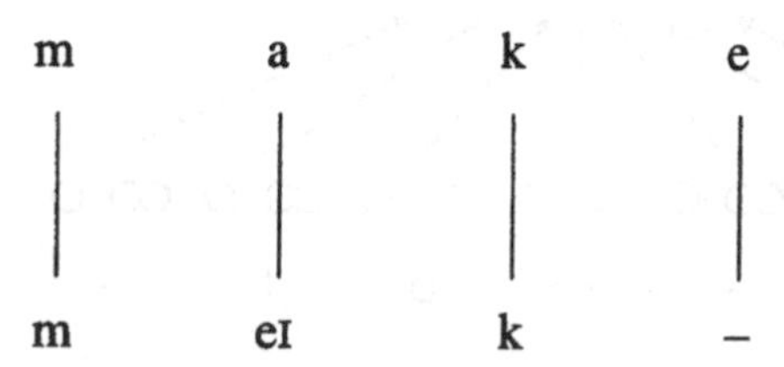

While the issue of alignment pervades approaches to automatic phonemisation – both rule-based and data-driven – it should be obvious from this example that it is all but impossible to specify a canonical set of correspondences for all the words of English, on which all experts could agree. For instance, why should we use the single-letter correspondences *a* → /eɪ/, *k* → /k/ and *e* → /–/ as above, rather than the composite *ake* → /eɪk/ which captures the noncontiguous marking by the final *e*? As we shall see, data-driven techniques typically require letters and phonemes in their training dataset to have been prealigned. An issue then arises as to whether alignment should be manual or learned/automatic. Methods for automatic alignment are described by Van Coile (1990), Luk and Damper (1992, 1993a) and Black, Lenzo, and Pagel (1998) among others.

Those techniques that apparently do not require prealignment generally perform an alignment as an integral part of the learning process or yield a *de facto* alignment as a by-product. The nub of the problem is associating letters with phonemes, so some form of alignment is inescapable.

3.4. NETtalk: A Data-Driven Solution

We now turn to considering a very well-known data-driven approach to automatic phonemisation: NETtalk as described by Sejnowski and Rosenberg (1987). Although this was not the first noteworthy attempt at applying machine learning techniques to the problem – the decision tree method of Lucassen and Mercer (1984) should probably be given this distinction – the work of Sejnowski and Rosenberg (S&R) is truly seminal for many reasons. Not least, they were the first to apply the then-emerging paradigm of back-propagation neural networks (Rumelhart, Hinton, and Williams 1986) to a sizable and important learning problem. Indeed, as mentioned earlier, the conversion of English text to phonemes has become a benchmark problem in the machine-learning community.

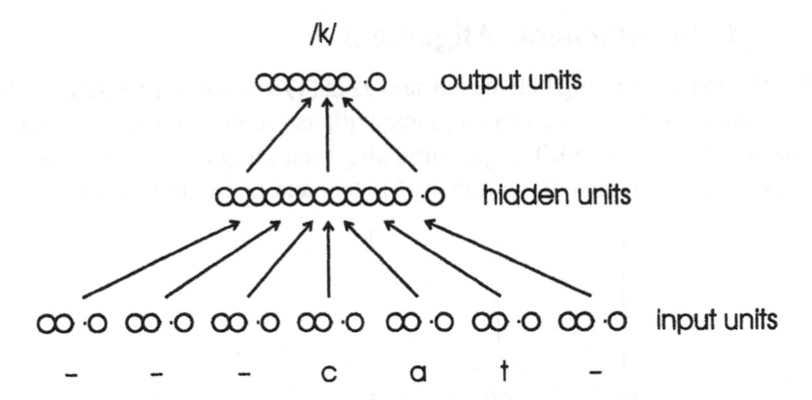

Figure 1.3. Architecture of NETtalk. A 'window' of 7 letters is fed to the (203) input units then via the hidden units (most usually 120 in number) to the 26 output units. The central letter in the input window is transliterated into its equivalent phoneme. In this case, the initial letter *c* of *cat* receives the pronunciation /k/. *After Sejnowski and Rosenberg (1987).*

NETtalk is a multilayer feed-forward network: Its basic structure is shown in Figure 1.3. This is not the place to review the extremely well-developed field of neural computing. Excellent introductory texts are Anderson (1995), Bishop (1995) and Hassoun (1995), with Cherkassky and Mulier (1998) giving useful background on the general topic of learning from data, and Dietterich (1997) giving pointers to current research directions. Suffice to say that a neural network consists of units (artificial 'neurons') connected by weighted links. Each unit computes a weighted sum of its input values ('activations') and passes the result through a nonlinear activation function (often sigmoidal), to produce an output which is fed on to other units. During the learning or training phase, connection weights are adjusted according to some algorithm (such as back-propagation) in a search for a desired, overall input-output mapping.

Input coding. The input consists of a window of an odd number of letters, where the central letter is the 'target' and the letters to left and right provide the context. The input text steps through the window letter-by-letter. NETtalk is trained to associate the target letter with its correct pronunciation – here a single phoneme. Thus, the translation is one-to-one (i.e. transliteration) and requires the training data to have been previously aligned: This was done manually by one of the authors (Rosenberg). In the usual case where the orthographic representation is longer than the phonemic one, S&R added null phonemes so as to effect a one-to-one alignment (as in our example of *make* above, which would be rendered /m/, /eɪ/, /k/, /–/). In the (rarer) case when the phoneme string is the longer of the two, S&R invented new 'phonemes' (e.g. they use /K/ to correspond to the letter *x* in *sexual*, which is actually sounded as the two phonemes /ks/).

The standard NETtalk (i.e. the architecture for which S&R collected the most complete results) has a 7-letter window. However, S&R also studied smaller and larger windows in the range 3 to 11 letters: Generally, performance improved with window size. Each input character is represented by a sparse (1-out-of-n) code of 29 bits – one for each of the 26 letters plus 3 additional bits for punctuation marks and word boundaries. Thus, the number of input units i in the standard net is $7 \times 29 = 203$.

Output coding. NETtalk uses 21 "articulatory features" to represent the (single) phoneme output. Examples of such features are voicing, place of articulation (labial, velar, ...) and tongue height. S&R also added 5 additional output units to represent stress and syllable boundaries. Hence, the number of output units o is $21 + 5 = 26$.

While targets during training can (in principle) easily be specified by binary codings from the set of 2^{26} possible outputs, actual outputs obtained in use are continuously graded in the range (0, 1). Even if these were thresholded to give 'hard' (0, 1) values, very few would correspond exactly to legal codings, since 2^{26} massively exceeds the cardinality of the phoneme set. So how do we know what the obtained output phoneme is? S&R's solution was to compute the inner product of the output vector with the codes for each of the possible phonemes. The phoneme that exhibited the smallest angle with the output was chosen as the "best guess" output.

Hidden units. Various numbers of hidden layers and units (h) were studied. Specifically, single hidden layers with $h = 0$, 80 and 120 units were trained and tested, together with one net with two hidden layers each having 80 neurons. Performance increased with additional units and with an additional hidden layer. However, the most comprehensive results are presented for the case of a single hidden layer of 120 units.

With this latter network, the number of connections (excluding variable thresholds) is:

$$(i \times h) + (h \times o) = (203 \times 120) + (120 \times 26) = 27{,}480$$

To this must be added the small number ($h + o = 146$) of variable thresholds. Thus, the number of adjustable weights (free parameters of the model) is approximately 28,000.

Performance. S&R trained and tested NETtalk on a (transcribed) corpus of continuous informal speech, and on words in *Miriam Webster's Pocket Dictionary* of American English. We will consider the dictionary train/test scenario only.

The 1,000 most commonly occurring words were selected from the total of 20,012 in the dictionary. For the net with no hidden units, performance was asymptotic to 82% "best guess" phonemes correct after some tens of passes through the training set and when tested on the training set. The best performance achieved with 120 hidden units on this 1,000-word corpus was 98%.

Generalisation is a crucial aspect of any data-driven technique (Wolpert 1995) because the whole purpose of learning is to be able to deal appropriately with novel cases which, by definition, are absent from the training dataset. Unfortunately, S&R never tested the generalisation ability of NETtalk on an entirely unseen dataset. Rather, generalisation ability was assessed by testing the net with 120 hidden units, trained to asymptote on the 1,000-word corpus, on the full 20,012-word dictionary. The result was 77% phonemes correct. With continued learning (and testing) on the complete dictionary (an even less satisfactory way of assessing generalisation power), performance reached 85% phonemes correct after one complete pass and 90% correct after 5 passes.

Discussion. NETtalk was highly influential in showing that data-driven techniques could be applied to the large, difficult problem of text-phoneme conversion with a degree of success. By S&R's use of artificial neural networks at a time when these were receiving heightened interest, and by quantifying performance on a fairly large test set, NETtalk won popular precedence over earlier approaches like that of Lucassen and Mercer (1984). (The latter authors tested their decision-tree method on just 194 words.) Cognitive scientists and psychologists made observations such as "NETtalk exhibits several striking and unanticipated analogies with real brains" (Lloyd 1989, p. 99). However, the following points should be emphasised:

- The training data were manually prealigned, so simplifying the learning problem significantly.

- The generalisation power of NETtalk was never properly assessed on a totally unseen test set.

- Scoring was in terms of phonemes correct only. Although this appears reasonable at first sight, we have argued (Damper et al. 1999) that it gives an inflated and insensitive view of performance. For instance, an apparently good figure of 90% phonemes correct predicts – on the assumptions of independent errors in the transliteration of each letter, and an average of 6 letters per word – a words correct score of 0.9^6, or just 53%. If the phonemes correct figure falls marginally to 89%, this implies a reduction to less than 50% words correct. Thus, performance should be specified in terms of *words* correct.

- NETtalk is concerned with learning mappings between sequences but a feed-forward network is only capable of learning mappings between static patterns. The reason that NETtalk can solve the problem at all is because the 7-letter window provides left and right context for the central letter. That is, NETtalk implicitly uses the idea of a time-to-space unfolding. Considering the example input *cat* as in Fig. 1.3, each of the patterns corresponding to each of the central letters *c*, *a*, and *t* is treated as entirely unrelated. There is no exploitation of the fact that these three patterns are actually derived from the same word. In an attempt to utilise this sort of information, Adamson and Damper (1996) used a recurrent neural network (with feedback connections and, hence, memory of previous inputs) on the problem, but with mixed results.

The interest engendered by NETtalk can be gauged by the fact that, within a year of S&R's paper appearing, an improved version was published. This was called NETspeak.

3.5. NETspeak– a Re-implementation of NETtalk

In 1987, McCulloch, Bedworth, and Bridle produced a multilayer feed-forward network for text-phoneme conversion called NETspeak. Although intended primarily as a re-implementation of NETtalk, it extended S&R's work in various ways:

1. The impact of different input and output codings was studied.
2. Generalisation on entirely unseen words was tested.
3. The relative performance of separate networks for the translation of common and of uncommon words was assessed.

Training was again by back-propagation on manually-aligned data, but using a different dictionary.

Input and output codings. These were thought to be important in that "... an appropriate coding can greatly assist learning whilst an inappropriate one can prevent it" (McCulloch, Bedworth, and Bridle 1987, p. 292). Like NETtalk, NETspeak uses a 7-letter input window through which the input is stepped. However, the NETspeak input coding is more compact (2-out-of-n, 11-bit). So, the number of input units is $i = 77$, compared to 203 for NETtalk.

The first 5 bits of the 11 indicate which of 5 rough "phonological sets" the letter belongs to while the remaining 6 identify the particular character. In place of NETtalk's $o = 26$ "articulatory features" plus stress and syllable markers, NETspeak uses $o = 25$ output features. NETspeak used $h = 77$ hidden units. Hence, the number of adjustable weights (including thresholds) is:

$$(i \times h) + (h \times o) + h + o = (77 \times 77) + (77 \times 25) + 77 + 25 = 7{,}956$$

Generalisation. NETspeak was trained on 15,080 words of the 16,280 words in a (previously-aligned) pronouncing dictionary (*The Teacher's Word Book*). After three passes through the training set, performance on these 15,080 words was 87.9% "best guess" phonemes correct. However, performance on new (unseen) words was only slightly worse at 86%, indicating (the authors claim) excellent generalisation. The authors do not say how many words were in the unseen test set; presumably, it was $16{,}280 - 15{,}080 = 1{,}200$.

Common versus uncommon words. NETspeak was subsequently trained on a frequency-weighted version of the dictionary. This was done by replicating words in the training set in appropriate proportions. McCulloch, Bedworth, and Bridle expected that the network would have more difficulty learning this training set because of the higher proportion of more common (and therefore, presumably, irregularly-pronounced) words. This turned out not to be the case, although performance on common words was reported to improve while that for "very regular words" deteriorated.

The authors then trained two separate networks – one on common words and the other on uncommon words. Each net was tested on an appropriate dataset, e.g. the net trained on common words was tested on held-out common words (i.e. not in training set). The corpora used were frequency-weighted and the numbers of (nonblank) characters in the two training sets and in the two test sets were equal. Results were "quite surprising" in that the network trained on the common words did better. This was unexpected since they reasoned that the pronunciation of these words was less regular, i.e. the mappings were "more complicated".

Additional light is thrown on this matter by Ainsworth and Pell (1989), who trained a neural network modelled on NETspeak on a 70,000-word dictionary divided into "regular" and "irregular" words according to whether pronunciation was correctly predicted by the Ainsworth (1973) rules or not. As one would expect, they found much better performance on the regular words than on the irregular words (asymptotic to 95.9% and 83.6% "best guess" phoneme scores on unseen words respectively). Thus, it seems that McCulloch, Bedworth, and Bridle's intuition that the more common words have less regular pronunciations may be incorrect. This is offered as a cautionary tale against the uncritical incorporation of tacit 'knowledge' in data-driven systems.

3.6. Pronunciation by Analogy

While it would be inappropriate to attempt an exhaustive review of data-driven approaches to automatic phonemisation, nor does space allow this, it is important to avoid the impression that neural networks are the only or the best means of learning about pronunciation. One very attractive approach is to use the hidden Markov model (HMM) techniques that have proved so valuable in speech recognition (Rabiner 1989; Knill and Young 1997). Early work in this vein is that of Parfitt and Sharman (1991) with more extensive, recent work from Donovan and Woodland (1999). In this subsection, another very promising alternative is outlined – pronunciation by analogy (PbA).

PbA is essentially a form of similarity-based (also known as case-, instance- or memory-based) reasoning (Stanfill and Waltz 1986; Stanfill 1987, 1988). It is a kind of 'lazy' learning (Aha, Kibler, and Albert 1991; Aha 1997), meaning that the prior training phase is minimised and compression of the training database is avoided. This is in stark contrast to NETtalk and NETspeak, where the prior training using back-propagation is computationally intensive and the method attempts to compress the training dictionary into a small set of connection weights. That is, translation of letters to phonemes uses the network's connection weights to find a pronunciation for an unseen word, rather than the dictionary itself. Adopting a very different approach, PbA treats the system dictionary as a primary resource not only for looking up the pronunciation of lexical entries but also for inferring a pronunciation for novel words. The underlying idea is that a pronunciation for an unknown word is derived by matching substrings of the input to substrings of known, lexical words, hypothesising a partial pronunciation for each matched substring from the phonological knowledge, and assembling the partial pronunciations to form a final output. To this extent, we implicitly reject the contention of Miller (1981), mentioned in Section 3.1, that $\text{Pronunciation}(x) + \text{Pronunciation}(y) \neq \text{Pronunciation}(x + y)$. Many times, this equality certainly will hold and PbA aims to exploit this situation.

An early example of PbA was PRONOUNCE described by Dedina and Nusbaum (1991). The basic PRONOUNCE system consists of four components (Fig. 1.4): the lexical database, the matcher which compares the target input to all the words in the database, the pronunciation lattice (a data structure representing possible pronunciations), and the decision function, which selects the 'best' pronunciation among the set of possible ones. This latter decision function is heuristic and its selection has a major impact on performance (Sullivan and Damper 1993; Damper and Eastmond 1997; Marchand and Damper 2000). The necessity to specify heuristic decision functions is arguably one of the weaknesses of the method.

In spite of this, we have obtained very good results with this technique (see Evaluation subsection below). Indeed, there are good theoretical and empirical reasons to favour analogy. For instance, Pirrelli and Federici (1995, p. 855) write: "... the space of analogy is ... eventually more accurate than the space of rules, as the former, but not the latter, is defined by the space of already attested base objects". There are, however, other advantages. Analogy does

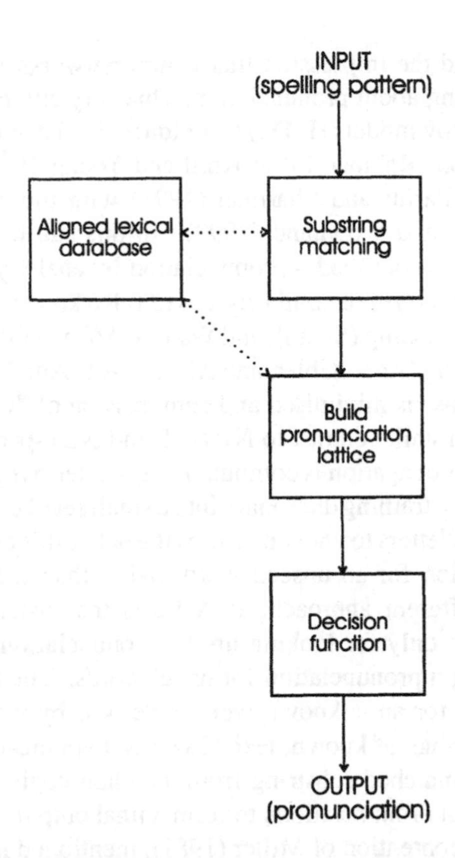

Figure 1.4. Schematic of Dedina and Nusbaum's PRONOUNCE. Note the way that pronunciation by analogy blurs the distinction between dictionary matching as a primary strategy and a back-up letter-phoneme strategy as depicted in Fig. 1.2.

not use a fixed-size window on the input text nor commit to use of specific units (graphemes, phonemes, syllables, words ...). Instead, input/output mappings are modelled together in variable-size chunks, so handling long-range dependencies easily and naturally. In particular, Daelemans, van den Bosch, and Zavrel (1999, p. 38) write: "empirical results strongly suggest that keeping full memory of all training instances is at all times a good idea in language learning". Finally, the 'no free lunch' theorems (Wolpert and Macready 1995, 1997) tell us there is no general, 'best' method for machine-learning problems. They provide a basis for believing that we should prefer "appropriateness-to-task over uniformity of method" (cf. van Santen and Sproat 1998) when selecting a learning technique, with all the evidence pointing to analogy as highly appropriate to our task.

Table 1.1. Words correctly pronounced using a variety of text-phoneme conversion techniques.

Elovitz et al. rules	25.7% out of 16,280
NETspeak	46.0% out of 8,140 unseen
	54.4% out of 16,280 seen
Tree method (IB1-IG)	57.4% out of 8,140 unseen
Pronunciation by analogy	71.8% out of 6,280 unseen

3.7. Evaluation

As mentioned in the Introduction, a widespread misconception in speech synthesis has been the assumed superiority of rule-based over data-driven methods. To quote Klatt (1987, pp. 770–1):

> "... the performance *[of* NET*talk]* is not nearly as accurate as that of a good set of letter-to-sound rules (performing without use of an exceptions dictionary, but with rules for recognizing common affixes)."

Similarly negative pronouncements have been made at regular intervals. For instance, Sproat, Möbius, Maeda, and Tzoukermann (1998, p. 75) write:

> "... there is certainly no reason to believe that the results of training a decision tree, a decision list, an inferable stochastic transducer or a neural net will necessarily result in a system that is more robust or predictable on unseen data than a carefully developed set of rules."

Until very recently, no sound basis existed to resolve the question of the relative performance of rule-based and data-based methods in automatic phonemisation. Arguably then, there was "no reason to believe that ... " the latter offered any advantage only because the required empirical comparison had never been made.

This state of affairs was corrected when Damper, Marchand, Adamson, and Gustafson (1999) published a comparison of the performance of four representative approaches to automatic phonemisation on the same test dictionary (*The Teacher's Word Book*). As well as rule-based techniques, three data-driven methods were evaluated: pronunciation by analogy, NETspeak and a tree-structured method, IB1-IG (Daelemans, van den Bosch, and Weijters 1997; van den Bosch 1997). The rule sets studied were that of Elovitz et al. (1976) and a set incorporated in a successful TTS product (identity kept anonymous for commercial reasons). The latter had been the subject of extensive development over many years. Many subtle issues involved in comparative evaluation were detailed and elucidated. The reader is referred to the original paper for details.

Table 1.1 gives the results of the comparison. *The Teacher's Word Book* contains 16,280 words. The Elovitz et al. rules could be straightforwardly tested on the complete dictionary. Two of the techniques studied, NETspeak and IB1-IG, require a large dataset for prior training, disjoint from the test set. Hence, the dictionary was divided into two and results are reported on 8,140 unseen words. We also trained and tested NETspeak on the entire dictionary, giving an estimated upper bound on performance on the 16,280 words. Because PbA is a form of lazy learning in which there is no requirement for prior training, and so no need for disjoint training and test sets, it was possible to test performance on the entire dictionary. Each word was processed in turn by removing it from the dictionary and deriving a pronunciation for it by analogy with the remainder. This can be seen as a version of the n-fold cross-validation technique described by Weiss and Kulikowski (1991), as used by van den Bosch (1997, p. 54) to evaluate his text-phoneme conversion methods, with n equal to the size of the entire lexicon.

From the table, we see that the data-driven techniques outperform experts' rules in accuracy of letter-to-phoneme translation by a very significant margin, contrary to the many opinions expressed in the literature on this matter. However, the data-driven methods require aligned text-phoneme datasets. Best translation results are obtained with PbA at approximately 72% words correct, compared to something like 26% words correct for the rules. Performance of around 25% was also obtained for the highly-developed rules incorporated in a commercial TTS system. So much for the presumed superiority of rules! Generally, in accordance with the findings of van den Bosch (1997), the less compression there is of the dictionary data, the better the performance. Thus, PbA is better than IB1-IG which, in turn, is better than NETspeak. Finally, we take a best word accuracy of 72% as indicating that automatic pronunciation of text is not yet a solved problem.

In a large multinational collaboration, Yvon et al. (1998) evaluated eight different automatic phonemisation modules for French. This language exhibits many of the same difficulties as English for TTS synthesis (see Section 3.1) and some additional ones besides (Catach 1984; Belrhali 1995). Unfortunately, for a variety of reasons, the results of this elaborate study were not as useful as they might have been. The main shortcoming is that no details were given of the compared methods of letter-phoneme conversion, so that the performance results obtained give no comparative information on the effectiveness of different approaches. Further, it appears that the tests must have been made with the system dictionary in place. While this parallels the situation in actual use, the primary conversion technique of dictionary matching always gives 100% correct performance when it is successful, i.e. when the word is actually in the dictionary. Hence, dictionary matching is essentially uninteresting. Interest focuses on the default or back-up strategy which, accordingly, ought to be assessed without the dictionary present. Unless one knows the size of the dictionary, the nature of the default strategy, and how many dictionary 'hits' are obtained in conversion, nothing much can be made of the performance figures.

These problems are exacerbated by Yvon et al.'s use of running text as test material. This choice was motivated by the high degree of context-dependent modification of the citation-form pronunciation which takes place at the sentence level in French. In their example of *les enfants ont écouté* (p. 394), liaison is compulsory between *les* and *enfants*, forbidden between *enfants* and *ont*, and optional between *ont* and *écouté*. Nonetheless, it is the nature of running text that very common words like *le*, *la*, *une* (whose citation-form pronunciations are trivially easy to get right by dictionary matching) are very highly represented. So if the treatment of these words can be got right at sentence level, this will bias the evaluation in the direction of high performance. Yvon et al. report results as phonemes-correct figures ranging from 97.1% to 99.6% (ignoring two anomalous results said to underestimate performance). They also report corresponding sentences correct figures, with performance ranging from 21.2% to 89.1%, emphasising the remarks above about the lack of sensitivity of a phonemes-correct measure. The latter gives an over-optimistic view of the state-of-the-art in automatic pronunciation generation, especially if (as seems to be the case here) most of the work is being done by dictionary look-up.

4. Prosody

As we said at the outset, the uptake of TTS technology is hindered by the quality and acceptability of the output from current systems. There is a general consensus in the research community that poor prosody is a significant factor in limiting speech quality. For instance, Tatham, Morton, and Lewis (2000, p. 1/1) comment on a representative set of synthesisers as follows:

> "... whatever the shortcomings of the segmental rendering ... the major fault lies with their prosodic rendering: rhythm, stress and intonation."

According to Granström (1997), the problem for intonation is that it remains poorly understood by speech scientists. The same could be said for other aspects of prosody. For instance, Wightman and Ostendorf (1994, p. 496) write "... the lack of a general consensus in these areas may be one reason why prosody has been under-utilized in spoken language systems". Also, prosody serves multiple functions in speech communication, many of them paralinguistic rather than linguistic. This lack of phonetic and linguistic knowledge impacts negatively on our ability to build speech systems with appropriate prosody. Taylor (2000) challenges Granström's analysis, pointing out that formalisms such as hidden Markov models and *n*-gram language models, which have proved so successful in large-vocabulary automatic speech recognition, are rudimentary – even downright incorrect – realisations of phonetic and linguistic knowledge. So "instead of fundamental research holding back the application of intonation, it is the lack of a suitable model which is robust, easily trainable,

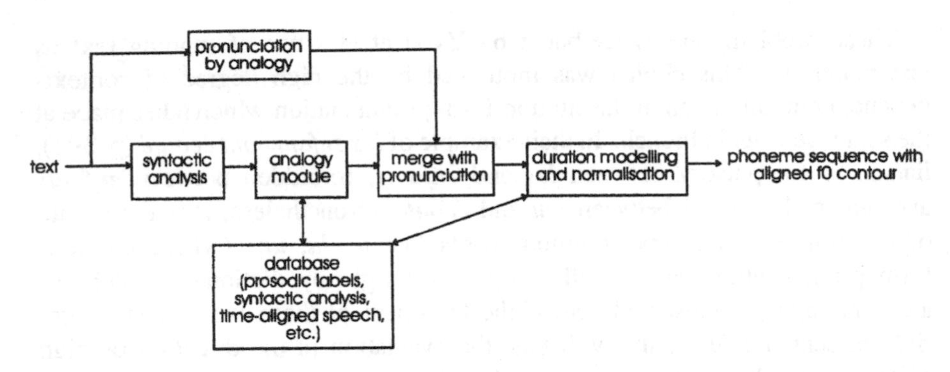

Figure 1.5. Schematic of a possible approach to analogical learning of prosody.

and amenable to statistical interpretation" (Taylor 2000, p. 1697). Bailly (1997, p. 160) goes so far as to write: "One could claim that this lack of theoretical background is the key to the success of sub-symbolic approaches in speech technology"! This is, of course, yet another instance of the 'knowledge' versus 'statistics' schism. As previously stated, it is unwise to see the two as mutually exclusive alternatives. Rather, at least some prior knowledge should be employed to structure and constrain data-driven models, so easing the central problem of training/learning and rendering our models linguistically-relevant, and thereby 'transparent'. This perspective is consistent with the recent authoritative commentaries of van Santen and Sproat (1998), Spärk Jones, Gazdar, and Needham (2000), Rosenfeld (2000) and others.

As far as the generation of prosody for TTS synthesis is concerned, the state of the art remains the use of hand-crafted rules (e.g. Malfrère, Dutoit, and Mertens 1998) rather than solutions learned from data – reflecting the fact, mentioned earlier, that the problem is much harder than automatic phonemisation. Over the last decade, however, an identifiable literature has been slowly accumulating on the topics of automatic prosodic labelling and/or acquisition of aspects of prosody (e.g. Bailly, Barbe, and Wang 1992; Emerard, Mortamet, and Cozannet 1992; Traber 1992; Black and Taylor 1994; Daelemans, Gillis, and Durrieux 1994; Hirschberg and Prieto 1994; Jensen, Moore, Dalsgaard, and Lindberg 1994; Ross and Ostendorf 1994; Taylor and Black 1994; Wightman and Campbell 1994; Wightman and Ostendorf 1994; Ross 1995; Fujisawa and Campbell 1998.) Many questions arise when adopting this approach. What aspects of prosody (pitch accent, lexical stress and tone, timing, duration, etc.) should be included? How can we derive usable prosodic information from text input in which such information is at best impoverished? What are the most appropriate learning techniques? How can success be evaluated? And having decided what prosody is appropriate, and assuming concatenative synthesis is used, how do we select units from the speech database to achieve this?

Given the considerable success enjoyed with pronunciation by analogy (Section 3.6), and the fact that lazy learning in general and analogy in particular seem to work well for many natural language processing tasks (Skousen 1989; Jones 1996; Daelemans, van den Bosch, and Zavrel 1999), it is surely worth considering analogy for prosody generation too. Thus far, this has not been tried in TTS synthesis except that Blin and Miclet (2000) have proposed a very similar idea. However, their work remains largely preliminary. McKeown and Pan (2000) have described memory-based prosody modelling for concept-to-speech synthesis, but this is a considerably easier problem. These latter authors offer the opinion that the approach "... will not work well if the input is unrestricted text unless there is a huge pre-analyzed, speech inventory available."

This cautionary message may be unduly pessimistic. Provided reasonably powerful methods are at hand for adjusting durations, fundamental frequency contours etc. of the speech segments selected by the analogy process, the amount of necessary labelled data may not be too great. Figure 1.5 shows a possible scheme for this. There are undoubtedly performance trade-offs to be considered here. To quote Bailly (1997, p. 158): "The best way to preserve the intrinsic naturalness of the sounds of the reference database *[in concatenative synthesis]* is to leave most parts of the speech untouched".

5. The Synthesis Module

During the early development of text-to-speech synthesis in the late 1970's and 1980's, for the reasons outlined above, the knowledge-based or rule-based approach was assumed mandatory. Following the convincing demonstration of Holmes (1973) that the formant model of speech production was adequate to yield highly natural sounding output – essentially indistinguishable from human speech – the majority of TTS systems were based on formant synthesis (e.g. Klatt 1980; Allen, Hunnicutt, and Klatt 1987). However, Holmes's demonstration was the result of 'copy synthesis' – meticulously careful hand-crafting of the control inputs to his parallel formant synthesiser to match examples of natural speech. The problem was that practical systems, that had to derive the control inputs from text in real time without knowing in advance what was to be said and without expert intervention, produced speech that was very far from natural sounding. Typically, the output was described by listeners as 'robotic' with levels of quality and intelligibility significantly below natural speech (Pisoni, Nusbaum, and Greene 1985; Logan, Greene, and Pisoni 1989; Duffy and Pisoni 1992). Certainly, no one would ever confuse it with human speech!

In spite of Holmes's demonstration of the adequacy of the model, formant synthesis attempts to do something that is clearly highly ambitious in our current state of knowledge, viz. produce speech *entirely* artificially. While formant synthesis remains a thoroughly valid and useful line of research, a less ambitious approach is to try to synthesise new messages from prerecorded natural speech which, accordingly, retains some – if not most – of the naturalness of the original. In particular, the speaker identity is retained. Since the new messages are produced by concatenating units selected from a large speech database, such concatenative synthesis only really became practical with the relatively recent availability of cheap computer memory, but is now widely used. To quote Campbell (1998, p. 19):

> "Trends in concatenative synthesis . . . are towards bigger unit databases, shifting the knowledge from the synthesiser into the data. By accessing the data directly, rather than modelling its variation by rule, we open up the possibility of encoding richer types of speech variation, such as dialectal, emotional or laryngeal differences without requiring explicit control over their realisation."

The main problem with concatenative synthesis is the audible discontinuities that occur where the individual units abut (Harris 1953). This can be reduced by having available a large variety of units which in turn implies a large recorded database and sophisticated means for selecting among them. According to Olive (1997), important research topics include:

- the nature of the recorded database (nonsense words, words in carrier sentences, natural running text, etc.);
- the size and type of units stored;
- concatenation algorithms (selection of units, adjustment at boundaries, etc.);
- the coding of the speech database.

Space does not allow us to develop these issues here; we content ourselves with a few pointers to the literature.

For coding, the aim is to maintain good speech quality and intelligibility while reducing storage requirements. The most popular technique has been linear prediction coding (LPC) (Makhoul 1975; Markel and Gray 1976; Varga and Fallside 1987). With increasing availability of computer memory, however, coding for data reduction is becoming less important. More recently, pitch-synchronous overlap-add (PSOLA) synthesis (Charpentier and Stella 1986; Moulines and Charpentier 1990) has become popular as a way of producing speech output. Its advantage lies in the way that 'pitch' (fundamental frequency) can be easily modified during synthesis. PSOLA is independent of any particular coding strategy and gives best output when no data reduction is used at all. The time-domain version, TD-PSOLA (Hamon, Moulines, and Charpentier 1989), is extremely simple and offers good quality and intelligibility. Like the formant model, TD-PSOLA is capable of essentially perfect copy synthesis

(Dutoit 1997, p. 256). Macchi, Altom, Kahn, Singhal, and Spiegel (1993) have studied the effect of different coding methods on the intelligibility of the speech output. They found that residual-excited linear prediction (RELP) provided higher intelligibility than PSOLA for voiced consonants, which were assumed to be more sensitive to coding methods and pitch changes than vowels. This is somewhat against the usual claim that PSOLA gives higher quality than LPC.

A variety of units has been used in synthesis by concatenation, including demisyllables (which seem to be most popular for the synthesis of German – Kraft and Andrews 1992), diphones (Charpentier and Stella 1986; Moulines and Charpentier 1990), a mixture of both (Portele, Höfer, and Hess 1997), context-dependent phoneme units (Huang et al. 1996; Plumbe and Meredith 1998) or non-uniform units (Nakajima and Hamada 1988; Sagisaki 1988; Takeda, Abe, Sagisaki, and Kuwabara 1989; Nakajima 1994). Selection schemes for choosing units, including ways of minimising mismatch at unit boundaries, have been extensively described (Dutoit and Leich 1993; Wang, Campbell, Iwahashi, and Sagisaki 1993; Dutoit 1994a; Hunt and Black 1996; Campbell and Black 1997; Conkie and Isard 1997).

6. Conclusion

This chapter has introduced the concepts of data-driven approaches to speech synthesis which will be developed throughout the remainder of the book. Inevitably, given the enormous literature that exists on machine learning and our interest in a particular application domain, the focus has been on speech synthesis from text and the motivation for using data-driven techniques, rather than on the techniques themselves.

Chapter 2

CONSTRUCTING HIGH-ACCURACY LETTER-TO-PHONEME RULES WITH MACHINE LEARNING

Ghulum Bakiri
Department of Computer Science, University of Bahrain

Thomas G. Dietterich
Department of Computer Science, Oregon State University

Abstract This chapter describes a machine learning approach to the problem of letter-to-sound conversion that builds upon and extends the pioneering NETtalk work of Sejnowski and Rosenberg. Among the many extensions to the NETtalk system were the following: a different learning algorithm, a wider input window, error-correcting output coding, a right-to-left scan of the word to be pronounced (with the results of each decision influencing subsequent decisions), and the addition of several useful input features. These changes yielded a system that performs much better than the original NETtalk. After training on 19,002 words, the system achieves 93.7% correct pronunciation of individual phonemes and 64.8% correct pronunciation of whole words (where the pronunciation must exactly match the dictionary pronunciation to be correct) on an unseen 1000-word test set. Based on the judgements of three human listeners in a blind assessment study, our system was estimated to have a serious error rate of 16.7% (on whole words) compared to 26.1% for the DECtalk 3.0 rule base.

1. Introduction

There have been many attempts to apply machine learning (ML) algorithms to solve the problem of mapping spelled English words into strings of phonemes that can drive speech synthesis hardware (see Damper 1995 for a comprehensive review). One of the best-known attempts was the NETtalk system, developed in 1987 by Sejnowski and Rosenberg (S&R). This chapter describes a line of research inspired by the NETtalk system. Our goal was to take the basic

R. I. Damper (ed.), Data-Driven Techniques in Speech Synthesis, 27-44.

approach developed in NETtalk and see how far we could extend it to create a high-performance letter-to-sound system for individual English words.

The chapter is organized as follows. In Section 2, we review the NETtalk formulation of the letter-to-sound task. Then, in Section 3, we describe a series of studies that were carried out to evaluate various design decisions. The resulting automatic pronunciation system is described in detail. Finally, we present an experimental evaluation of the learned pronunciation system in Section 4 before concluding with Section 5.

2. The NETtalk Approach

Here we give an overview of the approach to letter-to-sound conversion adopted by Sejnowski and Rosenberg (1987).

2.1. The NETtalk Dictionary

To apply ML algorithms to the automatic pronunciation problem, the first step is to construct a dictionary of words and their pronunciations. S&R developed a dictionary of 20,003 English words and their pronunciations by starting with Miriam Websters's Pocket Dictionary making some modifications to aid the machine learning process. The pronunciations in this dictionary are represented by two strings: one of phoneme symbols and one of stress symbols. Both strings are aligned one-to-one with the English word spelling. For example, the word *lollypop* is encoded in the dictionary as follows:

```
Word:            l o l l y p o p
Phoneme string:  l a l - i p a p
Stress string:   > 1 < > 0 > 2 <
```

There are 54 possible phoneme symbols (including word boundary and sentence-end period), and 6 stress symbols. The phoneme symbols are based on the ARPABET and, hence, they are very closely related to the phoneme symbols used in the DECtalk system (Digital Equipment Corporation 1985). Some modifications were required to maintain the one-to-one alignment. For situations where a single letter is actually pronounced as a sequence of two phonemes, S&R introduced a new phoneme symbol. For situations where a string of letters is pronounced as a single phoneme, that phoneme was aligned with the first letter in the string, and the subsequent letters were coded with the 'silent phoneme' (represented as /-/). Silent letters were also coded using the silent phoneme. Table 2.1 gives the meaning of each symbol.

The stress symbols indicate how much stress each syllable should receive and also locate the syllable boundaries. The meanings of the stress symbols are given in Table 2.2

Table 2.1. NETtalk phoneme symbols.

Phoneme	*Examples*	*Phoneme*	*Examples*
/a/	wAd, dOt, Odd	/E/	mAny, End, hEAd
/b/	Bad	/G/	leNGth, loNG, baNk
/c/	Or, cAUght	/I/	gIve, bUsy, captAIn
/d/	aDd	/J/	Jam, Gem
/e/	Angel, blAde, wAy	/K/	aNXious, seXual
/f/	Farm	/L/	eviL, abLe
/g/	Gap	/M/	chasM
/h/	Hot, WHo	/N/	shorteN, basiN
/i/	Eve, bEe	/O/	OIl, bOY
/k/	Cab, Keep	/Q/	Quilt
/l/	Lad	/R/	honeR, afteR, satyR
/m/	Man, iMp	/S/	oCean, wiSH
/n/	GNat, aNd	/T/	THaw, baTH
/o/	Only, Own	/U/	wOOd, cOUld, pUt
/p/	Pad, aPt	/W/	oUT, toWel, hoUse
/r/	Rap	/X/	miXture, anneX
/s/	Cent, aSk	/Y/	Use, fEUd, nEw
/t/	Tab	/Z/	uSual, viSion
/u/	bOOt, OOze, yOU	/@/	cAb, plAId
/v/	Vat	/!/	naZi, piZZa
/w/	We, liqUid	/#/	auXiliary, eXist
/x/	pirAte, welcOme	/*/	WHat
/y/	Yes, senIor	/^/	Up, sOn, blOOd
/z/	Zoo, goeS	/+/	abattOIr, mademOIselle
/A/	Ice, hEIght, EYe	/-/	silence
/C/	CHart, Cello	/_/	word-boundary
/D/	THe, moTHer	/./	period

Table 2.2. NETtalk stress symbols.

Symbol	*Meaning*
1	Syllable nucleus with primary stress
2	Syllable nucleus with secondary stress
0	Syllable nucleus with no stress
<	Nucleus of this syllable is to the left
>	Nucleus of this syllable is to the right
_	Non-letter (beyond start or end of word)

Table 2.3. Training examples computed by converting the word `lollypop` to a series of 7-letter windows.

Window ($\mathbf{x}$)	*Pronunciation* (y)
_ _ _ l o l l	l >
_ _ l o l l y	a 1
_ l o l l y p	l <
l o l l y p o	- >
o l l y p o p	i 0
l l y p o p _	p >
l y p o p _ _	a 2
y p o p _ _ _	p <

2.2. Inputs and Outputs

Most learning algorithms take as input training examples of the form $\langle \mathbf{x}, y \rangle$, where $\mathbf{x}$ is a fixed-length vector of feature values and y is a discrete-valued variable (the desired output class). Because English words have variable length, this poses a problem. The approach taken in NETtalk was to formulate the problem of pronouncing an entire word as the sequence of smaller problems of pronouncing each individual letter. Hence, one training example is created for each letter in the word. The $\mathbf{x}$ vector of features describes a 7-letter 'window' centered on the letter to be pronounced (and including the three letters before and the three letters after that letter). The y output is a pair of values: the desired phoneme symbol and the desired stress symbol. For example, the word *lollypop* is converted into 8 training examples as shown in Table 2.3.

In principle, there are 324 possible y values (corresponding to all possible combinations of phonemes and stresses). However, many of these combinations cannot occur in practice. Approximately 140 of these combinations occur in the NETtalk dictionary.

The learning method employed by S&R was the back-propagation algorithm for training feed-forward neural networks (Rumelhart, Hinton, and Williams 1986). To suit back-propagation, it was necessary to convert each input and output into a collection of binary feature values. Specifically, each letter was converted to a string of 29 binary features. There was one feature for each letter and three features to encode blank spaces between words and punctuation. The punctuation features were not used in this study, so only 27 binary features were strictly necessary. However, in most of our experiments we used all 29 binary features (the punctuation features always had the value 0). This produced a total of $(29 \times 7) = 203$ input features.

To convert the outputs to binary, S&R employed a 26-bit distributed output encoding. Each phoneme symbol was encoded using 21 bits, where each bit had an articulatory meaning. For example, six bits encoded the location of the articulation point in the oral cavity (i.e. alveolar, dental, glottal, labial, palatal, velar). Three bits encoded the vertical position of the articulation point (i.e. high, medium, low). Eight bits encoded other phoneme properties (i.e. affricate, fricative, glide, liquid, nasal, stop, tensed, voiced). The stress symbol was encoded as a string of five bits.

2.3. Training and Classification

S&R trained a large neural network to perform letter-to-sound conversion using the back-propagation algorithm. Once the network was trained, it could be used to pronounce a new word as follows.

First, the new word was converted into a sequence of windows. The input features for each window were given as input to the neural network, and the output of the neural network was a vector of 26 output activations (each between 0 and 1). The first 21 bits of this vector were compared with the 21-bit encodings of each of the 54 phoneme symbols, and the symbol whose 21-bit string was closest to the output (i.e. made the smallest angle with the output activation vector) was chosen as the output phoneme symbol. A similar procedure was followed for choosing the output stress symbol. The phoneme and stress symbols computed from each window were then concatenated to obtain the phoneme string and stress string for the whole word.

3. High-Performance ML Approach

Our research was initially motivated by the goal of comparing the performance of the back-propagation algorithm with alternative machine learning (ML) algorithms. We have performed several studies comparing back-propagation with algorithms for decision trees, radial basis functions, and various forms of nearest neighbor classifiers (Bakiri 1991; Wettschereck and Dietterich 1992; Dietterich, Hild, and Bakiri 1995). While these studies revealed interesting properties of the learning algorithms, they did not address the question of how well learning algorithms could do in comparison with non-learning approaches. To answer this question, we undertook the study described in this chapter.

For the purposes of evaluating various design decisions, we chose a small 2000-word subset of the NETtalk dictionary (at random, without replacement). We chose this subset because many learning algorithms are computationally expensive, and we wanted to make the experiments tractable. This 2000-word subset was further randomly partitioned into a 1000-word training set and a 1000-word test set. This 1000-word test set was used for all of the experiments reported here (with one exception – see Table 2.10).

3.1. Learning Algorithm

Training the back-propagation algorithm is difficult for two reasons: it is computationally expensive and the experimenter must choose several parameter values by trial and error. Specifically, it takes several hours to train a large neural network for the NETtalk task, even when restricted to the 1000-word training set. The experimenter must select the number of hidden units, the initial random weight values, the learning rate, the momentum coefficient, and the stopping point. While there exist automated methods for making these choices, they typically involve making multiple runs and employing cross-validation or held-out datasets. These increase the computational cost, often by orders of magnitude.

To avoid the above problems, we chose Quinlan's ID3 decision tree algorithm (Quinlan 1983, 1986) for all of our experiments. ID3 constructs a decision tree recursively, starting at the root. At each node, it selects, as the binary feature to be tested at that node, that a_i whose mutual information with the output classification is greatest (this is sometimes called the information gain criterion). The training examples are accordingly partitioned into those where $a_i = 0$ and those where $a_i = 1$. The algorithm is then invoked recursively on these two subsets of training examples, and halts when all examples at a node fall in the same class. At this point, a leaf node is created and labeled with the class in question. The basic operation of ID3 is quite similar to the CART algorithm developed by Breiman, Friedman, Olshen, and Stone (1984) and to the tree-growing method of Lucassen and Mercer (Lucassen 1983; Lucassen and Mercer 1984).

In our implementation of ID3, we did not employ windowing, χ^2 early stopping, or any kind of pruning. Initial experiments (Dietterich, Hild, and Bakiri 1990) showed that these stopping and pruning methods did not improve performance. We did apply one simple stopping rule to handle inconsistencies in the training data: if all training examples agreed on the value of the chosen feature, then growth of the tree was terminated in a leaf and the class having more training examples was chosen as the label for that leaf (in case of a tie, the leaf is assigned to class 0). Note that if all of the examples agree on the value of this feature, then the feature has zero mutual information with the output class (and since this was the best feature, every other feature must also have zero mutual information).

We applied ID3 to the initial NETtalk problem formulation. Unlike back-propagation, ID3 can only learn a single function at a time, whereas a neural network can have multiple output units. Hence, we run ID3 26 times, once for each of the output bits of the distributed output representation, and obtain 26 decision trees. When classifying a new example, we evaluate each of the 26 decision trees to assemble a vector of 26 bits. We then choose the output phoneme and stress using a slight modification of the procedure employed in

NETtalk. During training, we compute a list of all phoneme/stress combinations that occur in the training set. During classification, we compute the distance between the 26-bit result vector and the 26-bit encodings of each *observed* phoneme and stress pair. Ties are broken in favor of phoneme/stress pairs that occurred more often in the training data. This is called *observed decoding*. Previous studies have shown that observed decoding improves the performance of ID3 substantially, but that it is still not as good as the performance of back-propagation (Dietterich, Hild, and Bakiri 1995).

Table 2.4 summarizes the performance of ID3 applied using the NETtalk formulation of the letter-to-sound task. Performance is given in percentage correct at various levels of aggregation. The column labeled *Stress* gives the number of stress symbols correctly predicted on the test set. The *Phonemes* column gives the number of phoneme symbols correctly predicted on the test set. The *Letter* column gives the percentage of test letters where both the phoneme and stress symbols were correctly predicted, while the *Word* column gives the percentage of words for which every phoneme and stress symbol was correctly predicted.

Performance on the training set – not surprisingly – is nearly perfect, because the decision trees have been grown so that they correctly classify the training set. However, this does not mean that the trees have merely memorized the pronunciations of the words in the training set. On the test set, 12.5% of the (entirely new) words are pronounced completely correctly (i.e. every phoneme and stress is pronounced exactly as it appears in the NETtalk dictionary). The reason that the decision trees can generalize is that each tree only tests a small number of the available input features. The mean depth of the 26 decision trees is 29.3. This means that to determine any one bit of the output representation, on average, only a subset of about 29 features is examined. This is much less than the 203 features available. Hence, other 7-letter windows with the same values for this subset of relevant features will be classified the same way. The mean number of leaves of the 26 decision trees is 269.9. The ability of the decision trees to generalize depends on the granularity at which performance is measured. While the trees pronounce 81.3% of phonemes correctly, only 12.5% of entire words are correctly pronounced, because an error in any one phoneme or stress causes the entire word to be treated as mispronounced. In the remainder of this chapter, we will refer to this configuration of ID3 as the *base configuration*.

The table also shows the performance of back-propagation trained on the same training set and tested on the same test set. We see that back-propagation performs somewhat better than ID3. Nonetheless, the extremely long training times required for back-propagation forced us to adopt the very efficient ID3 for the research described in this chapter.

Table 2.4. Performance of ID3 using the NETtalk formulation on 1000-word training and test sets. The row labeled 'NETtalkģives the test set performance of back-propagation on a network with 160 hidden units and cross-validated early stopping (see Dietterich, Hild, and Bakiri 1995).

	% correct (1000-word dataset)				
Dataset used	*Level of Aggregation*				
for evaluation	*Word*	*Letter*	*Phoneme*	*Stress*	*Bit (mean)*
Training set	96.6	99.5	99.8	99.6	100.0
Test set	12.5	69.6	81.3	79.2	96.3
NETtalk	14.3	71.5	82.0	81.4	96.7

Table 2.5. Six alternative output configurations. Each cell gives the number of decision trees that need to be learned. For error-correcting output coding (ECOC) the number of trees depends on the length of the code(s) used. For ECOC with the separate phoneme and stress encoding, L_p is the length of the error-correcting code for the phonemes and L_s is the length of the code for the stresses.

	One-per-class	*Multiclass*	*ECOC*
Separate	60	2	$L_p + L_s$
Combined	126	1	L

3.2. Output Representation

After choosing the learning algorithm, the next design decision was to select an output representation. We considered six alternative output representations, as summarized in Table 2.5. The approached employed in NETtalk predicts the phonemes and stresses separately. We call this the 'Separate Phoneme/Stress' output representation. An alternative is to treat each observed combination of phoneme and stress as a distinct class. In fact, 126 such combinations are observed in the training data, so this can be implemented by defining a single output class variable with 126 distinct values. We call this the 'Combined Phoneme/Stress' method.

With either the Separate or the Combined Phoneme/Stress methods, there are three alternative ways of encoding the classes. The *one-per-class* approach defines a separate boolean function for each possible value of the class variable. In the Combined Phoneme/Stress method, this would involve learning 126 separate decision trees, one for each possible class. In the Separate Phoneme/Stress method, this would require 54 decision trees for phonemes and 6 decision trees for stresses. The one-per-class method is the method most commonly used in neural network problems. A drawback of the one-per-class method is that when the class of a new example is predicted, more than one of the decision trees may classify the new example as positive (i.e. more than one phoneme would be predicted). Such ties could be broken in favor of the more frequently-occurring phoneme.

The second encoding of the classes is the direct multiclass method. Unlike neural networks, decision trees can handle multiple classes easily. Each leaf of the decision tree is labeled with the value of the class variable that is to be predicted for examples which reach that leaf. This is the standard output encoding used in decision tree applications. With this representation, in the Separate Phoneme/Stress approach, only two decision trees need to be learned.

The third method is error-correcting output coding (ECOC) which was directly inspired by S&R's distributed output encoding. In the ECOC approach, each class, $k = 1, \ldots, K$, is represented by an L-bit binary codeword, C_k. The codewords are selected to form a good error-correcting code – that is, each pair of codewords differs in at least d bit positions, for some reasonably large value of d. A decision tree is learned for each bit position. The decision tree for bit position ℓ is trained to produce an output of 0 for each training example that belongs to a class y whose codeword has a 0 in the ℓ-th bit position (and to produce a 1 otherwise). To classify a new example, all L decision trees are evaluated to compute an L-bit binary string. The class k is chosen whose codeword C_k is closest to this binary string (where distance is measured by the number of bit positions in which the strings differ – the Hamming distance).

An alternative way of understanding the ECOC approach is the following. Suppose we take the K possible classes and partition them into two sets of approximately equal size, A and B. We could then train a decision tree to classify examples according to whether they belonged to set A or set B. If an example belongs to a class $k \in A$, then the decision tree should output a 0; otherwise, it should output a 1. Now suppose we repeat this process L times. Each time, we divide the classes into two sets: A_ℓ and B_ℓ. We learn a set of L decision trees.

To classify a new example, we classify it using each of these L trees. If decision tree ℓ outputs a 0, we give one vote to each class in A_ℓ, otherwise we give a vote to each class in B_ℓ. After all of the decision trees have been evaluated, the class with the largest number of votes is chosen as the output class.

There are two intuitions indicating why the ECOC approach should work well. From the error-correcting code perspective, we can view the decision trees as making 'noisy' predictions. The error-correcting code permits us to recover from errors in some of those predictions. Specifically, if fewer than $\lfloor (d-1)/2 \rfloor$ prediction errors are made by the individual trees, then the resulting class will still be correct. In practice, this permits roughly one quarter of the decision trees to make errors on each test example.

Another intuition is that ECOC is a form of committee or ensemble method, closely related to the methods of bagging (Breiman 1996a) and boosting (Freund and Schapire 1996; Quinlan 1996). According to this view, ECOC is an interesting method for constructing a group of diverse yet accurate classifiers.

Table 2.6. Results for output encoding methods for the Separate Phoneme/Stress approach. L is the length of the error-correcting code, and d is the minimum Hamming distance between each pair of codewords.

Configuration				*% correct (1000-word test set)*				
				Level of Aggregation				
Configuration				*Word*	*Letter*	*Phoneme*	*Stress*	*Bit (mean)*
Base				12.5	69.6	81.3	79.2	96.3
One-per-class				11.8	69.5	80.6	78.9	98.7
Multiclass				13.0	69.7	82.4	79.5	N/A
Phoneme		*Stress*		*Error-Correcting Output Coding*				
L	*d*	*L*	*d*	*Word*	*Letter*	*Phoneme*	*Stress*	*Bit (mean)*
10	3	9	3	13.3	69.8	80.3	80.6	90.8
14	5	11	5	14.4	70.9	82.3	80.3	91.0
21	7	13	7	17.2	72.2	83.9	80.4	91.2
26	11	13	11	17.5	72.3	84.2	80.4	91.0
31	15	30	15	19.9	73.8	84.8	81.5	91.6
62	31	30	15	20.6	74.1	85.4	81.6	92.0
127	63	30	15	20.8	74.4	85.7	81.6	92.4

Table 2.7. Results for output encoding methods for the Combined Phoneme/Stress approach.

		% correct (1000-word test set)				
		Level of Aggregation				
Configuration		*Word*	*Letter*	*Phoneme*	*Stress*	*Bit (mean)*
Base		12.5	69.6	81.3	79.2	96.3
One-per-class		8.7	66.7	76.4	74.5	99.5
Multiclass		13.5	70.8	81.1	78.3	N/A
L	*d*	*Error-Correcting Output Coding*				
63	31	20.3	74.3	83.8	80.3	87.4
127	63	22.3	75.5	85.2	81.5	87.8
255	127	22.4	75.5	85.2	81.6	87.8

Experimentally, the error-correcting output coding method has been shown to give excellent results with both decision tree and neural network learning algorithms (Dietterich and Bakiri 1995). We applied BCH code design methods (Hocquenghem 1959; Bose and Ray-Chaudhuri 1960; Lin and Costello 1983) to design good error correcting codes of various lengths for both the separate and combined phoneme/stress configurations.

Table 2.6 shows the results of the base configuration and the three alternative output encodings for the Separate Phoneme/Stress approach. Table 2.7 shows corresponding results for the Combined Phoneme/Stress approach. The first conclusion that can be drawn from these tables is that the ECOC method works very well. It clearly outperforms the other methods. The second-best method is the multiclass approach. The original S&R encoding comes next, and the one-per-class method gives the worst performance.

The comparison of the Separate and Combined Phoneme/Stress encodings is less dramatic. In general, however, the Combined approach works a little better, particularly at the *Letter* and *Word* levels of aggregation. Based on these observations, we decided to employ the Combined approach with error-correcting output codes in our final system. We also decided to use the 127-bit code (rather than the 255-bit code), because the two gave almost identical performance and the 127-bit code requires only half as much CPU time and memory to learn.

We performed various other experiments to determine how sensitive these results were to the exact assignment of codewords to classes and to the choice of training and test sets. These experiments showed that the results are the same regardless of how codewords are assigned to classes.

3.3. Input Representation

In parallel with our experiments on output encoding, we explored ways of improving the input representation for our system. One promising approach is to use the results of previous decisions when deciding how to pronounce a letter in a word. Specifically, suppose that we scan a word from left-to-right. After we predict the phoneme and stress of the first letter, we use this information as part of the input features to predict the phoneme and stress of the second letter. Then we use the results of the first two predictions as input to help predict the phoneme and stress of the third letter, and so on. One reason to expect this to work is that the stress patterns of English generally alternate between stressed and unstressed syllables. Hence, if the previous syllable was stressed, this could help the learning algorithm choose the correct stress symbol for the next syllable. This approach was first tested by Lucassen and Mercer (Lucassen 1983; Lucassen and Mercer 1984).

To construct training examples for this approach, we encode the letters, phonemes, and stresses of the three letters to the left of the target letter as well as three letters to the right of the target letter. Each training example has the following general form:

$$L_{-3}, L_{-2}, L_{-1}, L_0, L_{+1}, L_{+2}, L_{+3}, P_{-3}, S_{-3}, P_{-2}, S_{-2}, P_{-1}, S_{-1} \rightarrow P_0 S_0$$

where L_i is an element of the set of features representing the letter at position i (0 being the position of the target letter whose phoneme and stress we are trying to predict), P_i is an element of the set of features representing the phoneme at position i and S_i is an element of the set of features representing the stress at position i. During classification, we do not know the true value for $P_{-1}, S_{-1}, P_{-2}, S_{-2}, P_{-3}$, and S_{-3}, so we use the values computed by previous predictions.

We experimented with various methods for representing the phonemes and stresses. After performing several experiments, the following features were chosen:

- **Phoneme**: Each phoneme P_i is represented by 54 binary features, one for each possible phoneme symbol.
- **Tense**: Each phoneme P_i is also represented by a binary feature that is 1 if the phoneme is one of the symbols /a/, /e/, /i/, /o/, /u/, /A/, /O/, /W/, or /Y/. Otherwise it is 0. This encodes the difference between tense and lax vowels.
- **Stress**: Each stress S_i is represented by 6 binary features, one for each possible stress symbol.
- **Nucleus** and **Stressed**: Each stress S_i is also represented by two additional binary features. The first feature is 1 if the stress is one of /0/, /1/, or /2/, and 0 otherwise (i.e. this letter is a syllable nucleus). The second feature is 1 if the stress is /1/ or /2/, and 0 otherwise (i.e. this letter receives some stress).
- **Letter**: each letter L_i is represented by 29 binary features, one for each possible letter symbol (including blanks and punctuation).
- **Consonant** and **Non-Low-Vowel**: Each letter L_i is also represented by two additional binary features. The first feature is 0 if the letter is one of *a, e, i, o, u, y*, blank, or one of the punctuation symbols. Otherwise it is 1. The second feature is 0 if the letter is one of *e, i, o*, or *u*, and 1 otherwise.

We will call this the *extended* representation.

A question that arises when applying the extended representation is whether the word should be scanned left-to-right or right-to-left. This point was (briefly) investigated in the work of Lucassen and Mercer. The result was summarized by (Lucassen 1983, p. 11) as follows:

> "The *[system]* operates on a word *from left to right*, predicting phones in left-to-right order. This decision was made after preliminary testing failed to indicate any advantage to either direction. The left-to-right direction was chosen to simplify the interface to the linguistic decoder, as well as because of its intuitive appeal."

Our findings were entirely different. Table 2.8 shows the effect of the extended representation being applied left-to-right and right-to-left. (The outputs were encoded using the original S&R distributed output representation.)

Table 2.8. Performance of the Extended input representation applied left-to-right (L-R) and right-to-left (R-L) with and without 127-bit error-correcting coding (ECOC).

		% correct (1000-word test set)			
		Level of Aggregation			
Input Representation	*Direction of processing*	*Word*	*Letter*	*Phoneme*	*Stress*
Base		12.5	69.6	81.3	79.2
Extended	L-R	17.0	67.9	81.2	77.1
Extended	R-L	24.4	74.2	83.9	82.6
Extended + ECOC	R-L	32.2	78.1	86.8	83.7

We can see that the right-to-left order works much better (and, incidentally, yields smaller decision trees) than the left-to-right order. This result makes sense when one considers word pairs such as *photograph* and *photography*. In this case, the final *y* influences the pronunciation of the *a*. This then influences the pronunciation of the second *o*, which in turn influences the pronunciation of the first *o*. So even though the final *y* is more than 3 letters away from the first *o* and, hence, far outside the 7-letter window, the right-to-left scan can 'carry' the information from the end of the word back to the beginning. The table also shows that both forms of context dramatically improve performance compared to the base configuration.

After we developed the extended context, we checked to see how well it worked when combined with the ECOC method. The last line of Table 2.8 shows the performance of a 127-bit ECOC with the Combined Phoneme/Stress representation. We see that the combination of the extended context, right-to-left processing, and the error-correcting codes gives a substantial increase in performance.

We studied one other question concerning the input representation. All of our experiments thus far have employed a 7-letter window. What is the effect of using a larger window? A larger window can provide more useful information, but it can also provide more 'noise' in the form of irrelevant features. Also, the extended context may already be providing most of the benefit that could be obtained from a larger window. Table 2.9 compares the performance of the Extended Right-to-Left input representation (with the S&R output representation) for different window sizes. The table shows that larger windows perform slightly better, particularly at the level of whole words.

Based on these (and many other) experiments, we chose our final configuration to use a 15-letter window with the following input features:

- **Phoneme** (same as Extended context)
- **Tense** (same as Extended context)

Table 2.9. Effect of enlarging the window on the Extended encoding (scanned right-to-left).

Window Size	*% correct (1000-word test set)*				
	Level of Aggregation				
	Word	*Letter*	*Phoneme*	*Stress*	*Bit*
7	24.4	74.2	83.9	82.6	96.9
9	25.1	72.9	82.8	82.2	96.9
11	25.4	73.4	82.9	82.5	97.0
13	26.2	73.4	82.8	82.6	97.0
15	25.9	73.4	82.7	82.7	96.9

Table 2.10. Performance of the final configuration compared to the base configuration. Training and testing was on 1000 unseen words.

	% correct (1000-word test set)			
	Level of Aggregation			
Configuration	*Word*	*Letter*	*Phoneme*	*Stress*
Base[a]	12.5	69.6	81.3	79.2
Final[b]	37.3	83.9	86.9	91.6

[a]Evaluated on our original 1000-word test set.
[b]Evaluated on a disjoint 1000-word test set.

- **Stress**: The stress symbols /</, />/, and /0/ were replaced by the stress symbol /-/. This leaves only three stress symbols, so they were encoded as three binary features.
- **Stressed** (same as Extended context)
- **Letter** (same as Extended context)
- **Consonant** and **Non-Low-Vowel** (same as Extended context)

The output was encoded using a 127-bit error-correcting output code with the Combined Phoneme/Stress method. The performance of this method is given in Table 2.10. We can see that the modifications introduced above produced a very large improvement in performance.

4. Evaluation of Pronunciations

The purpose of this section is to describe an experiment comparing the performance of our best ML system with the performance of the DECtalk rule base.

4.1. Methods

DECtalk is an English text-to-speech synthesizer marketed by Digital Equipment Corporation. It contains hardware for synthesizing speech and software for mapping English words into strings of phonemes and stresses. At the heart of the software is a human-constructed rule base of letter-to-sound rules. This is augmented by a dictionary of exceptions – words that the rules mispronounce.

We trained our final ML configuration on all the 19,002 words in the NETtalk dictionary that are not part of our test set. (Tony Vitali of Digital Equipment Corporation kindly translated our 1000-word test set using the letter-to-sound rules of DECtalk version 3.0). So for each word in our test set, we had the DECtalk pronunciation and the ID3 pronunciation. In addition, we had the NETtalk dictionary pronunciation.

We performed an experiment with the help of three human judges. The experiment was designed as follows. First, the pronunciations produced by ID3 and by the NETtalk dictionary were converted into phoneme and stress strings suitable for playing through DECtalk. This was performed using a program call KLATTIZE written by Charles Rosenberg and distributed by Thinking Machines Corporation. If the pronunciations produced by ID3 or DECtalk exactly matched the pronunciation computed from the NETtalk dictionary, then the word was automatically considered correct.

Otherwise, the judge was asked to compare the pronunciation produced by the DECtalk rules and by ID3 with the NETtalk dictionary pronunciation. The judge was first shown the spelled word. Then the judge was played the NETtalk dictionary pronunciation followed by one of the proposed pronunciations, and was asked to give one of the following responses:

1 There is no difference between the two pronunciations;

2 There is a noticeable difference between the two pronunciations;

3 There is a serious difference between the two pronunciations; or,

4 The proposed pronunciation is better than the dictionary pronunciation. (Therefore, count the proposed pronunciation as correct.)

The order of the DECtalk and ID3 pronunciations was randomized so that the judge did not know which pronunciation was produced by which method.

4.2. Results

Each of the three judges performed this procedure independently (and on separate occasions). A response of 4 was considered equivalent to a response of 1, because both indicate that the word is correctly pronounced. The responses can be summarized by a three-digit string, such as 111 (all three judges gave a rating of 1), 112 (two judges rated 1 and one judge rated 2), and so on. Table 2.11 gives statistics showing the reliability of these judgements. The full experiment involved comparing two other ML systems as well as DECtalk

Table 2.11. Breakdown of the individual subject responses when evaluating correctness at the word level. Figures shown reflect the cumulative responses for all four systems compared.

Majority Response	*Individual Responses[a]*	*Actual Count*	*% of Total*
(1) No difference (23.7%)			
1	111	426	86.2
1	112	68	13.8
1	113	0	0.0
	Total:	494	(100%)
(2) Noticeable difference (34.6%)			
2	221	87	12.0
2	222	505	69.9
2	223	131	18.1
	Total:	723	(100%)
(3) Serious difference (41.7%)			
3	331	2	0.2
3	332	114	13.1
3	333	755	86.7
	Total:	871	(100%)

[a]Example: 332 means two subjects responded with 3 and the third responded with 2.

and the final ID3 configuration described above, and this table reflects all of those judgements.

The results are encouraging. Out of the 494 responses that were classified as 1 (i.e. no difference), 426 (86.2%) were by a unanimous decision. Similarly, out of the 871 responses that were classified as 3 (i.e. serious difference), 755 (86.7%) were by a unanimous vote.

The situation, however, is not so clear cut for responses classified as 2 (i.e. noticeable difference). Only 505 of the 723 such cases (69.9%) were by a unanimous decision. The rest (218 cases), were decided by a vote of two out of three. In 131 out of these 218 cases (60.1% of the time), the third judge who disagreed thought that difference was serious rather than noticeable, while in the remaining (39.9% of the 218 cases) the judge who disagreed thought that the difference was negligible.

It is interesting that there were *no* cases of the 113 response (two judges responding with 1, the third with a 3), only two cases of 331, and only one case of a 123 response (which was later changed to 222). Furthermore, all three judges agreed on 1,686 responses out of 2,088 (80.7%). These observations suggest that the results of the experiment are fairly robust against variations caused by differences between individual human judges.

The final results from the human judges are shown in Table 2.12. We can see that final ID3 configuration scores much better than the DECtalk rule base.

Table 2.12. Percentage of whole words correct as judged by majority vote of three human judges. A response of 2 indicated *noticeable difference*. A response of 3 indicated *serious difference*.

	Majority Response	
System	2 *or* 3	3
DECtalk rule base	55.7	73.9
ID3 (final configuration)	71.2	83.3

4.3. Discussion

The experiment provides clear evidence that the machine learning approach described in this chapter gives better results than the DECtalk rule base. However, there are two points to note, which suggest that these results should be interpreted with caution.

First, the error rates reported by our judges for DECtalk were higher than those reported elsewhere. It may be that our experimental setup, which required direct comparisons against a dictionary pronunciation was responsible for this, since the dictionary pronunciation is implicitly regarded as correct (even though we gave subjects the option of declaring the proposed pronunciation to be better).

This raises a second point. The ID3 system has been trained on other words from the NETtalk dictionary, so it can be expected to be better at mimicking the dictionary even if there are alternative legitimate pronunciations. If DECtalk produced those other pronunciations and they were seriously different from the NETtalk dictionary, then our study would count those as DECtalk errors.

The third point to note is that most of the experiments we performed during the design process used the same test set that was evaluated by the judges. Hence, information from this test set could have contaminated the learning process and yielded an overly optimistic rating for the ID3 pronunciations. In short, we may have overfitted the test set because of our extensive prior design experiments.

There are three arguments against this third point. First, given the size of the test set (1000 words, or 7,242 letters), overfitting is not very likely. Larger datasets are more difficult to overfit. Second, we replicated several of our design experiments on a different independent test set and obtained very similar results. Third, the authors of the DECtalk rule base probably also made use of some words from our test set when they were writing and debugging their rules. One advantage of computers is that you can explicitly train them on a fixed set of words and keep them ignorant of all other words in English. This is not possible with people. So in this sense, the DECtalk ratings are also potentially optimistic.

Despite these shortcomings in the experimental procedure, it is clear that our final ID3 system performs quite well. When evaluated according to how well it can exactly match the NETtalk dictionary pronunciations, our final system (trained on 19,002 words) correctly pronounces 64.8% of words, 91.4% of letters, 93.7% of phonemes, and 95.1% of stresses.

5. Conclusions

The task of mapping English text to sound is very difficult, because of the multiple sources of words in the English language and the shifts in pronunciation over the years. Because of this complexity, the letter-to-sound task is a good one for the development and testing of machine learning algorithms. The most important outgrowth of our experiments was the discovery and development of the error-correcting output coding method. We have demonstrated that this is a general method that can be applied to many application problems.

This chapter has described a machine learning approach that combined several general techniques (error-correcting output codes, sliding windows, and extended context) with some application-specific methods (special input features and the right-to-left scan). The result is a system that does a very good job of mapping English words to phonemes and stresses. An experimental evaluation by human judges provides evidence that this machine learning approach performs better than the hand-written rules in the DECtalk system. We hope that the techniques described here will be useful for other groups working on mapping problems in text-to-speech synthesis and for researchers studying similar difficult computational problems.

Chapter 3

ANALOGY, THE CORPUS AND PRONUNCIATION

Kirk P. H. Sullivan

Department of Philosophy and Linguistics, Umeå University

Abstract Reading aloud and text-to-speech synthesis share the commonality of taking a printed text and generating the acoustic correlate. This chapter starts from the premise that modelling the human solution for generating the pronunciation of a word previously unseen will result in more effective and accurate pronunciation modules for machine synthesis. The human and the text-to-speech system both have access to a lexicon: the mental lexicon in the first case and the system dictionary in the other. These are the data sources which can be mined to generate an unknown word's pronunciation. This chapter presents a computational approach to automatic pronunciation developed from a psychological model of oral reading. The approach takes the system dictionary – a frequency-tagged corpus – and uses analogy to generate the pronunciation of words not in the dictionary. A range of implementational choices is discussed and the effectiveness of the model for (British) English, German and Māori demonstrated.

1. Introduction

When confronted with a written word we, as literate human beings, rarely have a problem in assigning a pronunciation to that written word. Indeed converting the orthographic representation into an acoustic version of a word is almost a reflex action for the literate. The questions revolving around how this task is achieved are reflected most widely in the public arena as part of the political debate about how best to teach children to read – using phonics or the 'real' book technique. Interestingly, the arguments are paralleled here in both the psychological theories relating to the process of reading aloud and, more recently, the engineering solutions to the problem of computer-based text-to-speech synthesis. Reading aloud and computational text-to-speech conversion can be viewed as essentially the same problem. In the former, it is a human being who converts the text into an acoustic waveform and in the latter it is the computer.

R. I. Damper (ed.), Data-Driven Techniques in Speech Synthesis ,45-70.

Unfortunately, the conversion of a written word into a phoneme string is not always a simple process. In many languages, not least English, the conversion of text into some appropriate specification of the sounds of speech is not just a matter of assigning one phoneme to each letter. Indeed, the lack of correspondence between the writing system and the phonemic system is only too obvious when one has to write down or read an unfamiliar word in English. A frequently-cited example, which vividly illustrates this lack of invariance in the correspondence between orthographic and phonemic units in English, considers the initial letters of the words *cider*, *cat* and *kitten*. The initial letter *c* of *cider* is pronounced /s/, whereas the 'same' letter *c* is pronounced /k/ in *cat* – a one-to-many correspondence. Yet the initial letter *k* of *kitten* is also pronounced /k/, so that there is a many-to-one correspondence between letters *c* and *k* and phoneme /k/. A more complicated example is the pronunciation of the plural morpheme indicated by the letter *s*. Not only can this represent more than one individual phoneme (/s/ or /z/ as in *cats* and *dogs* respectively), it can also indicate a phoneme cluster (a regular noun ending with a sibilant forms its plural by adding the plural allomorph /ɪz/ as in *horses*). This illustrates the lack of one-to-one correspondence between single letters and single phonemes: a single letter can represent more than one phoneme. Further, contiguous groups of letters can correspond to a single phoneme as when the bigram *gh* is pronounced /f/ as in *enough*: such letter groups are called *graphemes* (Glushko 1981; Coltheart 1984). Yet in *doghouse* the *gh* bigram does not act as a grapheme. The fact that letter groups can sometimes act as graphemes, but do not always do so, further complicates the conversion of text into speech.

In the engineering solutions, the conversion process is traditionally achieved by one of two methods: looking up the input words in a dictionary, possibly with some form of morphological decomposition, or through the use of a set of grapheme-to-phoneme conversion rules. Most systems in reality use a combination of these methods; grapheme-to-phoneme rules with a small exceptions dictionary for the conversion of frequently-occurring words or a dictionary of words, or morphs, with a set of grapheme-to-phoneme rules for words not contained in the dictionary.

In recent years, it has become possible with the decline in the price of computer memory, to have increasingly large on-line dictionaries with which to convert the input text into phonemes, thereby overcoming many of the problems created by the lack of invariance in the relationship between the orthographic and the acoustic domains detailed above. As Coker, Church, and Liberman (1990) point out, the main motivation in moving to an entirely dictionary-based approach is accuracy. A dictionary is more accurate than grapheme-to-phoneme rules. Unfortunately for dictionary-based solutions, however, languages develop continuously with, for example, new words constantly entering a language and new acronyms being generated. Furthermore, texts frequently contain foreign place-names and surnames. So it is not possible to list all the words of a language (see Damper, Marchand, Adamson, and Gustafson 1999, Appendix A). Hence, even in an entirely dictionary-based approach to computational text-to-speech conversion, there is a need for some way to generate a pronunciation for those words not contained in the system's dictionary. The majority of systems use grapheme-to-phoneme rules for this purpose.

This solution parallels the position taken by psychologists such as Coltheart (1978) who favour the dual-route model of oral reading. They argue that there is one lexical route to pronunciation words the reader is familiar with (via their mental lexicon) and another, 'abstract' grapheme-to-phoneme route to read those words which are new and so cannot be found in the mental lexicon. Naturally, the arguments in favour of the dual-route theory are more complex than presented here: they include an explanatory ability of the theory to explain surface, phonological and developmental dyslexia (see Coltheart, Curtis, Atkins, and Haller 1993; Humphreys and Evett 1985). Thus, although dictionary methods are more accurate than grapheme-to-phoneme rules, such rules are an integral part of most dictionary-based systems. The words which are passed to the grapheme-to-phoneme rules by dictionary-based systems can be viewed as novel words – absent from a text-to-speech system's dictionary.

The approach posited and evaluated in this chapter is a radically different engineering solution. In contrast to earlier solutions, it recognises the central connection between human oral reading and computational text-to-speech synthesis. Previous technical solutions, while mirroring a psychological model of oral reading, were not inspired or guided by it.

This chapter initially discusses the importance of this connection to the engineering solution and the importance of the corpus as a resource for pronunciation. Thereafter, the analogy-based model of pronunciation developed by Sullivan and Damper (1990, 1992, 1993) is presented and its implications for text-to-speech synthesis application outlined in relation to a range of different implementations for (British) English, German and Māori.

2. Why Adopt a Psychological Approach?

As mentioned above the conversion of text into speech is performed efficiently by human beings. The task undertaken by a person reading aloud is the same as that a text-to-speech synthesiser attempts to achieve. There is, therefore, a natural connection between the development of text-to-speech systems and the development of psychological models of oral reading. Although novel-word pronunciation has been widely used as a means of assessing and developing psychological models of oral reading (e.g. Glushko 1979) and there is recognition that the development of such models can be assisted by computational modelling of reading (see Coltheart et al. 1993), there has been little, if any, inter-disciplinary research.

The models posited by experimental psychologists have only occasionally been considered by those developing automatic text-to-speech systems (Dedina and Nusbaum 1991; Sullivan 1992; Sullivan and Damper 1990, 1992, 1993; Yvon 1996a). Psychologists have become interested in computational models as they afford explicit control of variables, which cannot easily be achieved when experimenting with human subjects, rather than as a result of interest in the work conducted by speech technologists.

It was out of a recognition of the interconnections between the research in these disciplines that Sullivan and Damper (1989) considered the relevance of psychological model of oral reading to computational text-to-speech conversion. This early investigation forms the theoretical basis for the later computational model presented by Sullivan and Damper (1990, 1993) and Sullivan (1992). The model developed (hereafter the S&D model) aims not only to produce an effective automatic text-to-speech system, but also to reveal important aspects of the processes involved in human oral reading.

The S&D model is based on the experimental work of Glushko (1979) and Brown and Besner (1987). Glushko posited that human beings pronounce words not contained in their personal lexicon ('novel' words or pseudowords, i.e. non-words conforming to the spelling patterns of the language) by analogy with the entries in their orthographic lexicon – "assignment of phonology to non-words is open to lexical influence" (Glushko 1979). Brown and Besner, on the other hand, believe that their experimental results show that *phonological* similarity is the kernel of analogy. One important pronunciation-by-analogy system described in the literature prior to Sullivan and Damper's work is PRONOUNCE (Dedina and Nusbaum 1986, 1991) – but see also Byrd and Chodorow (1985). PRONOUNCE is based solely on orthographic similarity and is a rather simple system for American English. It was only tested on its ability to pronounce a subset of the pseudowords used by Glushko (1979); not on its ability to pronounce general text. The reported performance on the former task was very high, but attempts to duplicate Dedina and Nusbaum's results have, disappointingly, not proved successful (Damper and Eastmond 1996; Yvon 1996b).

Glushko's model contains no separate grapheme-to-phoneme route and thus clearly diverges from the dual-route theory. There is no demand for a separate route with a set of grapheme-to-phoneme rules to deal with 'novel' words. This is an obvious advantage when designing a machine text-to-speech converter. A fully operational text-to-speech synthesis system based on analogy would make it unnecessary to spend many hours of labour developing the grapheme-to-phoneme rule set. Dual-route theory does not specify the *canonical* set of grapheme-to-phoneme rules and, thus, does not aid the rule developer. One problem is that different pronunciations of the same 'novel' word are given by different people (see Sullivan 1992, Appendix B). So precisely which rule should be implemented?

Unfortunately, as Coltheart et al. (1993) and Damper and Eastmond (1996, 1997) point out: "pronunciation-by-analogy is under-specified: offering little meaningful guidance on the implementational choices which confront the programmer" (Damper and Eastmond 1996). The crucial question confronting any version of analogy theory is how to quantify the notion of *similar*. As Glushko (1981) wrote: "how similar is similar enough?" This is one of the issues which computational text-to-speech systems based on analogy have to confront.

Hence, we conclude that psychological models of oral reading are indeed relevant to the development of engineering solutions to the problem text-to-speech synthesis. In particular, useful approaches to grapheme-to-phoneme conversion can be inspired by theoretical developments in experimental psychology seeking to explain about how human beings read novel words aloud.

3. The Corpus as a Resource

The nub of analogy is the lexicon or, in the technical implementation, the system dictionary or corpus. In contrast to dual-route style implementations, the corpus is no longer a passive mechanism which is used to look-up pronunciation, but rather is an active source of information used to generate new pronunciations. The content of the corpus moves to the centre of the system design and becomes the system's primary resource. Computational implementations of analogy can address questions which can aid further development of the psychological model in respect of the size and content of the lexicon needed for effective performance. As stated above, analogy theory is under-specified and it is accordingly unclear whether it should operate upon the entire lexicon or on specialist (sub)lexicons depending upon the topic of the text.

Unfortunately for analogy based text-to-speech conversion by machine, the manner in which spoken language is encoded in orthography falls onto a deep/shallow continuum (Coltheart 1978; Liberman, Liberman, Mattingly, and Shankweiler 1980; Sampson 1985). In a shallow orthography (e.g. Finnish), there is a simple and direct relation between letters/graphemes and phonemes. In a deep orthography such as English, this relationship is complex and indirect as illustrated in the Introduction. Seidenberg (1985) found that for high-frequency words there was a difference between deep and shallow orthographies in the process of construction of word pronunciation. In relation to low-frequency words, Seidenberg does admit that it is possible for the detail of the recognition process to vary between different orthographic types and along the deep/shallow continuum. Investigations by Katz and Feldman (1981) further suggest the possibility of linguistic differences in coding preference between deep and shallow orthographies. They posit that the degree to which phonotactic constraints, visual and phonological codes and morphophonological knowledge are used in word-pronunciation construction from text is dependent upon the level of the orthography – whether the phonology is directly encoded (as in Finnish, for example) or not. Indeed, it is possible that the morpheme (and its attached meaning) is of more importance in deep orthographies than phonological information (Chomsky 1970).

These findings point to the difference between the degrees to which the various levels interact in the oral reading process changing along a deep/shallow orthographic continuum. The use of the constraints available from morphology and phonology alters from written language to written language; these differences may affect the way a novel word's pronunciation is generated.

The impact of corpus content on the performance of pronunciation-by-analogy is examinable through comparison of the results achieved when different corpora are used as the analogical database. In what follows, we will compare performance for the basic S&D model (presented in greatest detail in Sullivan 1992) but with a range of implementations all using a British English (Received Pronunciation) corpus. Similarly, the importance of the position of a language on the deep/shallow orthographic continuum will be examined by comparing results achieved by the same version of S&D model but using different English, German and Māori corpora. The German implementations are discussed in greatest detail in Sullivan (1992) and the Māori work in Sullivan (1995).

To recap, in the S&D model, the corpus is the central resource from which novel-word pronunciations are generated. Varying the design and content of the corpus allows the impact of this resource on performance to be evaluated within a single language. Further, using different corpora for different languages allows the effectiveness of pronunciation-by-analogy to be assessed as a function of position along the deep/shallow orthographic continuum.

4. The Sullivan and Damper Model

Figure 3.1 shows a schematic diagram of the full system which has, in light of the work of Brown and Besner (1987), both orthographic and phonemic routes to pronunciation. In a practical system, there must be some means of resolving conflicts between these two routes. To gain insight into the workings of the system, however, we consider the orthographic and phonemic outputs separately here: the conflict resolution module is not implemented. Also, at this stage, the phonemic analogiser is very simple – merely selecting the top-ranking pronunciation candidate (see later).

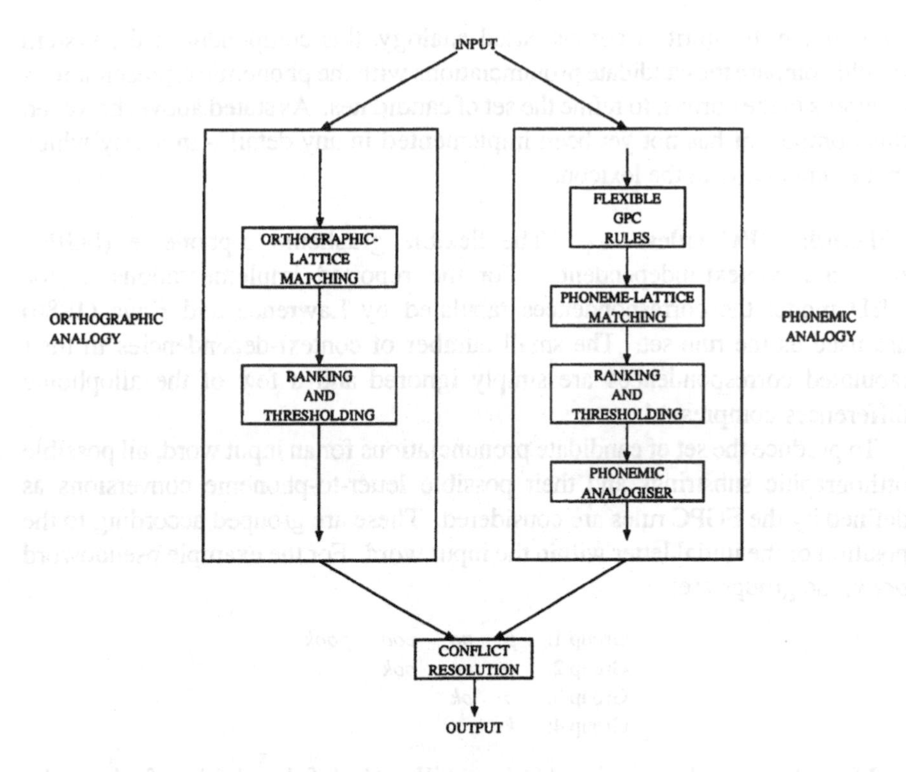

Figure 3.1. Schematic diagram of the Sullivan and Damper model of pronunciation-by-analogy, which applies the analogy process in both orthographic and phonemic domains. See text for details.

4.1. Analogy in the Phonemic Domain

Following Brown and Besner (1987), the basic notion of phonemic analogy is that a set of plausible, candidate pronunciations for an input word is generated and compared with the phonemic representations of known words in the lexicon. This set of candidates will ideally be exhaustive, covering all conceivable possibilities without regard to, for example, phonotactic constraints – to ensure that 'good' pronunciations are present somewhere in the set. Brown and Besner posit the use of 'flexible' GPC rules to achieve this. In the S&D model 'flexible' is taken to indicate that the rules are context-independent. 'Good' pronunciations are those having high similarity to known words, in terms of involving high-probability letter-to-sound mappings.

The function of lattice matching is to attach a numerical 'goodness' to each candidate pronunciation. This is referred to as the *confidence rating* after Brown and Besner. In principle, the set of candidates is then ranked according to these ratings and the ranked list thresholded, so that only the best N would be fed forward to the phonemic analogiser. In the scheme advanced by Sullivan (1992), to maintain the spirit of corpus-based analogy, this component of the system would compare the candidate pronunciations with the phonemic representations of words in the corpus, to refine the set of candidates. As stated above, however, this component has not yet been implemented in any detail – in a way which makes contact with the lexicon.

Flexible GPC rules. The flexible grapheme-to-phoneme (FGPC) rules are context-independent. For the reported implementations of the S&D model the correspondences tabulated by Lawrence and Kaye (1986) are used as the rule-set. The small number of context-dependencies in their tabulated correspondences are simply ignored and a few of the allophonic differences compressed.

To produce the set of candidate pronunciations for an input word, all possible orthographic substrings and their possible letter-to-phoneme conversions as defined by the FGPC rules are considered. These are grouped according to the position of the initial letter within the input word. For the example pseudoword *pook*, the groups are:

Group 1:	*p*	*po*	*poo*	*pook*
Group 2:	*o*	*oo*	*ook*	
Group 3:	*o*	*ok*		
Group 4:	*k*			

Most of these orthographic substrings will not be left-hand sides of rules in the FGPC rule-set. Ordering by length in this way permits the remaining substrings within a group to be disregarded as soon as the substring under consideration does not invoke a rule. The process then moves to the next group. For the above example, after successful conversion of *p* in Group 1, the attempted conversion of *po* will fail. (There is no FGPC rule for *po* and, therefore, none for any of the remaining letter clusters of Group 1). The next substring considered will be the first member (*o*) of Group 2. If a rule did exist for the cluster *poo*, then *po* would have transcribed (by rule) to *NIL*, rather than there being no rule.

Phoneme-lattice matching. Whenever a graphemic substring is successfully converted into phonemes, these are entered into a pronunciation lattice for the input word. The lattice contains information concerning phonemic outputs and the graphemic substrings producing each such output. Also represented is a numeric value reflecting the probability of that grapheme-to-phoneme mapping – the *preference values* (Sullivan and Damper 1993). The various different means of determining preference values which we have studied are detailed below (Section 6.1). Once all possible substrings of the input word have been processed in this way, candidate pronunciations can be generated. The enumeration of the possible paths through the lattice from start to end is conveniently done using a path algebra (Carré 1979).

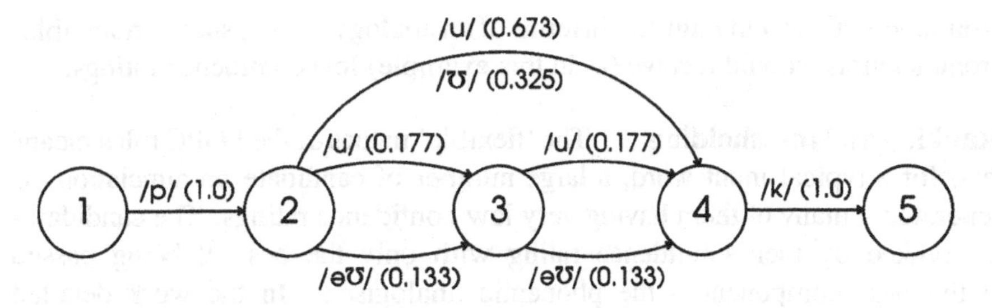

Figure 3.2. Simplified pronunciation lattice produced by phonemic analogy for the pseudoword *pook*. The nodes are the junctures between letters; the arcs therefore represent orthographic-to-phonemic mappings. Arcs are labelled with corresponding phonemic substrings and, in brackets, preference values (see text).

Figure 3.2 shows a much simplified illustration of the possible enumerated paths through the pronunciation lattice for the input *pook*, given the FGPC rules and their preference values (shown in brackets):

p	→	/p/ (1.0)	
o	→	/u/ (0.177)	as in *to*
	→	/əʊ/ (0.133)	as in *so*
oo	→	/u/ (0.673)	as in *moon*
	→	/ʊ/ (0.325)	as in *look*
k	→	/k/ (1.0)	

(Note that the preference values given here are those computed using Method 1 – see Section 6.1 below.)

The nodes represent the junctures between letters, i.e. node 2 is the 'space' between letter *p* and the first letter *o*. The arcs represent context-independent orthographic-to-phonemic mappings; they are labelled with corresponding phonemic substrings and, in brackets, preference values for that mapping. This example is simplified in that not all of the possible grapheme-to-phoneme correspondences are shown. For example, the correspondence *o* → /ɒ/ as in *cot*, along with others in the lexicon, is not considered. The candidate pronunciations corresponding to paths between nodes 0 and 4 in the illustrated lattice are /puuk/, /puəʊk/, /pəʊuk/, /pəʊəʊk/, /puk/ and /pʊk/.

A confidence rating is now attached to each each pronunciation (see Section 6.1). Various approaches for calculating these were investigated by Sullivan (1992); in the work discussed here, the confidence rating is simply calculating by taking the product of the preference values for the invoked correspondences to give confidence ratings of 0.0313, 0.0235, 0.0235, 0.0177, 0.673 and 0.325 respectively, for the above example pronunciations. Note that pronunciations are generated for later consideration which are 'impossible' in the sense that they violate the phonotactics of the language. For example, there are no words pronounced /CuuC/ in English. In terms of an analogy model, such 'impossible' pronunciations should receive (as in this example) low confidence ratings.

Ranking and thresholding. The 'flexible' nature of the FGPC rules means that, for a typical input word, a large number of candidate pronunciations is generated – many of them having very low confidence ratings. The candidates are ranked by their confidence rating with only the best *N* being passed to the next component – the phonemic analogiser. In the work detailed here, this component has not been implemented as originally envisaged by Sullivan (1992): we simply focus on the top-ranking pronunciation. For the example input *pook* above, the pronunciation /puk/ is at Rank 1 with confidence rating 0.673.

This simplification is somewhat against the spirit of pronunciation-by-analogy in that no consideration is given to the lexicon at this stage. Nonetheless, the contents of the lexicon do affect the numerical preference values and, thereby, the confidence values on which the rankings are based.

4.2. Analogy in the Orthographic Domain

The basic idea behind orthographic analogy is that "orthographic consistency in the spelling patterns of words is related to phonographical consistency in pronouncing these words" (Dedina and Nusbaum 1991). Thus, a novel input word has its pronunciation synthesised from the pronunciations of lexical entries with which it shares letters. While this could be done on a whole-word basis, computational efficiencies are gained by pre-compiling the knowledge in the lexicon into a set of letter-to-sound 'correspondences' – *analogy segments* in the terminology of Sullivan and Damper (1993). These are then used in lattice matching, much as in the initial stages of phonemic analogy. In fact, the initial stages of phonemic analogy differ from orthographic analogy only in the use and size of the analogy segments.

When pre-compiling the lexical knowledge, the orthography of each word is aligned with its corresponding phonemics using an algorithm similar to that described by Lawrence and Kaye (1986). An analogy segment is then defined as one or more consecutive alignments occurring within a word, including the whole word. For example, *green* → /grin/ contains the alignments: *g* → /g/, *r* → /r/, *ee* → /i/, *n* → /n/ and the analogy segments: *g* → /g/, *gr* → /gr/, *gree* → /gri/, *green* → /grin/, *r* → /r/, *ree* → /ri/, *reen* → /rin/, *ee* → /i/, *een* → /in/, *n* → /n/. Just as for phonemic analogy, preference values are computed for each analogy segment as described below (Section 6.1).

5. Parallels with Optimality Theory

Prior to discussing the possible methods for calculating the preference and confidence-rating values and the ways in which word boundaries can be considered, there is an interesting parallel to draw between the operation of the S&D model and the workings of *optimality theory*, the constraint-based theory of phonology developed by Prince and Smolensky (1993). Optimality theory's function *Gen* takes an input and generates the set of possible parses of the string into syllables. Thus, function *Gen* typically generates a large number of syllable structures which are "impossible" in the sense that they violate the syllable structure of the language. This is similar to the output of orthographic- and phoneme-lattice matching, where many pronunciations are generated for later consideration, which are "impossible" in the sense that they violate the phonotactics of the language. Thus, both S&D's model of pronunciation-by-analogy and Prince and Smolensky's constraint-based theory of phonology initially produce a large number of "impossible" candidate structures, which later do not become the favoured outcome.

Computational implementations of optimality theory (e.g. Tesar 1995) form a source of ideas for improvements to the computational implementation of the S&D model. For the interested reader, a simplified overview follows.

Optimality theory explores the idea that universal grammar consists of representational well-formedness constraints from which individual languages are constructed. Unusually in optimality theory, these constraints can be violated and can conflict resulting in "sharply contrary claims about the well-formedness of most representations" (Prince and Smolensky 1993). The resolution of these conflicts is achieved by an ordered ranking of the constraints. The constraints are applied in parallel to the set of candidate parses for a particular input and the parse which to the greatest degree satisfies the ranked set of constraints is considered to be the optimal candidate. This candidate, which has least violated the constraint set, is referred to as the most *harmonic*.

Figure 3.3(a) illustrates this point: the optimal candidate, A, violates both constraint X and Y. Candidates A and D violate the dominant constraint, X (where $X \gg Y$), to an equal degree, but to a lesser degree than do candidates B and C. Candidates B and C are therefore less harmonic than candidates A and D. Ultimately, candidate A is the most harmonic because it violates the dominated constraint, Y, less than candidate D.

For a given linguistic input, optimality theory proposes that *Gen*, by "freely exercising the basic structural resources of the representational theory" (Prince and Smolensky 1993), produces a candidate set which contains every possible parse of that input string, as detailed above. For a particular input string, then, *Gen* will always generate the same set of candidate parses, i.e. the result of applying *Gen* is language-independent. The constraints are, however, ranked

	Constraint X	Constraint Y
☞ Candidate A	**	**
Candidate B	***!	
Candidate C	***!	*
Candidate D	**	*****!

(a)

	Constraint Y	Constraint X
Candidate A	**!	**
☞ Candidate B	**	***
Candidate C	*!	***
Candidate D	*****!	**

(b)

Figure 3.3. (a) Constraint tableau for the constraint hierarchy $X \gg Y$. The degree of conflict is denoted schematically by the number of asterisks. The most harmonic candidate is marked with a pointing hand; the ! marks the crucial failure for each sub-optimal candidate and the shaded regions indicate cells which do not affect the choice of optimal candidate. (b) Constraint tableau for the reversed hierarchy $Y \gg X$. Here, constraint Y is applied first and, as there is only one winner, constraint X has no impact, so that the whole of the second column is now shaded.

in a language-specific manner, i.e. there is one ranking for Māori and another for English. Thus, the order of the constraints determines the output. For example, if the constraint hierarchy as applied in Figure 3.3(a) is reversed so that Y becomes the dominant constraint, $Y \gg X$, the optimal candidate is no longer A, but B (Fig. 3.3(b)). A schematic diagram of the process is shown in Figure 3.4.

6. Implementation

In this section, we outline the main implementation decisions taken in building the S&D model, and which will subsequently be investigated for their impact on performance.

6.1. Scoring Pronunciations

Preference values are intended to reflect the probability of individual letter-to-sound mappings (FGPC rules or analogy segments). Sullivan (1992) and Sullivan and Damper (1993) implemented two distinctly different ways of

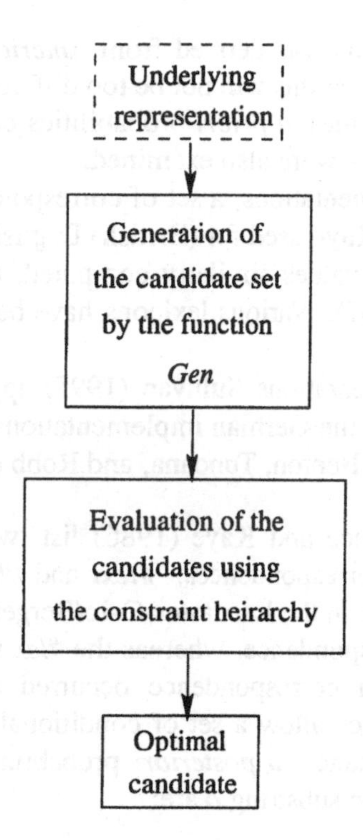

Figure 3.4. A schematic diagram of the processes of optimality theory.

determining preference values. The first of these gives a true probability whereas the second gives a probability-like value.

Method 1. A given orthographic substring o can map to a number of possible phonemic substrings p. Let the number of such correspondences be C. The *a posteriori* probability that orthographic substring o maps to phonemic substring p_j, given o, is estimated by the total number of $o \rightarrow p_j$ correspondences in the corpus, normalised by the total number of correspondences involving o:

$$P(o \rightarrow p_j|o) = \frac{\Sigma_{w=1}^{L} F_w N_w(o \rightarrow p_j)}{\Sigma_{w=1}^{L} F_w \Sigma_{i=1}^{C} N_w(o \rightarrow p_i)}$$

Here, $N_w(o \rightarrow p_j)$ is the number of $o \rightarrow p_j$ correspondences in word w, and F_w is the frequency of word w in the lexicon of size L.

For the English (RP) implementations, we use the correspondences tabulated by Lawrence and Kaye (1986) and an algorithm similar to theirs to align text and phoneme forms of words in computing these probabilities (Method 1 preference values). Word frequency information is taken from Kučera and Francis (1967).

(Note that this latter corpus was derived from *American* English newspaper usage but it was assumed that this will not be too different from British usage.) As well as *a posteriori* values, *a priori* probabilities conditioned on the total number of correspondences were also examined.

For the German implementations, a set of correspondences using the same criteria as Lawrence and Kaye used for (British) English were constructed and the Method 1 preference values similarly computed, taking word-frequency statistics from Meier (1967). Various lexicons have been used, as detailed in Section 7 below.

For the Māori implementations Sullivan (1995) applied the same criteria as for the development of the German implementations. The word-frequency statistics were taken from Benton, Tunoana, and Robb (1982).

Method 2. Lawrence and Kaye (1986) list two sets of statistics for each of their tabulated correspondences: *#text* and *#lex*. The *#text* statistic gives the number of words in the Lancaster-Oslo/Bergen Corpus (1978) which contain a particular correspondence, whereas the *#lex* value gives the number of times that a particular correspondence occurred in the Collins English Dictionary. These statistics allow a set of conditional mapping probabilities to be computed. For instance, *a posteriori* probabilities conditioned on the occurrence of orthographic substring *o* are:

$$P_{text}(o \to p_j|o) = \frac{\#text(o \to p_j)}{\Sigma_{i=1}^{C} \#text(o \to p_i)}$$

$$P_{lex}(o \to p_j|o) = \frac{\#lex(o \to p_j)}{\Sigma_{i=1}^{C} \#lex(o \to p_i)}$$

In principle, these equations provide a means of determining preference values. It should be noted that probability values based on *#text* ignore multiple occurrences of a correspondence in the same word while those based on *#lex* assume that all words have the same frequency of occurrence. In an attempt to produce preference values which might overcome these limitations of the *#text* and *#lex* statistics, Sullivan (1992) suggested simply amalgamating them by taking their product:

$$P_{prod}(o \to p_j|o) = \frac{\#text(o \to p_j)\#lex(o \to p_j)}{\Sigma_{i=1}^{C} \#text(o \to p_i)\Sigma_{i=1}^{C} \#lex(o \to p_i)}$$

Since implementations using P_{text} and P_{lex} values produced results which were consistently poorer than using the equivalent P_{prod} implementations, findings for the latter only (called Method 2) are discussed in this chapter.

Confidence ratings. While a preference value indicates the likelihood of an individual mapping, a confidence rating attempts to quantify the likelihood of a particular pronunciation for a whole word.

Many ways of obtaining a confidence rating have been investigated (Sullivan 1992). In the work reported here, this is done by taking the product of the preference values for the invoked correspondences as illustrated in Section 4.1 above for the example pseudoword *pook*. This simple approach seems to work reasonably well.

6.2. Treatment of Word Boundaries

Sullivan and Damper (1993) envisaged two possible ways of treating word-initial and word-final graphemes in computing preference and confidence-rating values, which we call *equivalent* and *distinct*. By *equivalent*, we mean that no significance is attached to the difference between word-initial and -final forms; *distinct* means they are considered to be different. Three 'varieties' were implemented as follows. Variety 1 treated word-initial and -final graphemes as equivalent, e.g. for the lexical entry *green*, scores are based on the alignment form *g* → /g/, . . . , *n* → /n/. Variety 2 considered these graphemes to be distinct. That is, the alignment form is *$g* → /g/, . . . , *n$* → /n/, where $ is the word delimiter. Finally, Variety 3 treated initial and final graphemes as both equivalent and distinct by including both types of mapping: the alignment form is *$g* → /g/, *g* → /g/, . . . , *n* → /n/, *n$* → /n/. All three varieties of alignment form were implemented for all three languages: English, German and Māori.

7. Corpora

As discussed in Section 3, the corpus is the key resource for pronunciation-by-analogy. It is from this resource that the pronunciation for a novel word is generated and chosen.

7.1. Corpus Construction

To investigate that impact of different lexical databases upon the effectiveness of a pronunciation-by-analogy system within a single language, Sullivan and Damper (1993) used three different English corpora. These were:

- the 800 words of Ogden's (1937) Basic English;
- the 800 most-frequent words of the Kučera and Francis (1967) corpus (KF800);
- the 3926 words in the Oxford Advanced Learners' Dictionary (Hornby 1974) – hereafter OALD – with a frequency of 7 or above in the Kučera and Francis corpus.

7.2. The Deep/Shallow Continuum

In order to compare the cross-language aspects of novel word reading, Sullivan and Damper (1993) constructed a German database consisting of the 800 most-frequent German words according to Meier (1967). This permitted a comparison of system performance between two Germanic languages lying at different points on the orthographic deep/shallow continuum. Sullivan and Damper's paper compared the resultant performance between the two languages as affected by the various implementation of their pronunciation-by-analogy system.

Subsequently, Sullivan (1995) constructed a dataset based on a 646 word Māori lexicon and the accompanying frequency statistics (Benton, Tunoana, and Robb 1982). This lexicon is a 'first basic word list' containing the most frequent words in the Māori language. This differentiates it from Basic English, where words are included on the basis of 'utility'. Ogden (1937) claims that with his 800 words one can express any concept.

Unfortunately, a strict comparison with the English and German implementations, based on their respective list of most frequent words, is not possible since the Māori list is not of the same length. However, as Māori lies at the shallow extreme of the deep/shallow continuum (see below), it still permits a useful comparison.

Māori is the indigenous language of Aotearoa/New Zealand and is the southernmost language of the Malayo-Polynesian family (Biggs 1961). Prior to the arrival of European settlers, it was the only language of Aotearoa/New Zealand, and was purely spoken. After many changes, the current alphabet for written Māori consists of the following graphemes: *a, e, i, o, u, h, k, m, ng, r, t, w*, and *wh*. The bigrams *ng* and *wh* are considered to be single graphemes corresponding to a single consonant in the same manner as the English bigram *gh* in *enough* is pronounced /f/. This small alphabet enables standard Māori, as taught in schools, to be spelled phonemically, resulting in a totally predictable word pronunciation. However, each of the five vowels can occur in either long or short forms, often leading to a minimal word pair. Vowel length is indicated in the orthography either by a doubling of the letter, e.g. *Maaori*, or by the addition of a macron, e.g. *Māori*. An example of a minimal pair is *mātau* or *maatau* (*to know*) and *matau* (*right*, as in the opposite of *left*). In the Sullivan (1995) implementations, long vowels are indicated by the double vowel, i.e. *ā* → *aa*.

The regular orthography of Māori also permits assessment of pronunciation-by-analogy systems without the problem of irregular word pronunciation. If the system fails for Māori, the cause of imperfect performance is likely to be fundamental, rather than a reflection of difficulties caused by the depth of the orthography (i.e., the complexity of the grapheme-phoneme relationship).

Consider, then, the corpora:

English: the 800 most-frequent words of the Kučera and Francis (1967) corpus (KF800);

German: the 800 most-frequent words of the Meier's (1967) *Deutsche Sprachstatistik*; and

Māori: the 646 words of Benton, Tunoana, and Robb's (1982) *The First Basic Māori Word List*;

From these corpora, some assessment can be made of how the effectiveness of pronunciation-by-analogy changes along the deep/shallow orthographic continuum. Further, given the highly regular orthography of standard Māori, we should be able to pinpoint inadequecies resulting from implementational shortcomings.

8. Performance Evaluation

The assessment and evaluation of the various implementations gives substance to the psychological theories upon which the S&D model is based. Until recently, with the advent of explicit computational models, analogy theory was seriously under-defined in terms of its detailed operation. The Sullivan and Damper implementations for English and German and the Sullivan implementations for Māori permit the evaluation of:

- the impact of implementational choices;
  - the method of determining the preference and confidence-rating values,
  - the treatment of word boundaries,
- the impact of the construction of the corpus;
- the impact of a language's position on the deep/shallow orthographic continuum.

The modules for English and German were assessed through the use of pseudowords. These were presented to native experimental participants who read them aloud. They were then transcribed into IPA. Different pronunciations were recorded for the same word from different participants. Sullivan and Damper (1993) defined a "correct" pronunciation for the assessment of their model as "one which was produced by any of the human subjects" because "a pseudoword does not necessarily have a *single* 'correct' pronunciation; people will read such words aloud differently". If this position were not adopted, one would in essence be denying the validity of some people's pronunciations. In scoring the output of the various implementations, the pronunciations at

Table 3.1. Percentage correct word pronunciation scores for the implementations using *a posteriori* Method 1 (i.e. $P(o \rightarrow p_j|o)$) preference values for English. Here, 'correct' means that a human-produced pronunciation appeared at the top of the confidence-rating ranking.

Database	*Analogical domain*	*Variety*	*% Correct*
Basic English	phonemic	1	53.7
Basic English	phonemic	2	46.3
Basic English	phonemic	3	58.8
Basic English	orthographic	1	58.1
Basic English	orthographic	2	60.3
Basic English	orthographic	3	61.8
KF800	phonemic	1	44.9
KF800	phonemic	2	47.8
KF800	phonemic	3	51.5
KF800	orthographic	1	50.7
KF800	orthographic	2	55.9
KF800	orthographic	3	53.7
OALD	phonemic	1	54.4
OALD	phonemic	2	54.4
OALD	phonemic	3	49.3
OALD	orthographic	1	60.3
OALD	orthographic	2	64.0
OALD	orthographic	3	64.7

Rank 1 in the confidence-rating rankings have been assessed. If the generated pronunciation is one of the pronunciations given by the human subjects, then the output is deemed to be correct. Otherwise, it is deemed to be incorrect. The Māori implementations were assessed through the use of words not contained in the corpus. In standard Māori, pronunciation is entirely predictable from the orthography. Thus, there is no need to generate novel pseudo-words and collect pronunciations from vernacular speakers of Māori as for the other two languages.

8.1. English

The English implementations were tested on the 131 pseudowords from Glushko (1979) plus two others (*goot* and *pome*) plus two lexical words (*cat* and *play*), together with the pseudohomophone *kwik*. This is the word-set used by Sullivan (1992). Glushko's pseudowords "were four or five characters long and were derived from monosyllabic words by changing one letter". 'Correct' pronunciations were obtained by asking 20 phonetically naïve readers to read these words aloud.

The results in terms of correct word pronunciations for the implementations using $P(o \rightarrow p_j|o)$ (Method 1 preference values) are shown in Table 3.1. Here, the treatment of word-initial and word-final graphemes as *both* equivalent

Table 3.2. Percentage correct scores for the English implementations using Method 2 (i.e. P_{prod}) preference values.

Database	*Procedure*	*Variety*	*% Correct (a priori)*	*% Correct (a posteriori)*
Basic English	phonemic	1	78.7	71.3
Basic English	phonemic	2	68.4	66.9
Basic English	phonemic	3	72.1	74.3
Basic English	orthographic	1	53.7	67.7
Basic English	orthographic	2	55.9	64.0
Basic English	orthographic	3	58.1	70.6
KF800	phonemic	1	59.6	58.8
KF800	phonemic	2	62.5	61.8
KF800	phonemic	3	58.8	61.8
KF800	orthographic	1	44.9	52.2
KF800	orthographic	2	47.8	57.4
KF800	orthographic	3	50.0	55.2
OALD	phonemic	1	58.8	56.6
OALD	phonemic	2	74.0	70.6
OALD	phonemic	3	57.4	58.1
OALD	orthographic	1	52.2	61.8
OALD	orthographic	2	53.7	62.5
OALD	orthographic	3	64.0	66.9

and distinct (Variety 3) resulted in the highest number of top-ranked correct pronunciations in the case of KF800 phonemic analogy (with 51.5% Rank 1 correct pronunciations) but not for orthographic analogy, where Variety 2 performed best (55.9% correct pronunciations). The overall best results for phonemic analogy are obtained using the 800 words of Basic English for Variety 3 (58.8% correct), and for orthographic analogy by OALD Variety 3 (64.7% correct). It is not unexpected that the implementations based on the 800 most-frequent words of the Kučera and Francis corpus performed least well. This lexicon contains all high-frequency words which are likely to be less regular in their pronunciation than the equally-sized 800 word-set of Basic English, which contains a cross-section of content and function words. Equally, it is not surprising that the best-performing orthographic analogy implementation was based on the OALD. This was the largest database and therefore contained the best cross-section of analogy segments.

Table 3.2 shows the results for the Method 2 P_{prod} implementations. Here, we have considered both *a posteriori* values, as described in Section 6.1 above, and *a priori* values conditioned on the total number of correspondences. The *a priori* Variety 1 implementation based on the 800 words of Basic English and phonemic analogy was the overall top-performing implementation with 78.7% correct. Orthographic analogy also performed generally better using

Table 3.3. Percentage correct scores for the German implementations using *a posteriori* Method 1 preference values. Here, 'correct' means that a human-produced pronunciation appeared at the top of the confidence-rating ranking.

Database	*Analogy*	*Variety*	*% Correct*
German	phonemic	1	47.0
German	phonemic	2	49.0
German	phonemic	3	47.0
German	orthographic	1	70.0
German	orthographic	2	80.0
German	orthographic	3	82.0

Table 3.4. Percentage correct scores for the German implementations using Method 2 preference values.

Database	*Procedure*	*Variety*	*% Correct (a priori)*	*% Correct (a posteriori)*
German	phonemic	1	42.0	42.0
German	phonemic	2	48.0	45.0
German	phonemic	3	46.0	42.0
German	orthographic	1	40.0	36.0
German	orthographic	2	75.0	62.0
German	orthographic	3	45.0	44.0

(*a posteriori*) P_{prod} values than using Method 1. A Variety 3 *a posteriori* implementation based on the 800 words of Basic English resulted in the overall best orthographic analogy performance (70.6% correct).

Unlike Method 1 preference values, with Method 2 values, the English orthographic analogy implementations generally perform worse than the phonemic ones. This is not the case for German (see below), where orthographic analogy again clearly outperformed phonemic analogy. The KF800 database produced the least successful of the English implementations. In contrast with English, the German implementations using Method 2 performed less well than those using Method 1. The top scores for both phonemic and orthographic analogy were produced by Variety 2 implementations using *a priori* P_{prod} values.

8.2. German

Using the same technique as Glushko used to create his English pseudowords, 100 German pseudowords were created. Then, 10 phonetically naïve (native) readers were asked to read these words aloud in order to obtain 'correct' pronunciations.

The results for the German implementations using Method 1 preference values are shown in Table 3.3 and the results for the Method 2 implementations are shown in Table 3.4. It is noticeable that the former are uniformly superior to the latter. For this language, orthographic analogy performed very much better than phonemic analogy when using Method 1 values. The top-ranked German pronunciations were produced by implementations which treated word-initial and word-final graphemes differently to their comparable KF800 implementations. The best phonemic analogy implementation (with 49.0% correct pronunciations) treated these graphemes as distinct (Variety 2) and the top-scoring orthographic analogy implementation (82.0% correct) treated these graphemes as both distinct and equivalent. For phonemic analogy, the KF800 and German implementations produced similar percentages of top-ranked pronunciations. Their respective best-performing implementations differed by only 2.5 percentage points, which is not statistically significant (see Damper et al. 1999, Appendix A). However, the situation is very different for orthographic analogy. The best German implementation produces 26.1 percentage points more top-ranked correct pronunciations than the best KF800 implementation. This score is 17.3 percentage points higher than the highest-scoring English implementation.

The pattern of results for the KF800 and German implementations are very different. In the case of German, orthographic analogy performed better than phonemic analogy, whereas for KF800 the opposite was true.

8.3. Māori

To provide a basis for comparison of the various Māori implementations 100 words from *A Name and Word Index to Nga Moteatea* (Harlow and Thornton 1986) were chosen, which were not also members of the 646 words of Benton, Tunoana, and Robb's (1982) *The First Basic Māori Word List*. Interestingly, no difference in the percentage of correct pronunciations was found between the outputs of the flexible grapheme-to-phoneme conversion module and the orthographic module when all other variables were left unaltered. This differs from the results produced for English and German (Sullivan and Damper 1993).

All the implementations based on Method 1 preference value produced 91% correct pronunciations (Table 3.5); the remaining 9% also produced correct pronunciations, yet they were ranked equal first with an incorrect pronunciation. Thus, they were scored as incorrect.

The Method 2 preference values followed a similar pattern (Table 3.6). For the *a posteriori* probabilities, both modules resulted in 91% correct pronunciations with the remaining 9% jointly correct/incorrect, as for Method 1. However, when the Method 2 preference values were calculated as *a priori* probabilities

Table 3.5. Percentage correct scores for the Māori implementations using *a posteriori* Method 1 preference values. Here, 'correct' means that a human-produced pronunciation appeared at the top of the confidence-rating ranking and that the pronunciation was not tied in first position.

Analogy	*Variety*	*% Correct*
phonemic	1	91
phonemic	2	91
phonemic	3	91
orthographic	1	91
orthographic	2	91
orthographic	3	91

Table 3.6. Percentage correct scores for the Māori implementations using Method 2 preference values.

Procedure	*Variety*	*% Correct (a priori)*	*% Correct (a posteriori)*
phonemic	1	100	91
phonemic	2	100	91
phonemic	3	100	91
orthographic	1	100	91
orthographic	2	100	91
orthographic	3	100	91

conditioned on the total number of correspondences, the percentage of correct pronunciations without a tied Rank 1 incorrect pronunciation increased to 100%.

8.4. Impact of Implementational Choices

It is clear from the range of these results that the exact nature of the implementation is of prime importance to the success of pronunciation-by-analogy. Yet the manner in which the variables examined affect the result is not clear and requires further investigation.

Preference values. Considering first the method used for determining preference values, Method 1 consistently results in better (or at least equivalent) results when the domain is orthographic. Method 2, on the other hand, produces better results in the phonemic domain. These are general tendencies, not universally found. For instance, Method 2 results in the highest scoring response for German but the implementation is in the orthographic domain.

Treatment of word boundaries. Here, the picture is far from clear. In the case of Māori, this variable had no effect on the results. For German, Variety 2 produced the best result in three of the four cases, although only marginally so.

For English, there was no clear superiority for any one variety, and this was true for both Method 1 and 2 implementations.

Corpus construction. The effects of varying the corpus have to date only been investigated for English. Three different corpora were constructed. Two of these were of the same size – 800 words. One (KF800) was constructed to include the most frequent words according to Kučera and Francis (1967); the other (Basic English; Ogden 1937) was selected to include 'useful' words and so contained a higher number of content words. By contrast, two of the three corpora were based solely on word frequency and differed only in size: KF800 and OALD. The former contained 800 words and the latter 3926 words. Overall, this permits both the importance of size and of selection criterion (frequency or 'usefulness') to be compared.

From the results for the various English implementations, it seems that content is more important than size of corpus. Yet content can be compensated for by size. The least successful corpus was KF800, containing the 800 most frequent words. The Basic English corpus performed, on average, as well as the much larger OALD corpus. Although it seems that content is more important than size, the detail of that content still has to ascertained. Equally, the point at which increasing the size of the corpus ceases to produce a performance improvement (if that is indeed the case) needs to be defined.

Position on the deep/shallow continuum. In this work, there were three corpora of the same size (800 words) and frequency-based method of construction – two for English and one for German. There was also a similar sized (646 words) Māori corpus constructed on the basis of word frequency. The three languages lie on different points on the continuum. Māori lies at the shallow end with German and English lying progressively nearer the deep end. So, these four corpora allow us to assess the effectiveness of pronunciation-by-analogy systems as we move along the continuum.

For the English implementations, Method 2 preference values outperformed Method 1 values. Using Method 2, phonemic analogy generally performed much better than orthographic analogy. For German, however, best performance was obtained for orthographic analogy. Indeed, the very best implementation (82.0% correct) produced a score 33 percentage points greater than the top-scoring phonemic analogy implementation (49.0%). Also in contra-distinction to English, Method 1 preference values did better than Method 2 values. The difference in the performance of the English and German implementations is possibly connected with the fact that the relationship between letter and sound is more direct in German than in English. In relation to the nature of German orthography, Dorffner, Kommenda, and Kubin (1985) point out "German spelling is fairly regular but pronunciation rules take knowledge of the internal structure for granted". Orthographic analogy may be capturing this 'internal structure'. The superiority of Method 1 over Method 2 for German could be accounted for by the different manner in which German orthography relates to speech.

For Māori, there is no difference in the success of orthographic versus phonemic analogy. All the pronunciation errors were caused by the failure to recognise the *wh* grapheme; it was transcribed as both /f/ and /wh/. The other bigram in Māori, *ng*, was always successfully transcribed, undoubtedly because (unlike *w* and *h*) there are no graphemes *n* and *g*. Interestingly, the *a priori* Method 2 implementations favoured the correct transcription, /f/, producing an unambiguous 100% correct transcription of the test words. This ability of the *a priori* implementations to deal more effectively with bigram pronunciation concurs with the better results produced by these implementations for English. For German, while it did not produce the overall best performance, the *a priori* Method 2 implementations performed as well, if not better than, the *a posteriori* Method 2 implementations.

Hence, there is an apparent commonality between English and Māori in that both perform best with *a priori* Method 2 preference values, whereas German performs best with Method 1 values. Oddly, it is the two languages at the different extremes of the deep/shallow continuum – English and Māori – that behave similarly. However, this could be due to implementational shortcomings. Intuitively, a totally regular orthography ought to analogise 100% correctly. The single ambiguity in the Māori orthography, which can produce the impossible phonemic string /wh/ as well as the correct phoneme /f/, highlights the necessity to address the question of multi-letter graphemes in future work.

It is clear, however, that analogy produces useful results, even if the detailed part played by implementational issues is not yet transparent. Yet, in the context of the results presented and the lack of implementational transparency, the conclusions for oral reading presented by Rosson (1985) seem apposite:

> "... it has been shown that both word-level and letter-level knowledge contribute to the pronunciation of words and non-words. And it has been shown that these two sources of knowledge interact in this process, such that when one source of information is strong, the effect of the other is considerably diminished."

9. Future Challenges

Many of the challenges which exist have been discussed in the previous section. These include the need for further, more rigorous examination of word-initial and word-final graphemes, together with a principled examination of the effect of, and relation between, size and content of the lexicon and how these interact with the position of the language on the deep/shallow orthographic continuum. Although these are important issues, they are also likely to be

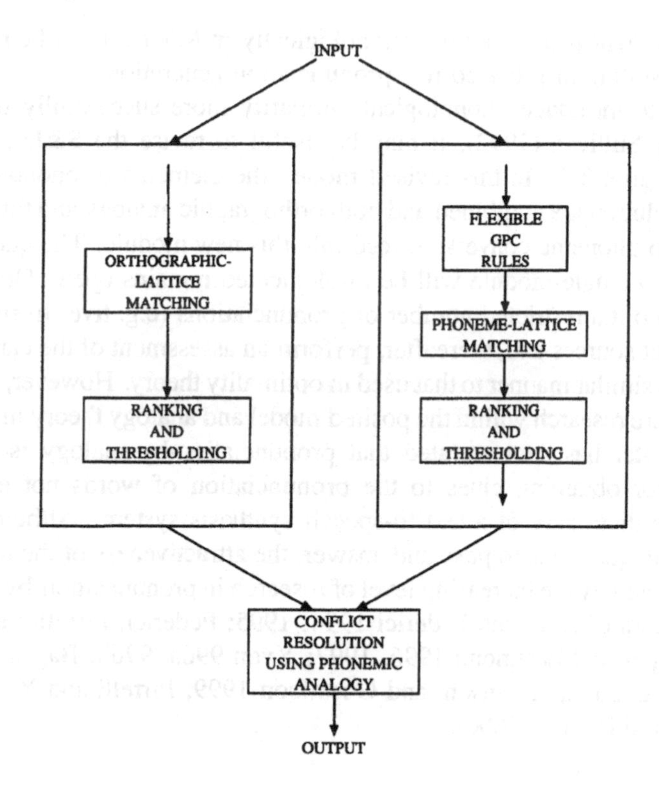

Figure 3.5. Schematic diagram of the revised Sullivan and Damper model.

influenced by the type of fundamental shortcoming revealed by the problem with bigrams in the Māori implementations. Thus, it appears important to concentrate on developing analogy techniques which are better able to deal with multi-letter graphemes.

A challenge for the S&D model, as it is depicted in Figure 3.1, is to implement phonemic analogy and conflict resolution more fully. The work to date has centered on orthographic analogy and the flexible grapheme-to-phoneme conversion required for phonemic analogy. Initial attempts at implementing a phonemic analogiser using whole-word pronunciations stored in the lexicon, as reported in Sullivan (1992), proved highly unsuccessful. Yet, there is evidence from the psychological literature that the phonological lexicon plays an active role in novel word pronunciation (Brown and Besner 1987). Further, the model requires phonological knowledge to limit possible outputs to those permitted by the phonological constraints of the language, except when the generated set of candidate pronunciations contains no candidates meeting these criteria. Such a process would have removed the source of error from the Māori implementation whereby both a correct and an incorrect pronunciation were tied in equal first place. Since /wh/ is not possible phonologically in Māori, it can be ruled out *ab initio*, resulting in 100% correct pronunciation generation.

In order to introduce phonological similarity more successfully than was achieved by Sullivan (1992), it may be useful to revise the S&D model as shown in Figure 3.5. In this revised model, the elements of phonology and conflict resolution are combined and both orthographic analogy and the flexible grapheme-to-phoneme conversion feed into this new module. The question of how this new single module will be implemented remains open. One option is to feed into it a restricted number of pronunciations (e.g. five) from each of the two input sources and, thereafter, perform an assessment of the candidates, perhaps in a similar manner to that used in optimality theory. However, this is an area for future research within the posited model and analogy theory in general.

This chapter has demonstrated that pronunciation-by-analogy is a viable technique for obtaining clues to the pronunciation of words not explicitly listed in the dictionary of a text-to-speech synthesis system. Although there remain many questions to pose and answer, the attractiveness of the technique is demonstrated by the increasing level of research in pronunciation-by-analogy in recent years (Pirrelli and Federici 1994, 1995; Federici, Pirrelli, and Yvon 1995; Damper and Eastmond 1996, 1997; Yvon 996a, 996b; Bagshaw 1998; Damper, Marchand, Adamson, and Gustafson 1999; Pirrelli and Yvon 1999; Marchand and Damper 2000).

Chapter 4

A HIERARCHICAL LEXICAL REPRESENTATION FOR PRONUNCIATION GENERATION

Helen Meng

Department of Systems Engineering and Engineering Management, The Chinese University of Hong Kong

Abstract We propose a unified framework for integrating a variety of linguistic knowledge sources for representing the English word, to facilitate their concurrent utilization in language applications. Our hierarchical lexical representation encompasses information such as morphology, stress, syllabification, phonemics and graphemics. Each occupies a distinct stratum in the hierarchy, and the constraints they provide are administered in parallel during generation via a probabilistic parsing paradigm. The merits of the proposed methodology have been demonstrated on the test bed of bi-directional spelling-to-pronunciation/pronunciation-to-spelling generation. This chapter focuses on the former task. Training and testing corpora are derived from the high-frequency portion of the Brown corpus (10,000 words), augmented with markers indicating stress and word morphology. The system was evaluated on an unseen test set, and achieved a parse coverage of 94%, with a word accuracy of 71.8% and a phoneme accuracy of 92.5% using a set of 52 phonemes. We have also conducted experiments to assess empirically (a) the relative contribution of each linguistic layer towards generation accuracy, and (b) the relative merits of the overall hierarchical design. We believe that our formalism will be especially applicable for augmenting the vocabulary of existing speech recognition and synthesis systems.

1. Introduction

We propose a unified framework which integrates a variety of relevant knowledge sources for speech synthesis, recognition and understanding. We conceive of a grand speech hierarchy with multiple levels of linguistic knowledge sources, grossly ranging from discourse, pragmatics and semantics at the upper levels, through the intermediate levels including prosody and stress, syntax, word morphology, syllabification, distinctive features, to the

R. I. Damper (ed.), Data-Driven Techniques in Speech Synthesis, 71-90.

lower levels of word pronunciations, phonotactics and phonology, graphemics (by "graphemes", we are referring to contiguous letters which correspond to a phoneme), phonetics and acoustics. The unified framework should encode not only the constraints propagated along each level of linguistic representation, but also the interaction among the different layers. From one perspective, the order of events in speech production is roughly simulated as we descend the hierarchy; while the reverse order as we ascend the hierarchy approximately models the speech perception process. Looking from another perspective, this unified body of linguistic knowledge should be applicable in speech generation/synthesis, recognition and understanding. Furthermore, improvements in the framework may be inherited by all three tasks.

To demonstrate the feasibility of our proposed framework, we have selected the test-bed of bi-directional spelling-to-phonemics and phonemics-to-spelling generation. Consequently, we will focus on the substructures in the grand hierarchy which are directly relevant to the English word. Our hierarchical framework has enabled us to formulate spelling-to-phonemics generation as a directly symmetric problem to phonemics-to-spelling generation, thereby achieving reversibility using a single system. This chapter focuses on the automatic generation of new word pronunciations, which is an important problem in speech synthesis, the focus of this book. Our results on the reverse task of phonemics-to-spelling generation can be found elsewhere (Meng, Seneff, and Zue 1994; Meng 1995).

This chapter is organized as follows. Section 2 begins with a brief survey of previous work in the area of letter-to-sound generation. Section 3 is a detailed description of our hierarchical lexical representation, followed by Section 4 which describes the *layered bigrams paradigm* for generation via parsing. Our pronunciation generation experiments, results and error analyses are reported in Sections 5, 6 and 7 respectively. Since we believe that the higher level linguistic knowledge incorporated in the hierarchy is important for our generation tasks, we will also provide an empirical assessment of: (a) the relative contribution of the different linguistic layers towards generation accuracy, and (b) the relative merits of the overall hierarchical design. These studies are based on spelling-to-pronunciation generation *only*, but we expect that the implications of the study should carry over to pronunciation-to-spelling generation. Results are reported in Section 8. The chapter will conclude with Section 9, which also presents some future directions.

2. Previous Work

A myriad of approaches have been applied to the problem of letter-to-sound generation. Excellent reviews can be found in Klatt (1987), Golding (1991) and Damper (1995). The various approaches have given rise to a wide range of letter-to-sound generation accuracies. Many of these accuracies are based on different corpora, and some corpora may be more difficult than others. Furthermore, certain systems are evaluated by human subjects, while others have their pronunciation accuracies reported on a per phoneme or per letter basis with respect to some lexical database which specifies correct pronunciations. Insertion errors or stress errors may be included in some cases, and ignored in others. There are also systems which look up an exceptions dictionary prior to generation, and the performance accuracies of these systems tend to increase with the use of larger dictionaries. For the above reasons, we should be careful when comparing different systems based on quoted performance values.

The approaches adopted for letter-to-sound generation include the rule-based approach, which uses a set of hand-engineered, ordered rules for transliteration. Transformation rules may also be applied in multiple passes to process linguistic units larger than the phoneme/grapheme, e.g. morphs. The rule-based approaches have by far given the best generation performance. A classic example of the rule-based approach is the MITalk system (Allen, Hunnicutt, and Klatt 1987), which uses a 12,000-word morph lexicon together with a morphological analysis algorithm as the major source of word pronunciation. If no analysis resulted, the word spelling was transformed to a string of phonemes with stress marks by a set of about 300–600 cyclical rules (Hunnicutt 1976). Using the MITalk rules alone gave word accuracies ranging from 66% to 76.5% for all phonemes and stress pattern correct (Hunnicutt 1980). However, the combination of the the morph lexicon, the analysis algorithm, and the set of ordered rules achieved a word accuracy of about 97%. Although the process of developing this combination is tedious and time consuming, the high performance level of 97% has not yet been matched by other more automated techniques.

Since the generation of rules is a difficult and complicated task, several research groups have attempted to acquire letter-to-sound generation systems through automatic or semi-automatic data-driven techniques. In the following we will provide a brief sketch. The goal is to provide as little *a priori* information as possible – ideally, only a set of pairings of letter sequences with corresponding (aligned or unaligned) phone sequences. Training algorithms are then used to produce a mechanism that is applied to predict the most 'promising' pronunciation. For example, the induction approach, which attempts to infer letter-to-sound rules from a body of training data, was adopted in Oakey and Cawthorne (1981), Klatt and Shipman (1982), Segre, Sherwood, and Dickerson (1983), Lucassen and Mercer (1984), Hochberg, Mniszewski, Calleja, and Papçun (1991) and Van Coile, Lyes, and Mortier (1992). Hidden Markov modeling was used in Parfitt and Sharman (1991) and Luk and Damper (1993c). Neural networks were used in the well-known NETtalk system (Sejnowski and Rosenberg 1987) and in Lucas and Damper (1992). Psychological approaches,

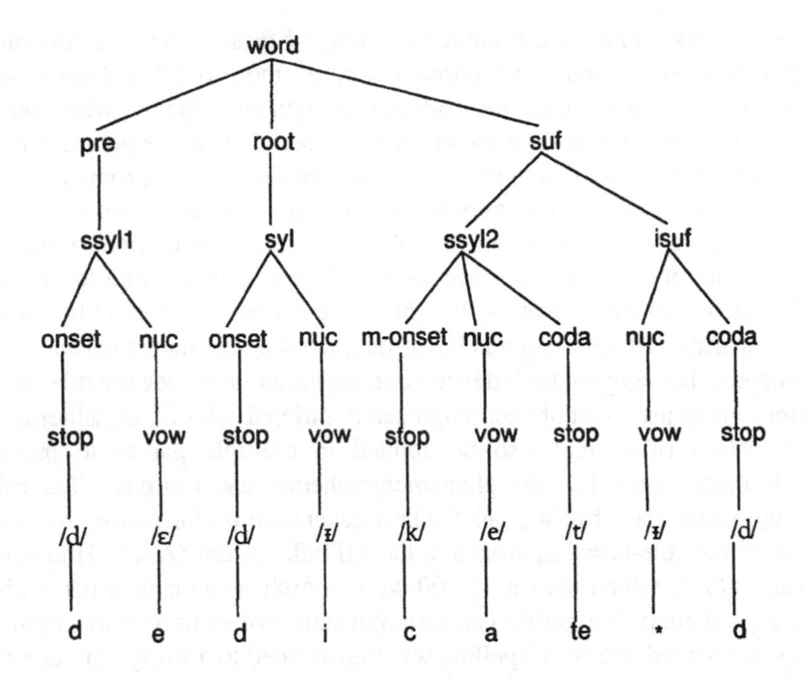

Figure 4.1. Lexical representation for the word *dedicated* – shown here in a parse tree format.

based on models proposed in the psychology literature, were used in Dedina and Nusbaum (1991), Sullivan and Damper (1993) and (Damper and Eastmond 1997). Case-based reasoning approaches, which generate a pronunciation of an input word based on similar exemplars in the training corpus, were adopted in Stanfill (1987), Lehnert (1987), Coker, Church, and Liberman (1990), Golding (1991) and van den Bosch and Daelemans (1993). Generally speaking, phoneme accuracies of the data-driven systems hover around the low 90 percentages. This roughly translates to 0.9^6 or 53% word accuracy, if we assume that an average word is six letters long, and the probability of pronouncing each letter correctly in a word is independent of the other letters.

3. Hierarchical Lexical Representation

We would like to formalize the relationship between spellings and phonemic pronunciations, by utilizing linguistic information which is of immediate relevance to the English word. Figure 4.1 illustrates the hierarchical lexical representation of the word *dedicated*, in the form of a parse tree. As we can see, the different knowledge sources embedded in the hierarchy are all substructures within the grand speech hierarchy mentioned in Section 1. As shown in Figure 4.1, they include:

1 top-level ([WORD] category, but may encode part-of-speech information);

2 morphs (prefix, root, suffix);

3 syllable stress (primary [SSYL1], secondary [SSYL2], reduced [RSYL]);

4 subsyllabic units (onset, nucleus, coda);

5 manner classes (stops, fricatives, vowels, etc.);

6 place and voicing classes encoded as phonemes;

7 letters (or graphemes).

The higher levels in the hierarchy encode longer distance constraints across the word, while the lower levels carry more local constraints. The layer of terminals nodes, which currently represent letters in a word spelling, can conceivably represent phones in the word pronunciation as well. As such, the task of letter-to-sound generation should involve deriving the higher levels in the hierarchy based on an input spelling, and subsequently generating a phone sequence while making use of all available hierarchical constraints. This process can also be used for sound-to-letter generation, with the exception of swapping the input/output specifications. Our current system does not involve phones, because our experimental corpus provides phonemic pronunciations only. However, if the terminal nodes were actually dual in nature, the hierarchical representation should be able to capture phonological rules between layers 6 (phonemes) and 7 (phones).

Although most of our labels are easily understood, a couple of special annotations that appear in Figure 4.1 should be explained. The graphemic 'place-holder' $*$ in layer 7 is introduced here to maintain consistency between the representations of the words *dedicate* and *dedicated*, where the letter *e* in the inflectional suffix [ISUF] *ed* is dropped when it is attached after the final silent *e* in *dedicate*. Also noteworthy is the special [M-ONSET] category, which signifies that the letter *c* should belong to the root *-dic-* (according to Webster's New World Dictionary, the root of *dedicated* is *-dic-*, which is derived from the Latin word *dicare*) but has become a moved onset of the next syllable as a result of syllabification principles such as the *maximal onset principle* and the *stress resyllabification principle*. The maximal onset principle states that the number of consonants in the onset position should be maximized when phonotactic constraints permit, and stress resyllabification refers to maximizing the number of consonants in stressed syllables.

4. Generation Algorithm

The paradigm for generation is one of probabilistic phonological parsing. Given the spelling of a new word, we employ a bottom-up, left-to-right

Table 4.1. A few examples of generalized context-free rules for different levels in the hierarchy.

word	→	[prefix] root [suffix]
root	→	stressed-syllable [unstressed-syllable]
stressed-syllable	→	[onset] nucleus [coda]
nucleus	→	vowel
nasal	→	(/m/, /n/, /ŋ/)
/m/	→	(*m, me, mn, mb, mm, mp*)

probabilistic parsing strategy to derive the tree representation, and simply walk the phonemic layer of the resulting tree to obtain the generation output.

We are adopting a technique that represents a cross between explicit rule-driven strategies and strictly data-driven approaches. Grammar rules are written by hand, and training words are parsed using TINA (Seneff 1992), according to their marked linguistic specifications. The parse trees are then used to train the probabilities in a set of 'layered bigrams' (Seneff, Meng, and Zue 1992). We have chosen a probabilistic parsing paradigm for four reasons. First, the probabilities serve to augment the known structural regularities that can be encoded in simple rules with other structural regularities which may be automatically discovered from a large body of training data. Second, since the more probable parse theories are distinguished from the less probable ones, search efforts can selectively concentrate on the high probability theories, which is an effective mechanism for perplexity reduction. Third, probabilities are less rigid than rules, and adopting a probabilistic framework allows us easily to generate multiple parse theories, which gives alternate pronunciations. Fourth, the flexibility of a probabilistic framework also enables us automatically to relax constraints to attain better coverage of the data.

4.1. Training Procedure

Our experimental corpus is derived from the 10,000 most frequent words appearing in the Brown corpus (Kučera and Francis 1967), where each word entry contains a spelling and a *single* unaligned phoneme string. We use about 8,000 words for training, and a disjoint set of about 800 words for testing. These test words are evenly distributed across the lexicon.

A set of training parse trees is first generated using the TINA parser with a context-free grammar. Each tree corresponds to a word in the training corpus. The training algorithm attaches probabilities to the sibling-sibling transitions in the context-free rules, which serve to incorporate linguistic knowledge into the parser. About 100 rules were written to parse the entire training corpus of 8,000 words into the format shown in Figure 4.1. As can be seen from the examples in Table 4.1, the rules are very general and straightforward. It should be noted that the terminal rules simply enumerate all possible spellings

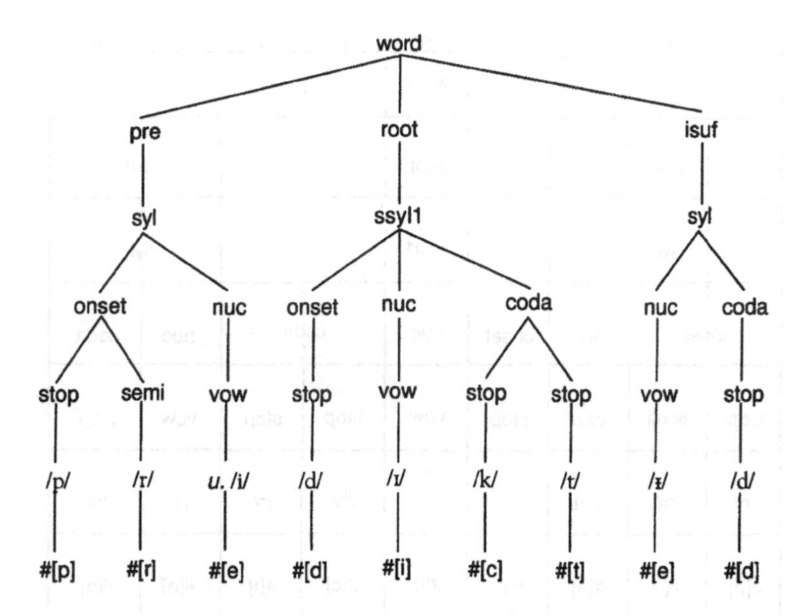

Figure 4.2. Lexical representation for the word *predicted* – shown here in a parse tree format.

that can associate with a given letter or letter sequence, without regard to context. Context conditions are learned automatically through the probabilistic training step.

Figure 4.2 shows the parse tree representation of the word *predicted*, and Figure 4.3 shows the representation in a format intended to elucidate the following discussion on parsing probabilities, which we name 'layered bigrams'.

The set of training probabilities is estimated by tabulating counts using the training parse trees. It includes bottom-up prediction probabilities for each category in the parse tree, and column advancement probabilities for extending a column to the next terminal. Hence, parsing proceeds bottom-up and left-to-right. The same set of probabilities is used for both letter-to-sound and sound-to-letter generation. The basic 4-tuple used in the analysis of the training data is:

1 left-history (LH)

2 left-sibling (LS)

3 right-parent (RP)

4 right-sibling (RS)

Using Figure 4.3 as an example, if we regard the current node (or RS) as the terminal node `[r]` in column 2, then the entire column 1

1	2	3	4	5	6	7	8	9
				word				
	pre			root			isuf	
	syl			ssyl1			syl	
onset		nuc	onset	nuc	coda		nuc	coda
stop	semi	vow	stop	vow	stop	stop	vow	stop
/p/	/r/	u. /i/	/d/	/ɪ/	/k/	/t/	/ɨ/	/d/
#[p]	#[r]	#[e]	#[d]	#[i]	#[c]	#[t]	#[e]	#[d]

Figure 4.3. Lexical representation for the word *predicted* – shown here in a layered bigrams format.

(`[WORD, PREFIX, UNSTRESSED SYLL, ONSET, STOP, /p/, [p]]`) constitutes its left-history. The left-parent and left-sibling are respectively the phoneme /p/ and the letter `[p]` in column 1, while the right-parent is the phoneme /r/ in column 2. Generally, the left and right parents may or may not be identical. Notice that in this example the left-parent is different from the right-parent. This 4-tuple derives from two context-free rules: /p/ → `[p]` and /r/ → `[r]`.

As a further example, let us consider columns 4 and 5 in Figure 4.3. Taking RS = `NUC` in column 5, then:

LH = `[WORD, ROOT, STRESSED SYLL, ONSET]` in column 4
LS = `ONSET`
LP = `STRESSED SYLL`
and RP = `SAME`

Notice that, in this case, the left-parent 'holds' (i.e. LP = RP). This is because this 4-tuple is derived from a single context-free rule – namely, `STRESSED SYLL` → `(ONSET) NUC (CODA)`. In other words, a 4-tuple encodes 'within-rule' constraints in a derivation if LP = RP. Otherwise, if LP ≠ RP, the 4-tuple corresponds to 'across-rule' constraints.

The set of probabilities that we compute is intended to capture the constraints provided by the context-free rules as well as other regularities inherent in the training parse trees but not explicitly specified. The set includes:

1. start terminal unigram, P_{StartUni} – this is the unigram probability over all the terminals that can start a word. In the letter-to-sound generation case, the start terminal is a grapheme, e.g. the letter *p* starts the word *predicted*, and the grapheme *ph* starts the word *philosophy*.

2. start column prediction probability $P_{\text{StartCol}} = P(\text{RP}|\text{RS}, \text{LH} = \text{START})$ – this is the bottom-up prediction probability given that we are at the start column of the word. The start column probability is the product of the start terminal unigram and all the bottom-up prediction probabilities in the start column, so that:

$$\begin{aligned} P_{\text{StartCol}} &= P_{\text{StartUni}} \times P(\text{RP}|\text{RS} = \text{start terminal}, \text{LH} = \text{START}) \\ &\quad \times \cdots \times P(\text{RP} = \text{WORD}|\text{RS}, \text{LH} = \text{START}) \end{aligned}$$

3. column advance probability P_{ColAdv} – this is the bigram probability over all the terminals than can follow the current column, that is $P(\text{RS} = \text{nextterminal}|\text{LH})$. The next terminal may be an END node.

4. column prediction probability P_{Col} – this is the bottom-up prediction probability conditioned on the left-history and the current (right-sibling) category. The layered bigrams have been modified to be driven entirely bottom-up here so that the 'within-rule' statistics and 'across-rule' statistics are shared. The bottom-up prediction probability $P(\text{RP}|\text{RS}, \text{LH})$ makes a prediction using the entire left-history as its left context. The 'column probability' is the product of the column advance probability and all the bottom-up prediction probabilities in the current column which we are trying to construct, so that:

$$\begin{aligned} P_{\text{Col}} &= P_{\text{ColAdv}} \times P(\text{RP}|\text{RS} = \text{current terminal}, \text{LH}) \\ &\quad \times \cdots \times P(\text{RP} = \text{SAME}|\text{RS}, \text{LH}) \end{aligned}$$

   Notice that we stop accumulating column prediction probabilities once we reach RP = SAME. This is because from then on the right-history merges with structures which are already in place in the left-history because of previous derivations from the context-free rules.

4.2. Testing Procedure

Given a test spelling or pronunciation, we attempt to construct a layered parse structure in a bottom-to-top, left-to-right fashion. A 'theory' is first proposed by constructing the start column and computing the corresponding start column probability, and then it is pushed on the stack to become a partial theory with a stack score. At each iteration of the algorithm, the stack is sorted, the partial theory with the highest stack score is popped off the stack, and advanced by one column. This advancement checks to see if the last column of the partial theory is valid top-down, and if it can reach the next terminal to follow. If either one of these conditions is not satisfied, the partial theory is eliminated. Otherwise, this newly expanded partial theory is pushed onto the stack with an updated stack score. The iteration continues until one or more complete theories (theories whose last column contains an END node) are popped off the stack.

It can be seen that the layered bigrams algorithm attempts to construct the entire parse tree structure based on some very local probabilistic constraints within adjacent layers. With the various different layers in the hierarchy, long-distance constraints are *implicitly* enforced. The layout of the hierarchical structure also enables us *explicitly* to enforce additional long-distance constraints through the use of filters. To illustrate this, consider the example of bisyllabic words with a [PREFIX]-[ROOT] morphology, where the nouns often have an UNSTRESSED-STRESSED pattern, while the verbs have a STRESSED-UNSTRESSED pattern (consider *permit*, *record*, etc.) There are also several known stress-affecting suffixes which tend to alter the stress contour of a word in a predictable way (consider *combine* versus *combination*). For these examples, we can build stress filters to eliminate any partial (or complete) theories which do not have the correct stress pattern. Similarly, it is possible to use morph filters to eliminate theories which contain illegal morph combinations. This work is presently under development.

Apart from allowing additional constraints to be enforced, the flexibility of the layered bigrams algorithm also enables us to relax some constraints. This can be done by using a backoff mechanism on the column advance probabilities to allow reasonable sibling-sibling transitions which did not occur in the training data because of sparse data problems. In this way, we are able to increase the coverage of the parser.

4.3. Efficient Search Algorithm

During the parsing process, the search for the best-scoring theory is guided by the stack score. Consequently, the evaluation function for computing the stack score is important. To avoid expensive computation to obtain a tight upper bound for the look-ahead score of a partial theory, we are currently using an evaluation function which invokes a score normalization mechanism. This mechanism aims at generating stack scores within a certain numeric range, and thus strives to achieve a fair comparison on the goodness of a partial path between the shorter partial paths and the longer ones. Scoring normalization may be accomplished by an *additive* correction factor in some cases, and a *multiplicative* correction factor in others. In our implementation, we use a 'fading' scheme as in:

$$\hat{f}(c) = \alpha \hat{f}(c') + (1 - \alpha)p(c', c)$$

where $\hat{f}(c)$ is the stack score from the [START] column to the current column c, c' is the column preceding c in the parse tree, $p(c', c)$ is the log-likelihood associated with extending the parse tree from c' to c, and α is some fading factor $(0 < \alpha < 1)$. The idea is to have the stack score carry short term memory, where the current node always contributes towards a certain portion of the stack score (according to some preset weight), while the remaining portion associated with the past gradually fades way. So the distant past contributes less to the stack score than the recent history, and the score tends to remain quite stable over time.

If multiple hypotheses are desired, the algorithm can terminate after a desired number of complete hypotheses has been popped off the stack. In addition, a limit is set on the maximum number of theories (partial and complete) popped. The complete theories are subsequently re-ranked according to their *actual* parse score (no fading). Though our search is inadmissible (Nilsson 1982, p. 76), we are able to obtain multiple hypotheses inexpensively, and our performance will be reported in the next section.

The parser was set to terminate after obtaining up to a maximum of 30 complete parse theories, or after the maximum number of theories (partial and complete) popped off the stack reaches 330, whichever happens first. These numbers are empirically chosen as a limit on the depth of the search. The 30 hypotheses are then re-ranked according to their actual parse score, and the performance accuracies reported below are based on the new set of rankings.

5. Evaluation Criteria

The two criteria which we use for evaluating letter-to-sound generation accuracies are similar to those used in the other systems reported previously.

1 Word accuracy – In the case of letter-to-sound generation, one can perform a match between a generated phoneme string and the reference phoneme string from the lexical entry of the word. Our experimental corpus provides only a *single* reference pronunciation per word. Generation is correct if there are no discrepancies between the two phoneme sequences. (In the case of sound-to-letter generation, a similar match is performed between the two letter sequences.) This is a strict evaluation criterion which does not permit alternate pronunciations for words, as any deviation from the reference string is regarded as an error.

2 Phoneme accuracy – To indicate the extent to which a generated spelling or pronunciation is correct, phoneme accuracy should be a good

> evaluation criterion to use for letter-to-sound generation. The generated string is aligned with the 'correct' string using a dynamic programming algorithm, which selects the alignment with the minimum number of insertion, deletion and substitution operations necessary to map the generated string to the reference string. The accuracy is computed by subtracting the sum of the insertion (I), deletion (D) and substitution (S) error rates from 100%:
>
> $$\text{accuracy } (\%) = 100 - (I + D + S)$$
>
> This evaluation criterion is the one adopted by the National Institute for Standards and Technology (NIST) in the USA (Pallet, Fiscus, Fisher, Garofolo, Lund, and Przbock 1994) for measuring the performance of speech recognition systems.

The phoneme accuracy evaluation criterion assumes that all discrepancies between the reference and generated strings have equal costs. This may not be a fair assumption because often a word has alternative pronunciations which are not provided by the lexicon. Moreover, certain confusions tend to be more acceptable than others. Vowel-vowel confusions in a reduced syllable, or confusions involving few differences in distinctive features are often tolerable. For example, one would probably allow the pronunciation for *proceed* to be transcribed as /proʊsId/ as well as /prʌsId/, but this /oʊ/–/ʌ/ confusion is unacceptable for the stressed vowels in *boat* and *but*. Therefore a better method of evaluation is to elicit opinions from human subjects. However, since our current emphasis is not on performance comparison with other systems, we have not undertaken the task of conducting human evaluation.

Although there are quite a few existing spelling-to-pronunciation systems, thus far there are no standardized datasets or evaluation methods employed. Evaluation criteria for letter-to-sound conversion that have previously been used include word accuracy (which may be based on human judgement), pronunciation accuracy per phoneme and pronunciation accuracy per letter. Errors in the generated stress pattern and/or phoneme insertion errors may be neglected in some cases. However, the phoneme accuracy measurement which we use above includes insertion penalties. To a certain extent, stress errors are also accounted for, since some of our vowel phonemes are stress-loaded, i.e. we distinguish between their stressed and unstressed realizations. In measuring pronunciation accuracy per letter, silent letters are regarded as mapping to a [NULL] phoneme. We believe that pronunciation accuracy per letter would generally be higher than per phoneme, because there are on average more letters than phonemes per word. To substantiate this claim, we tested on our training set, and measured the performance using *both* pronunciation accuracy per phoneme and per letter, based on the alignment provided by the

Table 4.2. Letter-to-sound generation experiments: Word and phoneme accuracies for training and testing data. Unparsable words are excluded.

Accuracy		*Top choice correct (%)*	*Top 5 correct (%)*	*Top 10 correct (%)*
train	word	77.3	93.7	95.7
	phoneme	94.2	–	–
test	word	69.3	86.2	87.9
	phoneme	91.7	–	–

training parse trees. Our results show that using the *per letter* measurement led to approximately 10% reduction in the quoted error rate. It should be kept in mind that throughout the chapter we will be quoting *per phoneme* results.

6. Results on Letter-to-Sound Generation

In letter-to-sound (spelling-to-phonemics) generation, about 6% of the test set was unparsable. This set consists of compound words, proper names, and other words that failed because of sparse data problems. Results for the parsable portion of the test set are shown in Table 4.2. The 69.3% word accuracy corresponds to a phoneme accuracy of 91.7%, where an insertion rate of 1.2% has been taken into account in addition to the substitution and deletion errors. Our phoneme accuracy lies within the low 90's percentage range of the automatic letter-to-sound generation systems described earlier. The word accuracies of the rule-based approaches, which are typically reported to be in the mid-80 percentage range (Klatt 1987, pp. 770–1) is considerably higher than our top-choice word accuracy, but comparable to our N-best accuracy with $N = 5$. This may suggest that we can seek performance improvement by means of better search procedures. Alternatively, we can try to improve performance by using more contextual information during parsing, or devise post-processes to select among the top few generated outputs.

Figure 4.4 is a plot of cumulative percent correct of whole word theories as a function of the N-best depth for the test set. Although 30 complete theories were generated for each word, no correct theories occur beyond $N = 18$ after re-sorting. Performance reaches an asymptotic value just beyond 89%.

7. Error Analyses

The cumulative plot shown (Fig. 4.4) reaches an asymptotic value well below 100%. The words that belong to the portion of the test set lying above the asymptote appear intractable – a correct pronunciation did not emerge as one of the 30 complete theories. We have grossly classified the errors into four categories:

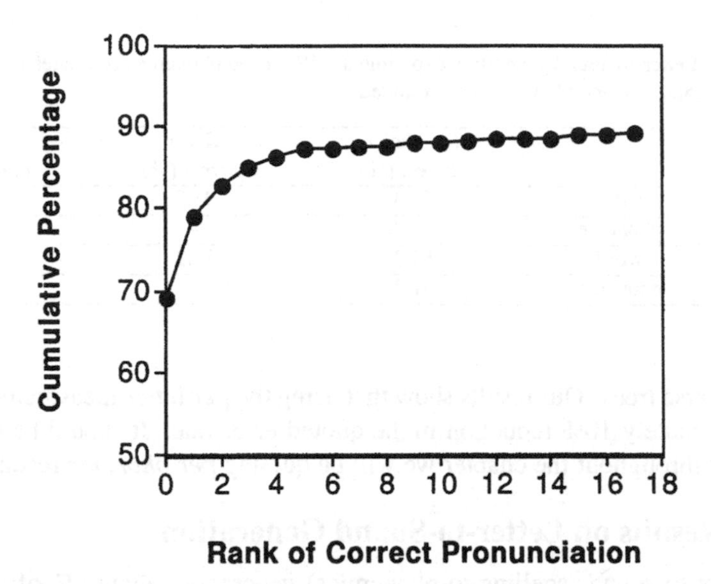

Figure 4.4. Letter-to-sound generation experiments: Percent correct whole-word theories as a function of *N*-best depth for the test set.

1 Generated pronunciations that have subtle deviations from the reference strings.

2 Unusual pronunciations due to influences from foreign languages.

3 Generated pronunciations which agree with the regularity of English letter-phoneme mappings, but which were nevertheless incorrect.

4 Errors attributable to sparse data problems.

Some examples are shown in Table 4.3.

Certain pronunciation errors, such as the generated pronunciation for (*acquiring*, /ɨkwɑʸrɨŋ/), may be considered by some as correct: likewise other examples such as /pæsyɨnet/ generated from *passionate* instead of /pæsyɨnɨt/, /kɨrtun/ for *cartoon* instead of /kɑrtun/, and /pɪpl/ for *people* instead of /pɪpbaril/. (We are using the 'underlying phonemic' form, and as a result the /ʃ/ phoneme is transcribed as /s y/ in *passionate*.) These cases can perhaps be rectified if alternate 'correct' pronunciations were available.

Further research may be devoted towards retrieving the 'correct' hypothesis from the *N*-best pool. In some of the words, different parse trees in the *N*-best list may give the same output spelling/pronunciation, but with different higher level linguistic analyses. One may attempt to sum the independent probabilities of the different parse theories with the identical phonemes, and re-rank the

Table 4.3. Some examples of generation errors.

	Category	*Spelling*	*Generated pronunciation*	*Correct pronunciation*
(1)	Subtle	acquiring	/ɨkwɑ^{ɣ}riŋ/	/ɨkwɑ^{ɣ}ɝɨŋ/
		balance	*correct*	/bælɨns/
		launch	*correct*	/lɔntʃ/
		pronounced	/prɨnɑ^{w}nst/	/pronɑ^{w}nst/
(2)	Unusual	champagne	/tʃæmpɨgnɪ/	/ʃæmpen/
		debris	/dibrɨs/	/dɨbri/
(3)	Regular	basis	/bæsɨs/	/besɨs/
		elite	/ɨlɑ^{ɣ}t/	/ɨlit/
		violence	*correct*	/vɑ^{ɣ}ɨlɨns/
		viscosity	/vɪskosɨti/	/vɪskɑsɨti/
(4)	Sparse	braque	/brækwi/	/bræk/

generated pronunciations. Alternatively, we can perhaps adopt a better stack criterion to target admissibility and curb search errors. Alternatively, we can eliminate systematic errors and refine generation outputs by post-processing with additional contextual information. An example of such technique is *transformational error-driven learning* (Brill 1992).

8. Evaluating the Hierarchical Representation

We believe that the higher level linguistic knowledge incorporated in the hierarchy is important for our generation tasks. Consequently, we would like to assess empirically: (1) the relative contribution of the different linguistic layers towards generation accuracy, and (2) the relative merits of the overall design of the hierarchical lexical representation. Therefore, we conducted two further studies. The first investigates the importance of each layer in our *hierarchical* framework, by observing how performance is affected by omitting the layer. The second compares our system with an alternative approach which uses a *single-layer* representation. These comparative studies are based on spelling-to-pronunciation generation *only*, although we expect that the implications of our study should carry over to pronunciation-to-spelling generation.

8.1. Investigating the Hierarchy

To explore the relative contribution of each linguistic level in the generation task, we conducted a series of experiments whereby an increasing amount of linguistic knowledge (in terms of the number of layers in the hierarchy) is omitted from the training parse trees. For each reduced configuration, the system is retrained and retested. In each experiment we measure:

1 Top-choice word accuracy on a disjoint (development) test set, where a word is considered correct when there is an exact match between the generated phoneme string and the *single* pronunciation provided by the lexical entry.

2 Perplexity, i.e. the average number of possible next letters, with the prediction based on the current letter.

3 Coverage of the test set, computed by subtracting the percentage of unparsable words from 100%. Unparsable words are those for which no complete parse is generated.

4 The number of parameters utilized by the system.

The top-choice word accuracy and perplexity measurements reflect the amount of constraint provided by the hierarchical representation, while coverage shows the extent to which the parser can generalize to account for previously unseen structures, by sharing training data across different layers in the hierarchy. The number of system parameters is a measurement from which one can observe the computational load, as well as the parsimony of the hierarchical framework in capturing and describing English orthographic-phonological regularities.

With the omission of each linguistic level, we expect to see two antagonistic effects on generation accuracy – the diminishing linguistic knowledge provides a decreasing amount of constraint for generation, which may induce degradation in performance. On the other hand, relaxing constraints brings about more sharing of training data across levels. This should help alleviate the sparse data problem and enhance wider coverage, which may potentially contribute to performance improvement.

The experimental results are tabulated in Table 4.4. Word accuracy refers to the percentage of the test set for which a correct pronunciation is generated from the word spelling. This is different from the word accuracy reported earlier, which is computed based on the parsable fraction of the test set. The number of system parameters in each case is rounded to the nearest hundred.

The advantages of using higher level linguistic knowledge for spelling-to-pronunciation generation can be gleaned from Table 4.4. Each layer in the hierarchical representation embodies one type of linguistic knowledge, and for every layer omitted from the representation, linguistic constraints are lost, manifested as a lower generation accuracy, higher perplexity and greater coverage. Fewer layers also require fewer training parameters.

Such phenomena are generally true except for the case of omitting the layer of broad classes (layer 5), which seems to introduce *additional* constraints, thus giving a higher generation accuracy, lower perplexity and lower coverage. This can be understood by realizing that broad classes can be predicted from

Table 4.4. Experimental results showing the relative contribution of the different layers in the hierarchical representation.

Omitted Layer	*Word Accuracy (%)*	*Perplexity*	*Coverage (%)*	*Number of Parameters*
NONE	65.4	8.3	94.4	32,700
2 (morphs)	60.4	9.3	95.7	24,700
3 (stress)	57.4	8.5	95.0	24,000
5 (broad class)	67.5	8.0	93.9	32,000
2 and 5	62.8	9.0	95.4	24,600
3 and 5	59.9	8.1	94.5	23,800
2, 3 and 5	56.4	9.1	96.1	17,300
2, 3, 4 and 5	51.1	10.1	97.1	14,800

phonemes with certainty. (These unity probabilities are, however, counted as system parameters. Broad classes may still serve a role as a fast match layer in recognition experiments, where their predictions might no longer be certain, because of recognition errors.) The inclusion of the broad class layer probably led to too much smoothing across the individual phonemes within each broad class.

8.2. Comparison with a Single-Layer Approach

We also compared our current hierarchical framework with an alternative approach which uses a single-layer representation. Here, a word is represented mainly by its spelling and an aligned phonemic transcription, using the [NULL] phoneme for silent letters. The alignment is based on the training parse trees from the hierarchical approach. For example, *bright* is transcribed as /braʸ NULL NULL t/. The word is then fragmented exhaustively to obtain letter sequences (word fragments) shorter than a preset maximum length. During training, bigram probabilities and phonemic transcription probabilities are computed for each letter sequence. Therefore this approach captures some graphemic constraints within the word fragment, but higher level linguistic knowledge is not explicitly incorporated. Spelling-to-pronunciation generation is accomplished by finding the 'best' concatenation of letter sequences which constitutes the spelling of the test word. Mathematically, let l denote the spelling of the test word, and s_i denote a letter sequence (or word fragment) with t_i being its most probable phonemic transcription. Furthermore, let S be a possible concatenation which constitutes l, i.e. $l = S = s_1 s_2 ... s_n$ which corresponds to the phonemic transcription $T = t_1 t_2 \ldots t_n$. The spelling-to-pronunciation generation process can then be represented as:

$$T = \arg\max_S \; P(T|S,l)\; P(S|l) = \arg\max_S \prod_{i=1}^{n} P(t_i|s_i,l)\; P(s_i|s_{i-1},l)^{\alpha}$$

Table 4.5. Experimental results for spelling-to-pronunciation generation using the single-layer approach.

Max. Word Fragment Length	*Word Accuracy (%)*	*Perplexity*	*Number of Parameters*
4	60.5	14.8	303,300
5	67.1	13.9	508,000
6	69.1	13.2	693,300

In the above equation, α is a weighting factor for the language score, and its value was optimized using the training data only. The underlying assumption is that the prediction of the next letter sequence depends on the current letter sequence only. This is purposely designed to maintain consistency with the layered bigram formalism, where the *current column* is used in the prediction of the *next column*. Another assumption is that the phonemic transcription of a letter sequence is independent of the context outside the letter sequence itself, so that each letter sequence is directly mapped to its *single* most probable phonemic transcription. In essence, the testing process involves a Viterbi search (Viterbi 1967; Forney 1973) which finds the highest-scoring segmentation for the spelling of a word. Lexical analogies are drawn within the context captured through the use of longer letter sequences. Eventually a phonemic transcription is generated from the segmentation.

To facilitate comparison with the hierarchical approach, we use the same training and test sets to run spelling-to-pronunciation generation experiments with the single-layer approach. Several different value settings were used for the maximum word fragment length. We expect generation accuracy to improve as the maximum word fragment length increases, because longer letter sequences can capture more context. However, this should be accompanied by an increase in the number of system parameters due to the combinatorics of the letter sequences.

Results are shown in Table 4.5. The maximum word fragment length was set to 4, 5 and 6 respectively. The mean fragment length used for the test set (with maximum fragment length set at 6) was 3.7, while the mean grapheme length in the hierarchical approach was 1.2. There are no unparsable test words in this case, because the approach can always backoff to mapping a single letter to its most probable phoneme.

The highest letter-to-sound accuracy of the hierarchical representation coupled with an inadmissible search is obtained when the layer of broad classes is omitted. Some 67.5% of the test words were correct, which corresponds to 71.8% of the parsable test words. About 6% of the errors in the hierarchical approach are due to parse failure. We are exploring a number of backoff strategies to overcome this problem. This outperforms the single-layer

Table 4.6. Error examples made by the single-layered approach.

Word	*Segmentation*	*Pronunciation*
bubble	#B+UBB+L+E#	/b/ /ʌ/ /b/ /l/
suds	#S+UDS#	/s/ /d/ /z/
thigh	#T+HIGH+#	/t/ /h/ /$ɑ^y$/

representation coupled with a Viterbi search, when the maximum word fragment length set at 4 and 5, but lies slightly below that when maximum length is set at 6. However, the hierarchical approach is capable of reversible generation using about 32,000 parameters, while the single-layer approach requires 693,300 parameters for uni-directional spelling-to-pronunciation generation. To achieve reversibility, the number of parameters needs to be doubled.

Table 4.6 shows some erroneous outputs of the single-layered approach. These errors seem to be mostly a result of incorrect segmentation reflecting a lack of linguistic knowledge. For example, the letter sequence *th* at the beginning of a word like *thigh* should be a syllable onset which is often pronounced as /ð/ or /θ/, but not /th/. Another example is the word *suds*, in which the letter sequence *uds* is transcribed as /ʤ/. Here, an analogy is drawn with *clouds*, /$klɑ^w$ʤ/), but the letters *ou* which should together map to the syllable nucleus have been split. Such errors are precluded by the higher level linguistic constraints in the hierarchical framework.

9. Discussions and Future Work

Our current work demonstrates the use of a hierarchical framework, which is relatively rich in linguistic knowledge, capable of bi-directional spelling-to-phonemics and phonemics-to-spelling generation. This chapter focuses on the former task. It is important to note that the use of layered bigrams in our hierarchy can be extented to encompass natural language constraints (Seneff, Meng, and Zue 1992): prosody, discourse and perhaps even dialogue modeling constraints on top, as well as phonetics and acoustics at the bottom. As such this paradigm should be particularly useful for applications in speech synthesis, recognition and understanding.

In this chapter, we have illustrated the hierarchical lexical representation, especially designed to incorporate multiple levels of linguistic knowledge. We have also described out letter-to-sound generation algorithm, which is based on a probabilistic parsing paradigm. We have attained a generation performance of 71.8% word accuracy and 92.5% phoneme accuracy with a corpus coverage of 94%, based on a disjoint training and test set of 8,000 and 800 words respectively. This result is comparable to the performance previously adopted by other automatic approaches.

Additionally, we have conducted two studies to assess the importance of higher level linguistic knowledge in our generation tasks. The first study shows that as different layers are omitted from the training parse trees, linguistic constraints are lost, manifested as a decline in generation accuracy, and an increase in perplexity and coverage. The converse is true when the layer of broad classes is omitted – generation accuracy improved, while perplexity and coverage decreased. This exception may be cause by the fact that broad classes can be predicted from the phonemes bottom-up with certainty, and their inclusion may have led to excessive smoothing in the subsequent predictions in the upper levels of the hierarchy.

The second study compares the hierarchical parsing framework with a non-linguistic approach. This alternative approach requires a one-to-one letter-to-phoneme mapping, creates a record of all possible word 'fragments' up to a maximum fragment length. Generation involves a Viterbi search on all the ways a word can be pieced together by word fragments, using a bigram fragment score and a phonemic transcription score. Comparative experiments showed that the two approaches achieves comparable performance for a predefined maximum fragment length, but the hierarchical approach requires 20 times fewer parameters. This indicates that the hierarchy helps impose structure on the lexical representation, which becomes a more parsimonious description of English graphemic-to-phonemic mappings.

In addition to the the ability of generation new word pronunciations for speech synthesis systems, our current *framework* may also be applied in a variety of other applications. These range from a lexical representation for large-vocabulary recognition, a clustering mechanism for fast match (Shipman and Zue 1982), to a low-perplexity language model for character recognition tasks, where our system gives a test set perplexity of 8.0 as contrasted with 11.3 from a standard letter bigram. In the near future, we plan to experiment with alternative search strategies in the layered bigrams framework, as well as to extend the coverage of the parser so as to address the sparse data problem.

Chapter 5

ENGLISH LETTER-PHONEME CONVERSION BY STOCHASTIC TRANSDUCERS

Robert W. P. Luk
Department of Computing, Hong Kong Polytechnic University

Robert I. Damper
Department of Electronics and Computer Science, University of Southampton

Abstract This chapter describes the use of stochastic transducers to model and to perform the conversion of English word spellings to phonemic equivalents. Generic word structures can be described by a simple regular grammar which usually overgenerates, producing many candidate translations. The 'best' candidate is selected based on the maximum likelihood criterion and the stochastic translation is assumed to be a Markov chain. The initial grammar allows any input string to translate to any output string. A set of example translations is used to refine this grammar to a more specific form: the Kleene closure of letter-phoneme correspondences. These correspondences were inferred by segmenting the maximum likelihood alignment of example translations when a special type of transducer movement is found, or when a segmentation point is found in one of the two (orthographic or phonemic) domains. For efficient translation, the transducers were implemented as stochastic generalised sequential Moore machines so that useless intermediate states in translation and Markov probabilities can be eliminated and reduced, respectively. The current translation accuracy on a per symbol basis is 93.7% on a representative 1667 word test set selected from the Oxford Advanced Learners Dictionary.

1. Introduction

The English alphabet was based on sound. The first stable alphabet was established in the Anglo-Saxon period, according to Venezky (1965) and Scragg (1975). Invasion by the Normans, the quest for a more Latin-like script in the Renaissance, as well as historical sound changes in England,

R. I. Damper (ed.), Data-Driven Techniques in Speech Synthesis, 91-123.

have all contributed to divergence between the relatively simpler spelling-sound system of earlier times and that of modern English. Extensive borrowing of foreign ('loan') words has also added to the increasing complexity of spelling-sound relations over time. As a result of this history, English exhibits one of the most complex letter-to-sound systems of any phonetic-based alphabet (Abercrombie 1981).

Despite the complexity of English pronunciation and the lack of a very transparent relation to spelling, it has long been alleged that a few hundred rules are adequate to produce the phonemic equivalents of word spellings (Ainsworth 1973; McIlroy 1973; Elovitz, Johnson, McHugh, and Shore 1976; Hunnicutt 1976). The Ainsworth rules, for instance, were applied depending on:

1 the context (i.e. surrounding letters) of the currently transcribed letter (e.g. *a*) or letter substring (e.g. *ch*), and;

2 which of possibly several matching rules has the highest 'priority' according to some rule-ordering scheme.

Improving these rules is difficult because they interact considerably. Hence, it is hard to manage the addition and deletion of rules – to assess the kind of errors which arise and to devise corrective actions which will not have unforeseen side effects. Although seemingly high accuracy is reported at around 90% on a per letter or phoneme basis, this depends on the suitability of the rules for the specific domain of application. For example, rules developed to pronounce dictionary-like words will not perform well with proper nouns (Spiegel, Macchi, and Gollhardt 1989). Hence, there is a need to obtain rules automatically from examples of translation. Alternatively, different translation mechanisms such as the NETtalk and NETspeak neural networks of Sejnowski and Rosenberg (1987) and McCulloch, Bedworth, and Bridle (1987) respectively can be employed. These automatically estimate their (numerical) parameters from examples of translation without the use of explicit rules.

Opinion is divided in the literature about the precise relation of trained neural networks to rule sets. There are many published examples of techniques for rule extraction from trained networks, yet Seidenberg and Elman (1999) write: "training a network to categorise stimuli ... using explicit feedback does not cause it to formulate a rule". However, any computation can always be interpreted in terms of production rules – Post (1943) originally devised such rules as a formal model of universal computation. Much depends in this debate on how transparent learned 'rules' are to the linguist, and how close they are in spirit to the context-dependent formalism that linguists would recognise.

We believe that there is much to be gained from using explicit rules and learned numerical parameters together in a translation model for English letter-phoneme conversion. This gives a way of building in at least some very basic *a priori* linguistic knowledge which can guide and constrain subsequent automatic learning. Generic word structures (e.g. in terms of syllable) have been known for some time; and they can be specified by a handful of rules or regular expressions. However, constraints have to be imposed to restrict the overgeneration which leads to many candidate translations. Numeric parameters can serve to define the most plausible translation among these candidates. Stochastic transducers offer this possibility, restricting rules to be regular. (Actually, this is true of finite-state transducers only. We will generally omit the *finite-state* qualification since context will indicate when the transducer is not finite-state, e.g. push-down transducers.)

2. Modelling Transduction

We begin by describing the historical development of finite-state transducer (FST) methods in text and speech processing. Next, we introduce formal translation and focus on capturing a generic word structure by regular translation and by finite-state transduction. We extend finite-state transduction with statistical descriptions and show that, by state expansion, the stochastic transducer with Markov probabilities can be transformed into a probabilistic finite-state automaton (as in syntactic pattern recognition).

2.1. History

Lexical processing for compilers of computer languages is traditionally described by regular expressions and carried out by finite automaton (Aho, Sethi, and Ullman 1986). The extension of these well-developed lexical processing techniques to letter-to-phoneme conversion is not obvious because:

- lexical processing is usually concerned with the recognition of tokens rather than the translation of tokens;
- rules for defining translations of letter substrings are dependent on context (e.g. Ainsworth 1973), and it is not readily apparent that such context-dependent rules can be converted into regular expressions.

Johnson (1972) first pointed out that context-sensitive rules in generative phonology, as introduced by Chomsky and Halle (1968), can be simplified to regular expressions if certain conditions are met (notably that reapplications of the context-sensitive rules to their own outputs are disallowed). From the 1980s, workers like Koskenniemi (1984) and Kaplan and Kay (1994) began to describe lexical properties of natural language by FSTs (or regular translation). A lexical item has a surface and a deep form, for example:

surface form : *works*
deep form : *work* + *s* (present progressive)

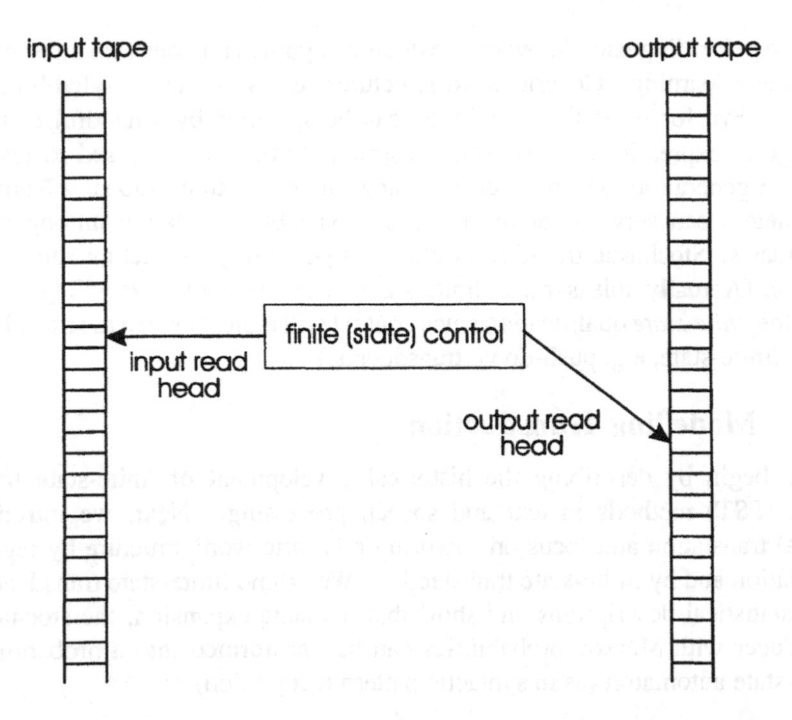

Figure 5.1. Representation of the processing of a finite-state transducer (FST) with an input and an output tape. The FST uses the input read head to read symbols from the input tape one at a time. When used for alignment or recognition, the output head reads symbols from the output tape; when used for translation, the FST writes symbols on the output tape. The input and output heads can only move forward along the input and output tapes, respectively.

Finite-state transduction can be used as a model of – and a practical technique for conversion between – these two forms of the lexical item.

Finite-state transducers are considered to have two tapes and a finite control (Figure 5.1). The transducer either reads one tape and outputs symbols on the other (translation), or it reads both tapes (recognition) and determines how input symbols correspond with output symbols (alignment). The finite control is generally defined by human experts in terms of context-dependent rules and their operators. These rules provide an adequate expressive language to model lexical phenomena (e.g. harmony in Finnish) across a variety of languages (e.g. English, Turkish and Korean). Kaplan and Kay (1994) showed how context-sensitive rules in generative phonology can be converted into context-dependent rules executed by FSTs.

In parallel with the early development of FST technology, dynamic time warping (DTW) (Sakoe and Chiba 1978) and hidden Markov model (HMM) methods (Jelinek, Bahl, and Mercer 1975) were found to be successful in automatic speech recognition. By the early 1990s, these came to be considered as *stochastic* versions of finite-state transducers (e.g. Deller, Proakis, and Hansen 1993). FSTs have now found a variety of applications in speech and language processing (Mohri 1997), from dictionary compression to efficient parsing.

2.2. Stochastic Grammars and Translation

Stochastic grammars (Gonzalez and Thomason 1978) were defined in pattern recognition to handle noisy or distorted input. Here, we extend these grammars for translation. A stochastic grammar, G_s, is a 5-tuple:

$$G_s = (N, S, R, \Sigma, \pi)$$

where N and Σ are the set of non-terminals and terminals respectively, R is the set of rewrite (or production) rules, $S \in N$ is the start symbol and π is the set of rule probabilities.

Permutational equivalence. For translation and alignment, Σ is the union of two disjoint sets: the input alphabet Σ_i and the output alphabet Σ_o. An equivalence relation called the *permutational equivalence* of two strings, denoted $\simeq$, holds if the precedence order of symbols over the input and output alphabet in the string is unaltered under permutation.

For example, if $\Sigma_i = \{a, b, c\}$ and $\Sigma_o = \{d, e, f\}$, then:

$$abcdef \simeq abdcef \simeq adbecf \simeq defabc$$

More formally, the $\simeq$-equivalence over Σ can be defined as:

$$\alpha \in \Sigma^* \simeq \beta \in \Sigma^* \text{ if } \mathrm{in}(\alpha) = \mathrm{in}(\beta) \text{ and } \mathrm{out}(\alpha) = \mathrm{out}(\beta)$$

Here, $(\mathrm{in} : \Sigma^* \rightarrow \Sigma_i^*)$ maps one string to another formed by deleting in the original string all symbols in Σ_o. Similarly, $(\mathrm{out} : \Sigma^* \rightarrow \Sigma_o^*)$ maps one string to another formed by deleting in the original string all symbols in Σ_i. So for the above example alphabets:

$$\mathrm{in}(abcdef) = \mathrm{in}(abdcef) = \mathrm{in}(adbecf) = \mathrm{in}(defabc) = abc$$

and:

$$\mathrm{out}(abcdef) = \mathrm{out}(abdcef) = \mathrm{out}(adbecf) = \mathrm{out}(defabc) = def$$

These functions are homomorphisms that can be defined (for single symbols) by the following rules:

$$\mathrm{in}(a \in \Sigma_i) \;=\; a$$

$$
\begin{aligned}
\text{in}(b \in \Sigma_o) &= \Lambda \\
\text{out}(a \in \Sigma_i) &= \Lambda \\
\text{out}(b \in \Sigma_o) &= b \\
\text{in}(a.b.c\ldots) &= \text{in}(a).\text{in}(b).\text{in}(c)\ldots \\
\text{out}(a.b.c\ldots) &= \text{out}(a).\text{in}(b).\text{in}(c)\ldots
\end{aligned}
$$

where '.' indicates concatenation. So, as an example:

$$\text{in(adbcef)} = \text{in}(a).\text{in}(d).\text{in}(b).\text{in}(c).\text{in}(e).\text{in}(f) = a.\Lambda.b.c.\Lambda.\Lambda = abc$$

The $\simeq$-equivalence naturally extends to strings over $(\Sigma \cup N)$ in which non-terminals in string s_1 must occur at the same position in the other string s_2. This is necessary because non-terminals can potentially produce strings over input, output or both alphabets. Thus, for correctness, the non-terminal cannot be moved unless proved otherwise (e.g. factorisation for context-free and regular translations). Formally, this extended $\simeq$-equivalence is defined as follows:

$$
\begin{aligned}
&\alpha \simeq \beta \qquad (\alpha, \beta \in \# \times (\Sigma \cup N) \times \#) \\
&\text{if } \forall i, \alpha[i] \in N \Rightarrow \beta[i] = \alpha[i] \\
&\text{and } \forall r, s\alpha[r, s] \in \Sigma^+, \alpha[r-1](N \cup \#), \alpha[s+1](N \cup \#) \\
&\qquad\qquad\qquad \Rightarrow \beta[r, s] \in \Sigma^+, \\
&\qquad\qquad\qquad\qquad \text{in}(\alpha[r, s]) = \text{in}(\beta[r, s]), \\
&\qquad\qquad\qquad\qquad \text{out}(\alpha[r, s]) = \text{out}(\beta[r, s])
\end{aligned}
$$

where # is the word boundary symbol that occurs at 0 and $|a| + 1$ positions in a string, i denotes a general position index in the string, and r and s denote a substring spanning these position indices.

(Note that later we will reserve '#' to denote the word boundary in the *orthographic* domain, and will use '/' for the word boundary in the *phonemic* domain.)

Alignment and translation as derivation. We define a relation $\stackrel{G}{\Rightarrow}$ called the derivation by the grammar G of the two strings over $(\Sigma \cup N)^*$. This relation is simplified to $\Rightarrow$ if the particular G is obvious from context. Derivation is simply the substitution of a substring ρ of a string $\alpha\rho\beta$ over $(\Sigma \cup N)^*$ by another substring σ if there is a rewrite rule $\rho \to \sigma$ in R. The transitive closure of a derivation is denoted by $\stackrel{G}{\Rightarrow}^*$ which is simplified to $\stackrel{*}{\Rightarrow}$ in the normal way. If there is a string w over $(N \cup \Sigma)$ and $S \stackrel{*}{\Rightarrow} w$, then w is called a sentential form.

The set of valid translations is defined as the language:

$$L(G) = \{\alpha\beta | \alpha\beta \in \Sigma^+, S \stackrel{*}{\Rightarrow} \gamma \simeq \alpha\beta\}$$

where $\alpha\beta$ and γ are *parallel* strings consisting of input and output symbols.

Alignment is the parse structure of an input string α and its corresponding output string β, defined as:

$$A_G(\alpha\beta) = \{\gamma | S \stackrel{*}{\Rightarrow} \gamma \simeq \alpha\beta\}$$

For translation, β is the unknown output, so that translation is defined as:

$$T_G(\alpha) = \{\beta | \alpha\beta \in \Sigma^+, S \stackrel{*}{\Rightarrow} \gamma \simeq \alpha\beta\}$$

This definition is more general and natural than formal (stochastic) transduction (Luk and Damper 1991, 1994, 1996) since elements in Σ_i (say a) or Σ_o (say b) can be considered as parallel strings of the form (a, Λ) and (Λ, b) respectively. Note that Σ is now considered to be an alphabet of *correspondences*, i.e. a parallel substring consisting of (each possibly empty) input and output parts. The length of a parallel string (a, b) is defined as the number of symbols in a plus the number of symbols in b which is the same as $|a.b|$. The parallel string (Λ, Λ) is identical to Λ.

Naturally, the Chomsky hierarchy in formal language theory can be extended for formal and stochastic alignment by imposing restrictions on the forms of the rewrite rules for the following classes:

Phrase Structure Grammar:

Rule Form:	$\alpha \rightarrow \beta$
Symbol type and constraints:	$\alpha, \beta \in (\Sigma \cup N)^*$
Corresponding transducer:	Turing transducer

Context-Sensitive Grammar:

Rule Form:	$\alpha \rightarrow \beta$
Symbol type and constraints:	$\alpha, \beta \in (\Sigma \cup N)^+$ and $\|\alpha\| \leq \|\beta\|$
Corresponding transducer:	Linear bounded transducer

Context-Free Grammar:

Rule Form:	$A \rightarrow \alpha$
Symbol type and constraints:	$A \in N$ and $\alpha \in (\Sigma \cup N)^+$
Corresponding transducer:	Push-down transducer

Regular Grammar (right linear):

Rule Form:	$A \rightarrow a.B$ or $A \rightarrow a$
Symbol type and constraints:	$a \in \Sigma$ and $A, B \in N$
Corresponding transducer:	Finite-state transducer

Automata for alignment and translation. Consider alignment (and recognition), in which there are no write operations. In this case, the transducers are closely equivalent to the automata which recognise the corresponding class of formal languages. But unlike the automata, which only read a single input tape, transducers read an input and an output tape (Figure 5.1). However,

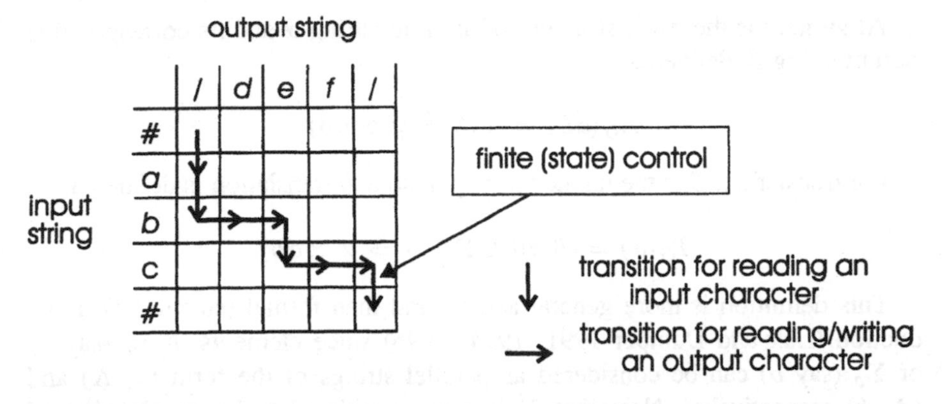

Figure 5.2. **The alignment of an input and an output string expressed as a path through a table. Vertical arrows represent movement of the transducer read head after reading a symbol from the input string. Horizontal arrows represent the movement of the transducer read head after reading a symbol from the output string. The alignment in the figure represents parsing the string *abdecf*, ignoring the word boundaries (# in the input domain and / in the output domain).**

this representation can be made more explicit to express the sequence of reading action (of an input or output symbol): the alternative is to represent the sequence of reading actions by the transducer as a path through a table as depicted in Figure 5.2. Reading an input symbol corresponds to moving the read head of the transducer downwards (i.e. vertical direction) and, likewise, reading an output symbol corresponds to moving the read head rightward (i.e. horizontal direction). Recognition terminates when the read head is at position $(|\alpha| + 1,\ |\beta| + 1)$ for the input string α and the output string β. The translation is valid (or accepted) if the state of the transducer at $(|\alpha| + 1,\ |\beta| + 1)$ is a final one (an acceptor state). A sequence of read actions from (0, 0) to $(|\alpha| + 1,\ |\beta| + 1)$ yielding an acceptor state is called a valid path or alignment. Each such path can be represented as a string of input and output symbols. The ordering of the symbols depends on the reading actions. The total number of valid paths in a table of size $|\alpha| + 1 \times |\beta| + 1$ is ${}^{|\alpha|+|\beta|}C_{|\alpha|} = {}^{|\alpha|+|\beta|}C_{|\beta|}$. The path in Figure 5.2 is represented as the string *abdecf*, which is permutationally equivalent to $\alpha\beta$ (i.e. *abdecf* $\simeq$ *abcdef*).

For alignment, the computational complexity of the transducer is the same as that of the corresponding automaton. The proof of equivalence between the language defined by the class of grammar and aligned by the corresponding type of transducer is based on existing knowledge:

1 Turing machines are equivalent in computational power no matter if they have a single input tape or multiple tapes (two input tapes for alignment by transducer);

2 Linear bounded transducers (LBTs) are the same as the Turing transducers except that the tape to store the sentential form is restricted to a linear bound on the length of the input (sum of the input and output strings). The equivalence between the language recognised by a linear bounded automaton (LBA) and that defined by a context-sensitive grammar is based on that idea that the LBA performs derivation and an exact match between the sentential form and the input string after each derivation. For LBT, two modifications of matching are necessary because of $\simeq$-equivalence: matching the left-hand side of the rewrite rules with the sentential form; and matching the sentential form with the input and output strings (i.e. $\alpha\beta$). These modifications do not increase the computational power of the LBT over the LBA because they amount to adding self-transitions of output symbols to states of the finite control and to simple control of the two read heads of the input and output strings;

3 The push-down transducer and finite-state transducer can align and translate (simple) context-free and regular translations respectively according to Hopcroft and Ullman (1979).

For translation, the computational complexity of the transducers corresponding to different classes of language may differ from their corresponding automata. The difference is that the output string is provided in the alignment, whereas the output string is unknown for translation. Effectively, translation of an input string $\alpha = a_1.a_2...a_n$ can be considered as the intersection of the regular translation $r(a) = \Sigma_o^*.a_1.\Sigma_o^*.a_2.\Sigma_o^* \ldots a_n\Sigma_o^*$ with the translation set $L(G)$. That is:

$$T_G(\alpha) = L(G) \cap r(\alpha)$$

(Although the intersection between the any class in the Chomsky hierarchy with any regular set is closed, this needs to be proved for formal translation because of $\simeq$-equivalence. The notable case is that intersection between two regular translations is not closed.)

Stochastic alignment and translation. Here, stochastic alignment and translation are defined by the maximum likelihood (ML) criterion such that, for alignment:

$$A_{G_S}(\alpha\beta) = \arg\max_r \{p(\gamma^r) | S \overset{*}{\Rightarrow} \gamma^r \simeq \alpha\beta\}$$

where r indexes the different parses of γ. If there are n rule applications in the derivation of γ^r, then by the chain rule:

$$\begin{aligned} p(\gamma^r) &= p(R_0, R_1, \ldots, R_n) \\ &= p(R_0) \times p(R_1|R_0) \times \cdots \times p(R_n|R_0, \ldots, R_{n-1}) \end{aligned}$$

where R_k is the selected rewrite rule at the kth iteration. For independent and Markov assumptions, we have, respectively:

$$p(R_k | R_{k-1}, \ldots, R_0) = p(R_k) \in \pi \qquad \text{(independent)}$$
$$p(R_k | R_{k-1}, \ldots, R_0) = p(R_k | R_{k-1}) \in \pi \quad \text{(Markov)}$$

If the independent and Markov chains are homogeneous, then the probability of γ^r can be simplified to, respectively:

$$p(\gamma^r) = \prod_k p(R_k)$$
$$p(\gamma^r) = p(R_0) \prod_k p(R_k | R_{k-1})$$

Although stochastic grammars in pattern recognition generally use the independent assumption, for regular grammars the Markov assumption is more relevant since the sentential form must end with a single non-terminal (say A) and a non-terminal determines the possible subset of rules to apply (i.e. those with the non-terminal A on the left-hand side). We will use this assumption henceforth.

Stochastic translation can be defined in terms of stochastic alignment as:

$$T_{G_S(\alpha)} = \{\beta | \alpha\beta \in \Sigma^+, \alpha\beta \simeq A_{G_S}(\alpha\beta)\}$$

3. Stochastic Finite-State Transducers

A stochastic finite-state transducer is a 7-tuple:

$$\Gamma_s = (Q, \delta, \lambda, \Sigma, F, i, \pi)$$

where Q is a set of states, i is the start state, $F \subseteq Q$ is a set of final states, $\Sigma = (\Sigma_i \cup \Sigma_o)$, $\delta : Q \times \Sigma \rightarrow Q$ is the state transition function, $\lambda : Q \times \Sigma \times Q \rightarrow \Sigma$ is the output function and the π are the state transition probabilities.

3.1. Transducer Topology

Finite-state transducers can only carry out translation and alignment based on regular rewrite rules. Although this is the most limited descriptive class in the Chomsky hierarchy, these rewrite rules can describe generic word structures in terms of prefix, infix and suffix position orderings as in the following regular expression:

word $\Rightarrow$ [prefix].stem$^+$.[inflectional-suffix].derivational-suffix* as well as syllabic structures:

$$\begin{aligned} \text{stem} &\Rightarrow \text{syllable}^+ \\ \text{syllable} &\Rightarrow \text{[onset] rhyme} \\ \text{rhyme} &\Rightarrow \text{peak [coda]} \end{aligned}$$

where [] means optional, + means repeat one or more times and $*$ means repeat zero or more times. These regular expressions can be converted into a finite-state transducer as in Figure 5.3(a). The transitions have labels which can be expanded into a set of parallel strings of the form $x.y$ where x and y are letter and phoneme substrings respectively. For instance, the label [prefix] could be a set of word prefixes with their phoneme strings (e.g. [prefix] = {(*re.*/ri/), (*un.*/ʌn/), (*in.*/ɪn/)}). A problem is that the sets of strings for onset, peak and coda are not usually readily available from dictionaries. Although the number of phoneme substrings for these sets is limited by phonotactics, deriving them manually is labour-intensive because there can be large numbers of letter substrings corresponding to these phoneme substrings. Later, we will show how these sets can be inferred automatically from examples.

For efficient processing, ϵ-transitions should be eliminated by computing the ϵ-closure, producing the transducer in Figure 5.3(b). If there are ϵ-cycles, they have to be eliminated first before removal of ϵ-transitions. A non-sequential transducer is usually obtained after eliminating the ϵ-transitions. For example, from state 2 in Figure 5.3(b) we can move to state 2, 3, 4 or 5 after a peak is read. Subset construction (Aho, Sethi, and Ullman 1986, pp. 117–121) should be carried out to obtain the sequential version.

3.2. Statistical Modelling

Undoubtedly, regular translations and alignment would massively overgenerate using these rewrite rules. The amount of overgeneration can be reduced by statistical modelling allowing us to retain only the most likely translation or alignment instead of all possible translations or alignments. Better estimation of model probabilities should lead to reduced overgeneration, although this is not necessarily the case.

Earlier probabilistic (finite-state) automata had probabilities assigned to each state transition. The behaviour of the automaton was described as a Markov chain, making a sequence of state transitions before reaching the final state. For example, a probability could be assigned to the transition from state 2 to state 3 in Figure 5.3(a). Since there is only one transition, this probability must be one.

For our model, the statistical dependencies are (usually) between the previous letter substring and/or phoneme substring instead of the previous state. For

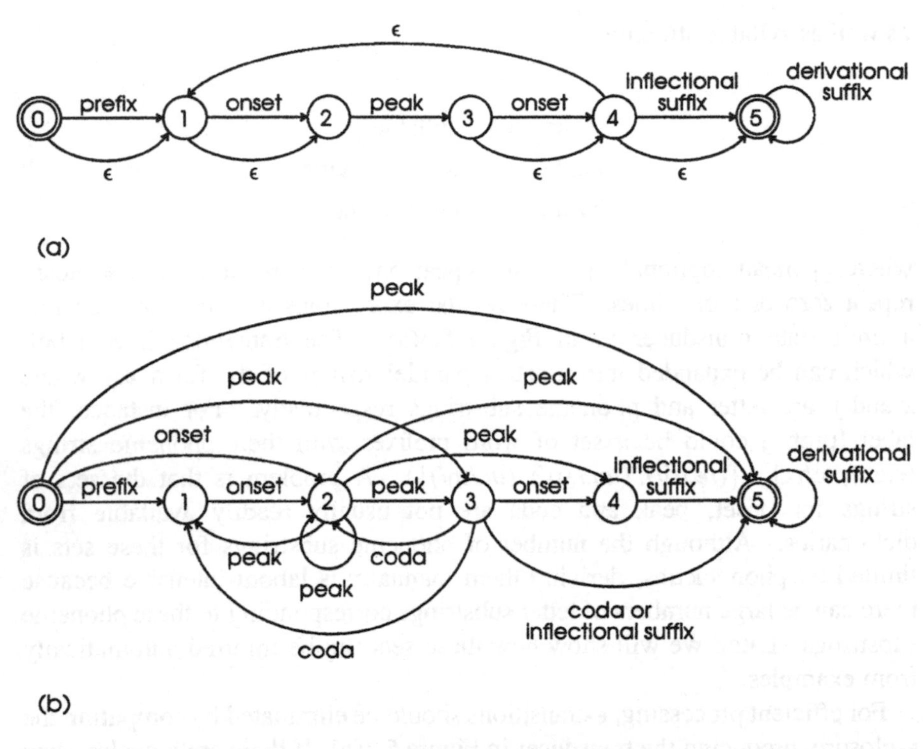

Figure 5.3. **A stochastic finite-state transducer representing the regular expression *word* in the text. Here, the regular expression stem expressed as a sequence of syllables is substituted into the regular expression *word*. The start and end states are 0 and 5 respectively. (a) shows the transducer with ϵ-transitions and (b) is the corresponding transducer without ϵ-transitions.**

example, the phonotactic constraint that the peak (*u.*/w/) must follow the given onset (*q.*/k/) is modelled by the conditional probability $p((u./w/) \mid (q./k/))$. This cannot be directly modelled by state dependency probabilities in the transducer of Figure 5.3(a) or (b). The reason is that state 2 is reached after the set of onsets (if non-empty) and not any particular onset with the particular peak (*u.*/w/). So we can only model, for example, the probability of state 3 after reading a peak conditional on state 2 after reading an onset. Even using a hidden Markov model does not improve the situation because the state dependencies remain the same.

To improve the statistical modelling, each state must associate with a letter and a phoneme substring. The dependencies between these substrings can then become the state dependencies of the transducer. For example, if $2.i$ represents reaching state 2 in Figure 5.3(b) after reading the onset (*q.*/k/) (with index i in the set of onsets) and $3.j$ is reaching state 3 after reading peak (*u.*/w/) (with index j in the set of peaks), then probability $p(3.i \mid 2.j)$ represents the probability that

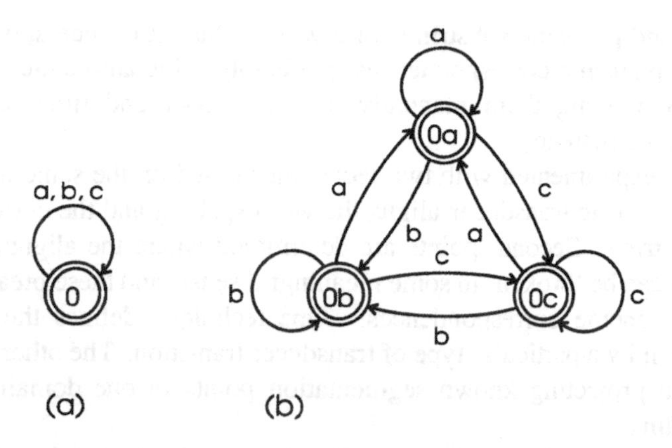

Figure 5.4. A single state transducer with self transitions in (a). The transducer in (b) is an expanded version of the transducer in (a) so that the single state 0 becomes the three states $0a$, $0b$ and $0c$.

the peak is (*u.*/w/) given that the onset is (*q.*/k/). Thus, each state in Figure 5.3(b) is a set of equivalent states and each label is a set of substrings.

Let us consider an example for clarification. Figure 5.4(a) shows a simple single-state transducer. State 0 represents a class of states $0a$, $0b$ and $0c$ of the transducer in Figure 5.4(b). The label of the transition from state 0 to itself the set of symbols a, b and c. Thus, after reading any number of a, b or c, the transducer remains in state 0. For the transducer in (b), since the three states are equivalent, as long as the transition is to one of the three states, the transducer is effectively making a self-transition as in (a). To enable Markov modelling, states $0a$, $0b$ and $0c$ represent the situation that the transducer has read symbol a, b or c, respectively. In this case, the transition probability $p(0.b|0.a)$ is simply $p(b|a)$. In this way, the transducer in (b) can be considered as a traditional probabilistic automaton as used in pattern recognition (Paz 1971). This kind of state expansion into equivalent states can be generalised for any bounded, variable-order Markov modelling because the number of equivalent states must be finite, because of bounded order and a finite alphabet (otherwise the transducer is not a finite control device).

4. Inference of Letter-Phoneme Correspondences

One of the problem with stochastic transducers is that letter-phoneme correspondences are not readily available. (Possible exceptions to this generalisation for English are Venezky 1965 and Lawrence and Kaye 1986 but their correspondences are not necessarily in the right form for our purposes.) Dictionaries give the word spelling and the corresponding phonemic string, instead of giving detailed specifications of correspondences between letter substrings and phoneme substrings of a word. Thus, it is necessary to obtain these letter-phoneme correspondences, preferably using automatic procedures rather than deriving them manually to save labour and time, as well as maintaining consistency.

We have experimented with two techniques based on the same idea. First, a simple stochastic transducer aligns the word spelling and the corresponding phonemic string. Second, points are determined where the alignment (as in Figure 5.2) can be 'broken' in some meaningful sense, and these breaks delimit the letter-phoneme correspondences. One technique defines the points of segmentation by a particular type of transducer transition. The other technique is based on projecting known segmentation points in one domain onto the other domain.

The use of a simple transducer to align input and output strings has both a theoretical and practical basis. In theory, any finite-transducer can be simulated by a single-state finite-state transducer with a self-transition at the start state. The process of inferring a formal grammar to describe parallel string generation can be considered as an expansion of this single-state transducer to more specific forms, depending on the available examples of translation. In practice, the single-state transducer is simple to implement and modify.

4.1. Stochastic Alignment

The simplest stochastic transducer M_1 for alignment and translation is the Kleene closure of the input and output alphabet, which covers all possible input and output strings. Its regular translation is:

$$L(M_1) = \#(\Sigma_i \cup \Sigma_o)^*\#$$

where, as before, # is the word boundary.

Figure 5.5(a) shows the transducer for the above regular translation. Consider an example where $\Sigma_i \cup \Sigma_o = \{a, b, c, d\}$. For Markov modelling, it is necessary to expand state 1 of the transducer in Figure 5.5(a) to the equivalent states 2, 3, 4 and 5 of the transducer in Figure 5.5(b). The implicit encoding of the states is that 2, 3, 4 and 5 represent the situations that the symbols a, b, c and d are read, respectively. Adding transitions from state 0 to states 2, 3, 4 and 5 would violate the encoding of these states because they may have read the boundary symbol # as well as a, b, c and d. This problem is simply resolved by adding an extra state 1 which indicates that the symbol read was #. The same encoding applies to state 6.

The state transition probabilities are estimated from training data. These probabilities can be viewed as entries in a table like that of Fig. 5.2. The estimation procedure makes an initial estimate of the probabilities from a set of training examples $\mathcal{E}$ (i.e. a set of word spellings and corresponding

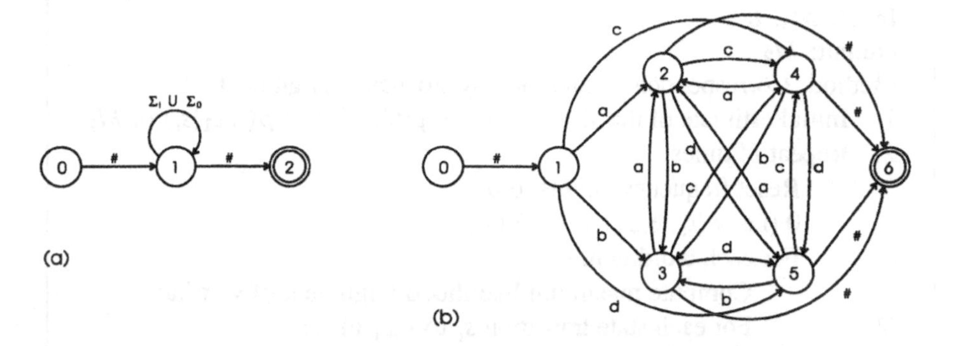

Figure 5.5. (a) A simple transducer M_1 that can perform translation of all possible input strings from the alphabet $\{a, b, c, d\}$. (b) The equivalent stochastic transducer M_2 after expansion of states based on Markov modelling.

phonemic strings). Next, the procedure repeatedly re-estimates the next set of probabilities by counting the state transitions in the maximum likelihood alignment of each word based on the current set of probabilities. Re-estimation continues for N iterations (typically 3 to 4) and terminates. Next, the maximum likelihood alignment of each word is found by dynamic programming (Bellman 1957). The segmentation points are defined to obtain the set of letter-phoneme correspondences. Algorithm SA summarises the inference procedure.

4.2. Diagonal Movement Segmentation

Transducers M_1 and/or M_2 produce alignments which consist of a sequence of alternating vertical and horizontal movements (or links) as in Figure 5.2. A letter-phoneme correspondence can be defined by one alternation of consecutive vertical (or Type 1) and horizontal (or Type 2) movements. For example, the letter-phoneme correspondences would be (*ab.*/de/) and (*c.*/f/#) in Fig. 5.2.

The intuitive interpretation is that vertical movements suggest that the strength of association between consecutive symbols of the input string is high whereas horizontal movements suggest that the association between consecutive symbols of the output string is strong. According to this interpretation, whenever there is a change of movement direction from vertical to horizontal, the input association is weak compared to the output association. For example, the association between *b* and *c* in the input string is not strong compared with # and *d*, and with *d* and *e*. Thus, the correspondences obtained should consist of a strongly-binding letter substring and a corresponding strongly-binding phoneme substring. However, this leaves out of account the strength of cross-domain binding between letters in the input substring and the corresponding phonemes in the output substring. In addition, the number of

Input: M_2, $\mathcal{E}$
Output: M_2
Method: Correspondence inference by stochastic alignment
1. Initial estimate of the state transition probabilities $p(s_{i+1}|s_i)$ of M_1
2. Repeat N times:
3. Reset frequency counts to 0
   0 (i.e. $f(s_i, s_{i+1}) = 0 \quad \forall(s_i, s_{i+1})$)
4. For each word w in $\mathcal{E}$:
5. Compute maximum likelihood alignment of word w
6. For each state transition s_i to s_{i+1} in w:
7. $f(s_i, s_{i+1}) = f(s_i, s_{i+1}) + 1$
8. Compute $p(s_{i+1}|s_i)$ using the frequency counts $f(s_i, s_{i+1})$
9. For each word w in $\mathcal{E}$:
10. Compute maximum likelihood alignment of word w
11. Determine the segmentation points on the alignment
12. Obtain letter-phoneme correspondences as alignment between consecutive segmentation points

Algorithm SA: for inferring letter-phoneme correspondences by segmenting the maximum-likelihood (stochastic) alignment of each example translation.

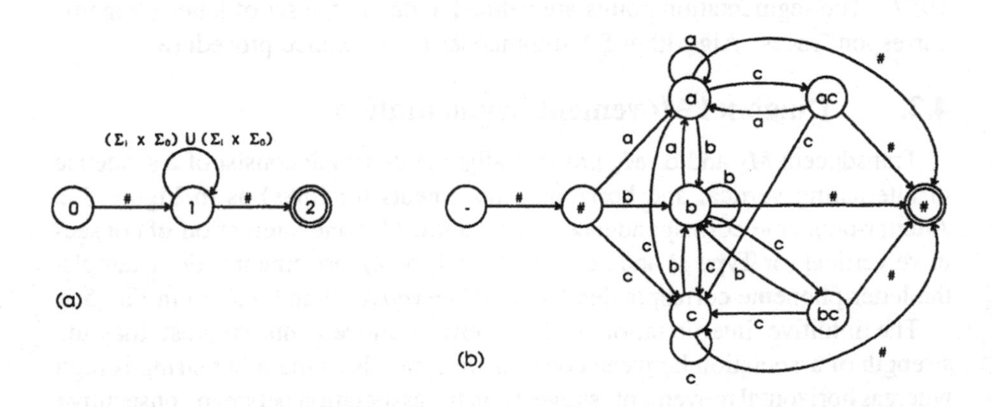

Figure 5.6. Simple (a) and equivalent forms (b) of the transducer with diagonal movement where $\Sigma_i = \{a, b\}$ and $\Sigma_o = \{c\}$.

vertical and horizontal movements may not necessarily pair up as in Fig. 5.2, leaving an output substring which is not associated with any input substring.

A remedy is to add a type of state transition that specifies how strongly input and output symbols are associated with each other. Such transitions take the form $(a./b/) \in (\Sigma_i \times \Sigma_o)$. This amounts to the transducer making

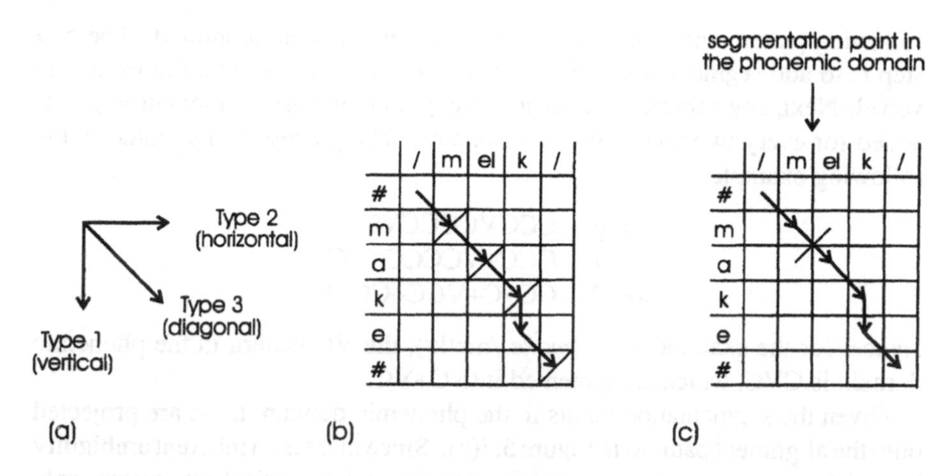

Figure 5.7. In (a), the three types of movements by the stochastic transducer are shown. Two methods are used to break the maximum likelihood alignment (represented as a sequence of arrows) and so infer correspondences. In (b), breaks are introduced at the end of Type 3 (diagonal) movements. In (c), a predefined VC pattern induces a segmentation of the phoneme string which is projected onto the alignment to break it. The segmentation is before a single vowel (as here) and after every two consecutive consonants (see text).

a simultaneous vertical and horizontal movement, which is effectively a diagonal (or Type 3) movement in the table of transducer movements, as shown in Figure 5.6(a). A form of transducer which can do this is depicted in Figure 5.6. Figure 5.7(b) shows how the maximum-likelihood alignment of the word (*make*, /meɪk/) is segmented based on Type 3 transitions. In this case, the inferred correspondences are (*m*, /m/), (*a*, /eɪ/) and (*ke*, /k/).

4.3. Projection Segmentation

The second approach is based on the intuition that vowels should be grouped with consonants appearing in the same syllable – in order to model vowel-dependency effects such as 'silent'-*e* affixation (Gimson 1980). The approach therefore predefines vowel/consonant (VC) patterns which induce segmentation boundaries when matched to the phoneme string. We work with the phoneme string since this is by definition closer to the word's pronunciation than is the letter string. For instance, the letter *y* could correspond either to a consonant or to a vowel phoneme as in the examples (*yes* /jɛs/) and (*easily*, /izəli/).

Because of computational load, we restricted the various different vowel-consonant (VC) patterns used to predefine segmentation points in the phonemic domain. Two types of VC pattern were experimented with, namely: vowel followed by a consonant sequence (likely to be the rhyme of a syllable) and consonant sequence followed by a vowel (or diphthong). The number of consonants in the pattern was also limited to at most two. For example, the following pattern shows how segmentation points (+) are identified. The first step is to add segmentation points between each consonant and the following vowel. Next, any consonant sequence longer than two has segmentation points added for every two consecutive consonants. The process is illustrated in the following example:

String: CCCVCCCCVC
Step 1: CCC+VCCCC+VC
Step 2: CC+C+VCC+CC+VC

Hence, for the example word (*make*, /meɪk/), the VC pattern in the phonemic domain is CVC, which is segmented into C+VC.

Given the segmentation points in the phonemic domain, these are projected onto the alignment paths as in Figure 5.7(c). Since there is an inherent ambiguity in projecting segmentation points onto a horizontal or vertical movement, only diagonal movements along the segmented points in the phonemic domain are allowed in the dynamic programming. Since this process leads to a large number of inferred letter-phoneme correspondences in total, correspondences which are a concatenation of others (i.e. they are decomposable) are eliminated.

5. Translation

The stochastic transducer finds the most-likely translation of an input string. The transducer reads an input symbol and produces the output symbol. In a direct implementation, the stochastic transducer finds all possible translations of the input string and their corresponding probabilities. Since the number of output strings can grow exponentially with the length of the input string, the direct implementation is not practical.

Linear processing speed can be achieved by determinising the transducer. However, unlike a finite-state automaton, not all transducers can be determinised. Fortunately, it is now known that every non-deterministic finite-state transducer can be simulated by a deterministic one (Mohri considered this as the trivial application of transducer composition) such that the set of output strings of the non-deterministic transducer is the same as the output string of the deterministic one up to some morphic mapping. Such determinisation is carried out by an extended version of subset construction but this procedure produces some 'dead' states as output, as well as some states that can be merged to save storage. By appealing to the Kleene closure of the inferred correspondences, a more efficient form of determinisation can be carried out.

For statistical modelling, expansion of states can transform the stochastic transducer into a probabilistic automaton as in pattern recognition. However, addition of new states would make determinisation and computation of the output string more costly in terms of both storage and speed because the number of state transitions would be increased as well as the number of states. Thus,

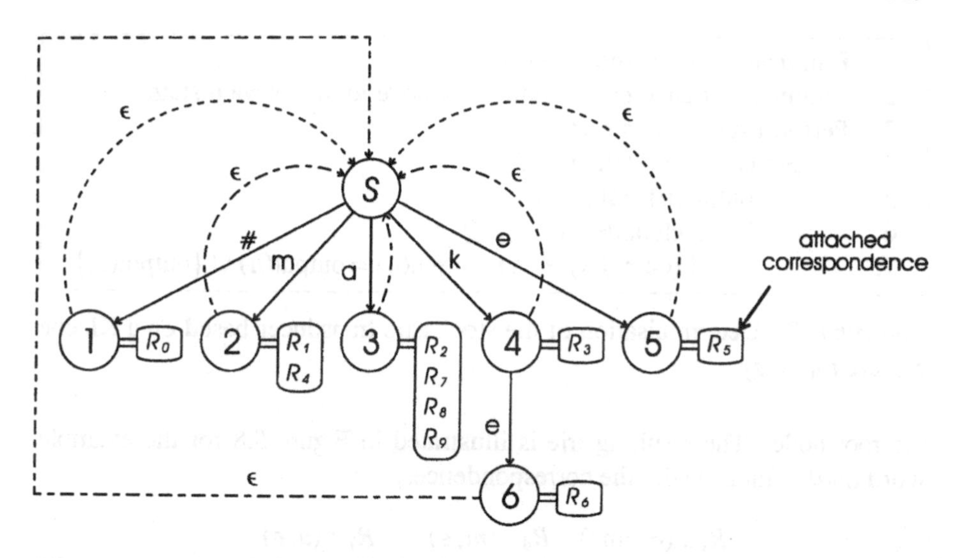

Figure 5.8. The trie constructed for the set of correspondences in (1). The correspondences attached to a state s_k are the union of all the correspondences attached to the trie nodes that belong to s_k. Correspondences enclosed in a box are attached to each node s_k if they have the same letter substring and ϵ-transitions are drawn as dotted lines.

it is preferable to compute the maximum likelihood translation and alignment using the original Markov probabilities between substrings instead of the state transition probabilities of the expanded stochastic automaton.

5.1. Determinisation

Given a set of letter-phoneme correspondences Σ, we want to find a stochastic transducer that can efficiently compute the most likely translation by concatenating these correspondences. We encode this problem so that the transducer becomes a finite automaton and standard subset construction can be used. Since we are matching the set of letter substrings ρ of the correspondences with the input string α_i, we can consider them as the pattern strings $P = \rho \in \Sigma_i^+ | \mu \in \Sigma_o^+, \rho.\mu \in \Sigma$ to match with the text string α_i in the Aho-Corasick algorithm (Aho and Corasick 1975). Thus, we need to construct a digital trie (Fredkin 1960) representing the set of letter substrings.

Whenever a substring is matched, the corresponding phoneme substrings must be output. Therefore, these phoneme substrings of correspondences are attached to nodes representing a successful match of the corresponding letter substring. In addition, once the letter substring is matched, matching must begin again from the root of the trie because the output should be a sequence of (concatenated) matched correspondences. This can be modelled by augmenting ϵ-transitions from nodes representing successful match of letter substrings to

1. Eliminate ϵ-cycles (none here)
2. Eliminate ϵ-transitions (add the root node to the relevant state)
3. Perform subset construction
4. For each new state $n \in Q_D$:
5. output(n) = $\emptyset$
6. For each node $s \in n \subseteq Q_N$:
7. if (output(s) $\neq \Lambda$), output(n) = output(n) $\cup$ {output(s)}

Algorithm D: Determinisation of the stochastic transducer based on a Kleene closure topology.

the root node. The resulting trie is illustrated in Figure 5.8 for the example word (*make*, /meɪk/) with the correspondences:

$$\begin{array}{lll} R_1 : (m, /\mathrm{m}/) & R_4 : (m, \epsilon) & R_7 : (a, \epsilon) \\ R_2 : (a, /\mathrm{eI}/) & R_5 : (e, \epsilon) & R_8 : (a, /\mathrm{ə}/) \\ R_3 : (k, /\mathrm{k}/) & R_6 : (ke, /\mathrm{k}/) & R_9 : (a, /\mathrm{æ}/) \end{array} \quad (1)$$

where ϵ is the null phoneme string. The special correspondence R_0: (#, /) is reserved for word boundaries. (Note that the orthographic word delimiter # is not counted in $|\alpha_i|$.) We can treat the digital trie with augmented ϵ-transitions as a non-deterministic finite automaton. The determinisation procedure is specified in Algorithm D.

The Aho-Corasick algorithm can convert the digital trie to a deterministic finite automaton (DFA) with a tight space bound equal to the number of nodes, $|\tau|$ in the digital trie. However, this bound cannot be applied to our digital trie with ϵ-transitions because the regular expression, R_{AC}, for matching substrings in the Aho-Corasick algorithm is different from ours, R_T:

$$\begin{aligned} R_{AC} &= \Sigma_i^*.\Sigma.\Sigma_i^* \\ R_T &= \Sigma^+ \end{aligned}$$

Fortunately, the number of states in the deterministic finite automaton is τ in practice, as discussed in Luk and Damper (1998, p. 221) and shown in the results of Section 6 later. The reason is that the inferred correspondences cover all possible single-letter substrings which effectively makes $R_T = R_{AC}$. If they did not, there would be no effective default match for some letters. A simple proof is to partition Σ into $M \cup \Sigma_i$ so that:

$$\begin{aligned} R_T &= \Sigma^+ \\ &= \Sigma^*.\Sigma.\Sigma^* \end{aligned}$$

$$= (M \cup \Sigma_i)^*.\Sigma.(M \cup \Sigma_i)^*$$
$$= \Sigma_i^*.\Sigma.\Sigma_i^*$$
$$= R_{AC}$$

where $(M \cup \Sigma_i)^+ \subseteq \Sigma_i^+$ since M is a set of letter substrings that is a subset of Σ_i^+.

5.2. Computing Translation and Alignment

The maximum likelihood translation is found by dynamic programming, i.e. the Viterbi algorithm (Viterbi 1967; Forney 1973). The algorithm works from left to right through the input string (i.e. increasing x). At each symbol position of the input string α_i, the matched correspondences R_y are emitted by the stochastic transducer. For each matched correspondence at x, the cumulative path value in a table – or trellis – with entries $T(x, y)$ is found by recursion. Assuming a first-order Markov model, the relevant recursive equation is:

$$T(x, y) = \max_t [T(s, t) + \log p(R_y | R_t)]$$

with $T(0, 0) = 0$. Here, t denotes the index of any matched correspondence at position $s = x - |\rho_y|$, and (x, y) and (s, t) are bounded as:

$$0 \leq x \leq |\alpha_i| + 1$$
$$0 \leq y \leq |\Sigma|$$

A link is made between state (x, y) and the state (s, t) that gives the maximum value of $T(x, y)$. These links can either be stored along with the table entry $T(., .)$, separately in another table, or otherwise. When the iteration has terminated, the ML translation is found from the most likely state sequence, by tracing the links from right to left along α_i starting from $(|\alpha_i| + 1, 0)$.

This process is illustrated in Figure 5.9, which shows the form of the constructed table for the word *make* using the correspondences given earlier in (1). In this example, the ML translation is /m.eɪ.k / (where, again, '.' denotes concatenation) corresponding to the (left-to-right) state sequence $\{R_0, R_1, R_2, R_6, R_0\}$.

When re-estimating the probabilities, it is necessary to find the ML alignment, which is the same as for ML translation. The only difference is that the phonemic equivalent of the letter string is known, so that the input is not simply a letter string, but an array $I(., .)$ of size $(|\alpha_i| + 2) \times (|\beta_i| + 2)$.

A three-dimensional table with entries $A(x, y, z)$ is used to store intermediate results as in Figure 5.10, where x and y denote a particular position in the input array $I(., .)$ and z denotes a particular correspondence R_x matched at $I(x, y)$ in the right-to-left and bottom-to-up direction (i.e. backwards). The algorithm starts with $x = 0$, and increments y from 0 to $|\beta_i|$ for every position x.

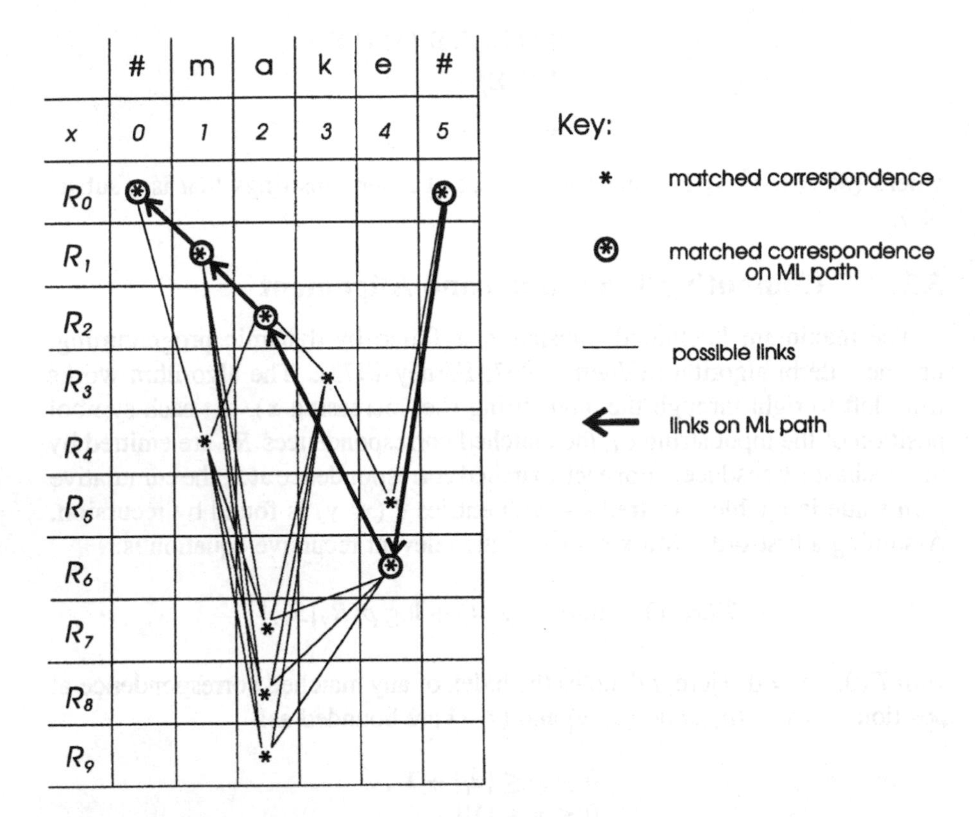

Figure 5.9. The two-dimensional trellis used to find the maximum likelihood (ML) path for the word make given the correspondence listed in (1). The matched correspondences are the states of the trellis, the possible links are the allowed state transitions in the trellis, and the ML path is found by dynamic programming, giving the pronunciation /m.eɪ.k/.

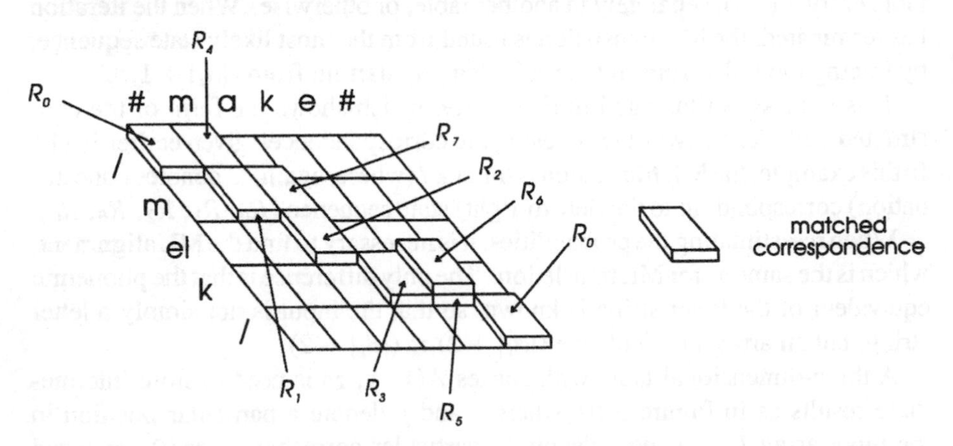

Figure 5.10. The maximum likelihood alignment is found using the three-dimensional table illustrated here for the word (*make*, /meɪk/). Possible links are omitted for clarity.

The ML alignment for R_z at (x, y) is defined (for the Markov model) according to the recursive equation:

$$A(x, y, z) = \max_y [A(r, s, t) + \log p(R_z|R_t)] \quad \begin{cases} 0 \le x \le |\alpha_i| + 1 \\ 0 \le y \le |\beta_i| + 1 \\ 0 \le z \le |\Sigma| \\ A(0, 0, 0) = 0 \end{cases}$$

where t indexes a particular correspondence, $r = x - |\rho_z|$ and $s = y - |\mu_z|$. A link is made between (x, y, z) and that (r, s, t) which yields the ML path value at $A(x, y, z)$. The ML alignment of the word is found by tracing the links back from $A(|\alpha_i| + 1, |\beta_i| + 1, 0)$.

5.3. Efficiency of Transduction

In this subsection, we analyse the complexity of the transduction to show that a straightforward implementation of Viterbi decoding has quadratic time complexity in the number of state survivors, ψ, which have to be considered. By precompiling the set of correspondences into a deterministic finite-state automaton, as above, we can reduce the number of state survivors which need to be considered, and so increase the speed of transduction considerably.

Computational complexity. The following discussion only identifies the major iterations of the algorithm where the problem size is dependent on the number of correspondences and the length of the input string. Additive constants are not included, but multiplicative constants are.

The algorithm iterates for every position x along α_i. At each of these $|\alpha_i| + 2$ positions (except the first), we suppose that it is necessary in the worst case to examine for matches all the $\psi = |\Sigma| + 1$ correspondences. In essence, the process is one of matching with each δ_y in the *backwards* direction the letter substring in the input ending with the letter at x. So, the number of iterations for matching grows in the worst case as:

$$I_M \sim |\alpha_i|\psi L \tag{2}$$

where L is the length of the longest letter substring in Σ.

For each successful such match, we then consider possible links to the previously-successful matches in column $(x - |\delta_y|)$ of the trellis in Figure 5.9, before computing the locally-optimal move to (x, y) which is then stored as the actual link for that cell. Hence, for each cell of the table, we have to examine all correspondences, and the number of iterations, I_L, for computing the links grows in the worst case as:

$$I_L \sim |\alpha_i|\psi^2 \tag{3}$$

In addition, to the matching of correspondences, the algorithm has to clear all the table entries $T(.,.)$ for the next translation. Hence, the number of iterations for clearing, I_C, is quadratic in the worst case:

$$I_C \sim |\alpha_i|\psi \tag{4}$$

So, combining equations (2)–(4), the total CPU time is estimated as:

$$\begin{aligned} T &\sim C_M I_M + C_L I_L + C_C I_C \\ &\sim C_1|\alpha_i|\psi^2 + C_2|\alpha_i|\psi \end{aligned} \tag{5}$$

where the Cs are appropriate constants and L has been absorbed into C_2.

Considering a word W_i as input, $|\alpha_i| << \psi$ in practice so that the complexity could be regarded as quadratic in the number of correspondences. In a text-to-speech system, however, the Viterbi algorithm has to be applied for every word in the input text, and the number of words is indeterminately large. In many situations – such as obtaining correspondences and estimating their transition probability distributions – the algorithm is applied for every word in a large training corpus. Accordingly, there is a case for regarding the complexity of matching as cubic, which tells against practical deployment of the algorithm. This motivates the development of the fast implementation using a deterministic finite-state automaton (DFA).

Since the table used to store the entries $T(.,.)$ has dimensions $|\alpha_i| + 2$ and ψ, and the number of possible state-transition probabilities is ψ^2, the required storage space S in the worst case is quadratic:

$$S \sim C_3|\alpha_i|\psi + \psi^2$$

where C_3 is a constant whose size depends upon the storage requirements for the table entries $T(.,.)$, which are cumulative log probabilities, possibly also including the locally-optimal link.

Measurement model. The obvious way to measure the speed-up is to compare CPU times for the fast (using a DFA) and straightforward implementations of stochastic transducers. Our attempts to measure speed-up directly in this way, however, were quickly abandoned in view of the very long execution times for the latter (even with small sets of correspondences and dummy transition probabilities). Instead, we decided to predict the speed-up indirectly.

The time complexity depends on the number of states emitted (i.e. correspondences considered as state survivors, or possible contributors to the ML path) per input symbol read. In the straightforward Viterbi implementation, we consider all correspondences at each letter position x linking from $x - |\rho_t|$ with t ranging over ψ. This is performed for each of the ψ potential state survivors at x, so that the number of computations is ψ^2. With the fast implementation, however, the number of state survivors at x reduces to some value, f say, since only relevant states survive. Further, if we use a linked-list implementation of the Viterbi trellis, instead of considering all ψ possible linking values of t, we again only need to consider f states. Hence, the total number of computations for optimal linking reduces to f^2.

For different input strings, f varies and therefore it is treated here as a statistical variate. Our approach is to seek a statistical measure of f, the number of correspondences associated with the states of the DFA, which is highly correlated with the observed CPU time. The correlation should allow us to predict the CPU time needed for ML translation and alignment, given that we know certain characteristics of the set of correspondences. The statistical measure considered here is the mean number of correspondences associated with the states of the DFA, $\overline{f}$.

Let $f_{k,\Sigma}$ be a random variable that denotes the number of states emitted per symbol read – where k is an index of the symbol and Σ indexes the set of correspondences employed. Considering both ML translation and alignment, the total number of symbols involved in the computation (Fig. 5.10) is:

$$n = \sum_i (|\alpha_i| + 2)(|\beta_i| + 2)$$

where i is the index of all words in the training and/or test sets. We regard translation and alignment as sampling processes of size n from an underlying distribution of the number of states emitted per symbol. The distribution is dependent on the particular set of correspondences used, which yield different means and variances for $f_{k,\Sigma}$.

According to equation (5), the total CPU time for the straightforward implementation is a quadratic in the number (ψ) of states emitted. Hence, we model the CPU time for the fast implementation as a quadratic in the (reduced) number of states emitted. Taking account of an additive constant A:

$$T_\Sigma = C_1 \sum_k f_{k,\Sigma}^2 + C_2 \sum_k f_{k,\Sigma} + A \tag{6}$$

where k ranges from 1 to n.

Since n is large in practice:

$$\sigma_\Sigma^2 \sim \frac{\sum_k \left(f_{k,\Sigma} - \overline{f_\Sigma}\right)^2}{n}$$

$$= \frac{\sum_k f_{k,\Sigma}^2 - 2\sum_k f_{k,\Sigma}\overline{f_\Sigma} + \sum_k \overline{f_\Sigma}^2}{n}$$

$$\sim \frac{\sum_k f_{k,\Sigma}^2 - \sum_k \overline{f_\Sigma}^2}{n} \qquad \text{assuming} \sum_k f_{k,\Sigma}\overline{f_\Sigma} \sim \sum_k \overline{f_\Sigma^2}$$

$$\text{and} \quad \sigma_\Sigma^2 \sim \frac{\sum_k f_{k,\Sigma}^2}{n} - \overline{f_\Sigma}^2$$

Assuming further that the standard deviation does not vary significantly across different sets of correspondences, $\sigma_\Sigma = \sigma$, then:

$$\sum_k f_{k,\Sigma}^2 = n\left(\sigma^2 + \overline{f_\Sigma}^2\right)$$

Substituting this and $n\overline{f_\Sigma} = \sum_k f_{k,\Sigma}$ into equation (6) yields:

$$T_\Sigma = C_1 n \overline{f_\Sigma}^2 + C_2 n \overline{f_\Sigma} + A' \qquad (7)$$

where $A' = A + C_1 n \sigma^2$ (i.e. another additive constant).

In this work, we use the mean number of emitted states per symbol, $\overline{f_\Sigma}$, as the predictor of the speed-up according to equation (7). To examine its adequacy, we fit quadratic polynomials using regression analysis to the total CPU time spent in translation and alignment and examine the obtained correlation.

6. Results

Three statistical models (independent, Markov and hidden Markov) were used for ML translation and alignment for the approximately 20,000 words in the Oxford Advanced Learner's dictionary. This was repeated 3 times (i.e. 60,000 ML translations and alignments) to reduce the sensitivity of the results to extraneous operating-system conditions.

Table 5.1 shows various statistics and measures associated with eight sets of correspondences studied as well as the speed-up factor Ω. As identified in the table, these sets of correspondences were:

Ains: the manually-compiled context-dependent rewrite rules of Ainsworth (1973) with the context-dependency simply ignored.

Elo: the manually-compiled context-dependent rewrite rules of Elovitz et al. (1976), again with the context-dependency ignored.

LK: the manually-compiled correspondences tabulated by Lawrence and Kaye (1986). In the small number of cases where the correspondences were context-dependent, the context-dependency was again ignored.

Table 5.1. Complexity measures for different sets of correspondences and their associated automata. See text for explanation of symbols and specification of sets of correspondences.

Set	ψ	$\lvert Q\rvert = \tau$	$\overline{f_\Sigma}$	λ
Ains	161	86	5.77	195
Elo	311	300	6.40	687
LK	571	254	9.08	1788
DD	343	202	11.4	512
HMM	1083	646	18.9	2566
LK+Aff	856	686	9.33	3918
DD+Aff	658	626	12.0	1775
GST	1173	28	41.9	742

DD: the correspondences automatically inferred by the dynamic programming and delimiting method of Luk and Damper (1992).

HMM: the set inferred on the basis of hidden Markov statistics by Luk and Damper (1993b).

LK+Aff: the Lawrence and Kaye correspondences with added affixes as in Luk and Damper (1993c).

DD+Aff: the DD set with added affixes as above.

GST: the correspondences automatically inferred by the generalised stochastic transducer method by Luk and Damper (1993b).

In all cases, the number of states of the DFA, $|Q|$, was equal to the number of nodes, $|\tau|$, of the trie, as mentioned in Section 5.1 above.

Best-fit quadratics were obtained for the three statistical models by quadratic regression based on equation (7), i.e. treating the CPU time as a measure of T_Σ. Our regression analysis shows that the quadratic term in the equation dominates the linear term. For instance, the Markov statistical model yielded:

$$T = 0.6499\overline{f_\Sigma}^2 + 0.086\overline{f_\Sigma} + 56.2$$

Luk and Damper (1998, p. 223) detail statistical tests which reveal that the C_2 coefficient can be ignored in relation to C_1 (here it is about 8 times smaller), but the constant term A can be significant when $\overline{f_\Sigma}$ is small. Hence, a reasonable estimate of the speed-up is:

$$\lambda = \frac{C_1 n \psi^2 + A}{C_1 n \overline{f_\Sigma}^2 + A}$$

taking $C_1 = 0.6$ and $A = 60$. The λ values in Table 5.1 range between 195 and 3918, indicating that a very good speed-up can be expected from use of the fast DFA implementation, although the exact gain is dependent upon the set of correspondences used.

6.1. Performance Measures

We want to evaluate the prediction by stochastic transducers of the pronunciation of a word given its spelling. Hence, the main determinant of performance is pronunciation accuracy. However, letter-phoneme alignment of words when both spelling and pronunciation are known is an important component of the training process. For this reason, we include alignment performance in the evaluation.

There are four performance measures:

1 % align is the percentage of words in the set that can be aligned. (There is, of course, no guarantee that any set of inferred correspondences is sufficient to align all words). Note that this is not a measure of correct alignment, since such an alignment is not well defined.

2 % words correct is the word translation accuracy evaluated according to the number of distinct words ND in the set.

3 % symbols (aligned) is the symbol translation accuracy computed as the Levenshtein distance (Levenshtein 1966) between the obtained and correct phoneme strings, according to the number of words that can be aligned.

4 % symbols (total) is the symbol translation accuracy computed as the Levenshtein distance between the obtained and correct phoneme strings, according to the number of words NT in the set.

6.2. Training and Test Data

All words used for training and evaluation are transcribed as British Received Pronunciation, with a phoneme inventory of size 44 as used in the Oxford Advanced Learners Dictionary (Hornby 1974). The primary stress symbol was ignored: in effect, we leave the problem of stress assignment for later study. The dictionary was split into two disjoint parts: the training data (set A) with 18,767 distinct words, and an unseen test set (B) with 1667 distinct words. With this training set, of the order of 2500 distinct correspondences were inferred by the VC-pattern segmentation method described in Section 4.3 above. These were compacted in a set of approximately 1800 undecomposable correspondences. Many fewer correspondences (just over 1000) were inferred by the diagonal movement segmentation method described in Section 4.2, and these were used uncompacted.

Table 5.2. Number of words in the training and test material. Word set A is training data but all sets A to F are used as test data. See text for further specification.

	A	B	C	D	E	F
Total words, N_T	20,704	1667	174	1694	479	134
Distinct words, N_D	18,767	1667	165	1681	462	134

Using the test data, we can measure the extent to which regularities of pronunciation are captured by the particular stochastic transducers. To determine if the transducers have wider applicability, we tested them with proper nouns (Hornby 1974) – country names (set C), surnames (D) and forenames (E) – as well as novel, or pseudowords (set F) (Glushko 1979; Sullivan and Damper 1993) produced by changing the initial consonant clusters of high-frequency, monosyllabic dictionary entries.

Table 5.2 shows the size of each set, in terms of the total number of words and the number of distinct words. Not all words in the dictionary had a unique pronunciation. For instance, *eager* appears as /iːgə(r)/, i.e. with /r/ variable – attempting to cover both rhotic and non-rhotic dialects. In such cases, both pronunciations were included in the training data with either considered a correct output.

6.3. Performance Comparison

Here we compare performance on the 4 measures enumerated above for the 2 different ways of inferring the correspondences. To provide a baseline for the comparison, we also include in the evaluation the set of correspondences manually derived by Lawrence and Kaye (1986), previously denoted LK, which we believe to have been handcrafted to a high standard (although for a different purpose). Transition probabilities for the LK correspondences were estimated from word set A as for the other correspondences. A small number of the LK correspondences is actually contest-dependent. For the purpose of this evaluation, we have simply ignored any such context dependency in the correspondences (although context effects are, of course, handled implicitly in stochastic transducers).

Figure 5.11(a) shows the % align performance measure evaluated for the three different sets of correspondences/probabilities on word sets A to F. Alignment can be achieved for about 99.7% of the dictionary words (sets A and B) for both inference algorithms (i.e. for both ways of inferring correspondences). By contrast, the manually derived LK correspondences perform less well for these words, failing to approach 100%. For the remaining word sets, alignment performance is rather variable. The general superiority of the LK correspondences on personal names (sets D and E) is not surprising in view

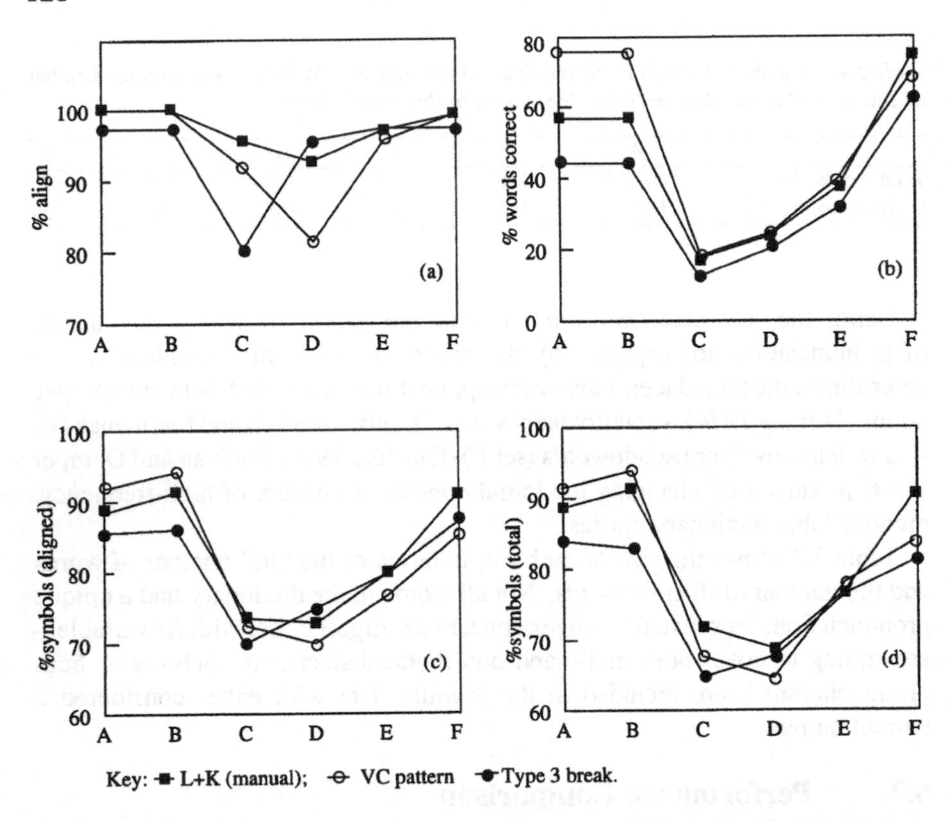

Figure 5.11. Comparison of stochastic transducer performance for word sets *A* to *F* with three different sets of correspondences: those inferred by breaking Type 3 transitions; those inferred on the basis of a predefined VC pattern; and the manually derived correspondences of Lawrence and Kaye. Performance measures are: (a) percentage of words which can be aligned; (b) word score for the distinct words; (c) symbol score for the alignable words; and (d) symbol score for the total number of words.

of the fact that Lawrence and Kaye (1986, pp. 160–165) explicitly considered words like *John*, *Byron*, *Michael*, *Lincoln*, ... in their work. The poor showing on country names is less understandable given that they also considered *Laos*, *Belgium*, etc.

Figures 5.11(a), (b) and (c) reveal that the correspondences inferred from VC patterns yield much better translation accuracy than those found by breaking at Type 3 transitions, for both training and unseen dictionary words (sets *A* and *B*). Thus, % words correct exceeds 70% for the VC pattern correspondences while it is less than 60% for the Type 3 break correspondences, and less than 50% for the LK correspondences. Similarly, both % symbols (align) and % symbols (total) exceed 90% by some margin for the VC pattern correspondences but are only around 90% for the Type 3 break correspondences, and are at around 80–85% for the LK correspondences. The two different sets of inferred correspondences both seem to capture the regularities of English pronunciation well as there is little difference in performance between the training and unseen data, sets *A* and *B* respectively, although the latter set is considerably smaller, of course.

As expected, the performance on proper nouns (sets *C*, *D* and *E*) is poor for both automatically-learned and manually-derived correspondences, indicating that the spelling-sound relations for these words are sometimes rather different from those for the dictionary entries. Hence, special techniques are required to derive pronunciations for unknown proper nouns in text-to-speech systems (e.g. Coker, Church, and Liberman 1990; Andersen and Dalsgaard 1994). The LK correspondences do as well as, or only slightly worse than, the automatically-inferred correspondences on these words, whereas they do much worse on the dictionary entries. This is no doubt a result of Lawrence and Kaye's explicit consideration of such words, as described above. Performance on novel (pseudo-) words (set *F*) is generally good, as one would expect from the small size of the set and the fact that words were directly derived from high-frequency, monosyllabic words.

The superior performance on the dictionary data of the correspondences inferred from VC patterns indicates the importance of supplying even quite limited prior knowledge about the structure of English words to the inference process. On the other hand, these correspondences give lower performance for pseudowords. A possible interpretation is that novel words might not embody the same kind of vowel/consonant structural information as the dictionary entries. It would be surprising, however, if this were the case: changing consonant clusters should not be too disruptive of VC patterning.

6.4. Errors

We have not so far examined and classified the kinds of errors produced by the current stochastic transducers. In previous work which included use of the hidden Markov model transducer (Luk and Damper 1993a), however, it was found that common classes of error were as follows:

1. Incorrect vowel changes in root morphemes due to silent-*e* markings and affixation.
2. Substitution of the high-probability schwa for lower-probability non-reduced vowels and diphthongs.
3. Vowel deletion (not surprisingly, as short strings are more probable).

The last two of these are understandable errors for a probabilistic system to make. To some extent, such errors can be ameliorated by requiring that the translated phonemic form of a word contains at least one non-reduced vowel. However, this is *ad hoc*, inconsistent with the stochastic transduction model, and would in any case only help with the shorter words.

The problems with affixation arise because the stochastic transducer has too little prior knowledge of morphemic structure. Some simple experiments were performed (Luk and Damper 1993c) in which we started with an explicit set of affix correspondences, used these to strip affixes (using the delimiting algorithm in Luk and Damper (1992) to yield (sometimes incorrect) roots and then applied the inference algorithm to these. This led to a statistically significant improvement in % words correct of some 10 percentage points (e.g. from 48.4% to 59.9% for word set *B* using the LK correspondences).

7. Conclusions

This chapter has reviewed the use of stochastic transducers for English letter-to-phoneme conversion. Stochastic transduction is related to formal language theory and statistical learning. That is, the spellings and pronunciations of English words are modelled as the productions of a stochastic grammar, inferred from example data in the form of a pronouncing dictionary, using a stochastic transducer to align word spellings and pronunciations. For efficient translation, these productions are converted into the finite control of a stochastic transducer to compute the ML translation. Although the formalism is in principle very general, restrictive assumptions are necessary if practical, trainable systems are to be realised, as Vidal (1994) says of finite-state transducers: "... in fact, no method is known so far to approach the learning of these devices". Since he is referring to algorithms that identify any finite-state transduction, stochastic transducers are also not learnable in this sense. This emphasises the necessity of making restrictive assumptions, even for finite-state models.

In the work reported here, we address the problem of learnability by assuming a restricted form of regular grammar whose language is the Kleene closure, Σ^*, of the set of correspondences Σ. This is sometimes called a universal grammar since it generates/accepts all strings in Σ^*. Such a restricted grammar would be of no practical interest were it not for the addition of rule probabilities to control overgeneration. Hence, the grammar is stochastic and its rewrite rules simply concatenate terminal symbols, which are identical to letter-phoneme correspondences. The rule probabilities are estimated from first-order transition counts in the training data. Thus, the formalism is a Markov model rather than a hidden Markov model. In early work (Luk and Damper 1993a), an HMM approach performed less well than the Markov approach adopted here when identically trained and evaluated. One possible reason for this is that there may be fewer distinct transitions from correspondence to correspondence (and so a higher and more statistically stable count for each one) than there are transitions between phoneme strings plus emissions to particular letter strings. Another possibility is that the Markov approach may avoid some of the problems of duration modelling with HMMs.

The current status of stochastic transducers has been defined by presenting our latest results on text-phoneme alignment and transduction for this (highly) simplified model. Best results are obtained if structural information concerning vowel-consonant patterning is supplied to the inference algorithm. We have compared stochastic transducer performance with automatically inferred and manually-derived correspondences. This has shown that the data-driven approach significantly outperforms the specific expert/knowledge-based approach of Lawrence and Kaye (1986). In interpreting this finding however, it must be remembered that the automatically-inferred correspondences were learned in a setting consistent with the transduction model, and this is not the case for the LK correspondences.

We have not yet evaluated the performance of the stochastic transduction model in comparison with other data-driven approaches to automatic pronunciation. Recently, Damper, Marchand, Adamson, and Gustafson (1999) compared typical linguists' rules and three different data-driven techniques on the same (approximately 16,000 word) dictionary. Although the comparison was complicated by practical issues such as the different transcription standards used by the rule-writers and the compilers of the dictionary, and the necessity to use some of the dictionary data for training the data-driven systems (meaning that it could not then be used as test data), a clear and enormous superiority of the data-driven techniques over the rules was found. For the data-driven methods, it appeared that there was a clear performance advantage to retaining the training data (i.e. the dictionary) in its entirety rather than attempting to compress it, for instance into a set of neural-network connection weights. This is consistent with the recent findings and arguments of Daelemans, van den Bosch, and Zavrel (1999) to the effect that "forgetting exceptions is harmful in language learning". From this perspective, we might expect that there would be some loss of performance with respect to analogy or memory-based methods, because the stochastic transduction approach attempts to learn a compact representation in terms of probabilities estimated from first-order transition counts in the training data. But this speculation remains to be confirmed.

Finally, we believe that stochastic transduction can be used with advantage at a variety of levels in text-to-speech conversion. Thus, as well as a performing letter-to-phoneme transduction as here, we could incorporate within our data-driven formalism (stochastic) morphemic and syllabic grammars based on sequences of pairwise transitions.

Chapter 6

SELECTION OF MULTIPHONE SYNTHESIS UNITS AND GRAPHEME-TO-PHONEME TRANSCRIPTION USING VARIABLE-LENGTH MODELING OF STRINGS

Sabine Deligne
IBM T. J. Watson Research Center, Yorktown Heights

François Yvon
Départment Informatique, ENST, Paris

Frédéric Bimbot
IRISA, Rennes

Abstract Language can be viewed as the result of a complex encoding process which maps a message into a stream of symbols: phonemes, graphemes, morphemes, words ... depending on the level of representation. At each level of representation, specific constraints like phonotactical, morphological or grammatical constraints apply, greatly reducing the possible combinations of symbols and introducing statistical dependencies between them. Numerous probabilistic models have been developed in the area of speech and language processing to capture these dependencies. In this chapter, we explore the potentiality of the multigram model to learn variable-length dependencies in strings of phonemes and in strings of graphemes. In the multigram approach described here, a string of symbols is viewed as a concatenation of independent variable-length subsequences of symbols. The ability of the multigram model to learn relevant subsequences of phonemes is illustrated by the selection of multiphone units for speech synthesis.

R. I. Damper (ed.), *Data-Driven Techniques in Speech Synthesis*, 125-147.

1. Introduction

Written or spoken utterances of natural language can be viewed, at some level of representation, as the result of a complex encoding process which maps a message into a stream of symbols. Depending on the level of representation chosen, these symbols correspond to linguistic units such as phonemes, graphemes, syllables, morphemes, words, or even higher-level groups (syntagms, rhythmic groups, etc.). Traditional accounts of these levels assume some kind of hierarchical organization, where each unit of a given level is composed of several units of the previous one. These linguistic units are of variable-length: a syllable contains a variable number of phonemes, a word a variable number of graphemes, and so on.

At each level of representation, specific constraints apply, greatly reducing the possible sequences of symbols: phonotactical constraints restrict the possible successions of phonemes, syntactical constraints rule out invalid sequences of words, etc. These constraints are responsible for a substantial degree of redundancy in symbolic natural language representations, such as word sequences or phonemic strings. For instance, in the phonemic transcription of a conversation, not all phonemes are equally likely, nor are their two-by-two combinations (bigrams), their three-by-three combinations (trigrams), etc.

This redundancy is partly captured by probabilistic models, among which the most popular and successful in the field of language engineering is certainly the n-gram model (Jelinek 1990). This model, however, relies on the assumption that the probability of any given linguistic symbol depends only on a fixed number n of its predecessors, with n being determined *a priori* and supposed constant over the whole stream of symbols. One undesirable consequence of this assumption is that when n is small, the long-term dependencies tend to be improperly captured by the model. Increasing n, however, increases exponentially the number of parameters required for the model, which makes their reliable estimation extremely difficult.

By contrast, the n-multigram model, initially proposed by Bimbot, Pieraccini, Levin, and Atal (1994), is based on the hypothesis that dependencies between symbols span over variable lengths, from 0 (independence) up to length n. Other work along these lines includes the variable-length sequence models proposed by Riccardi, Pieraccini, and Bocchieri (1996) or by Hu, Turin, and Brown (1997).

In this chapter, we report on a number of experiments conducted with this model in the domain of speech synthesis. In Section 2, we introduce the formalism of the n-multigram model, and provide theoretical results concerning the estimation of the model parameters. We then present (Section 3) a first application of the model, which consists of selecting automatically, from a corpus, variable-length speech units for concatenation synthesis systems. A multi-dimensional extension of the n-multigram model is then applied to the problem of converting an orthographic text into the corresponding sequence of phonemes (Section 4). We finally discuss in Section 5 the applicability of this model to other related tasks in the area of natural language engineering.

2. Multigram Model

This section introduces the theoretical background for the n-multigram approach. We first formulate the model in the most general case. We then address the problem of parameter estimation under the assumption that successive sequences are independent. A final subsection extends the one-dimensional model to the case where the observed stream comprises several parallel strings of symbols.

2.1. Formulation of the Model

Under the n-multigram approach, a string of symbols is assumed to result from the concatenation of shorter (sub-) sequences, each having a maximum length of n symbols. We denote as $\mathcal{D}_O = \{o_i\}$ the dictionary from which the emitted symbols are drawn, and by $O = o_{(1)}o_{(2)}\ldots o_{(T_O)}$ any observable string of T_O symbols. Let L denote a possible segmentation of O into T_S subsequences of symbols. We denote as S the resulting string of subsequences: $S = (O,\ L) = s_{(1)}, s_{(2)}\ldots s_{(T_S)}$. The dictionary of distinct subsequences, which can be formed by combining $1, 2, \ldots$ up to n symbols from $\mathcal{D}_O$ is denoted $\mathcal{D}_S = \{s_j\}$. Each possible segmentation L of O is assigned a likelihood (probability) value, and the overall likelihood of the string is computed as the sum of the likelihoods for each segmentation according to:

$$\mathcal{L}(O) = \sum_{L \in \{L\}} \mathcal{L}(O,\ L) \tag{1}$$

To illustrate, the likelihood of a 4-symbol string $abcd$ computed with a 3-multigram model, assuming that the subsequences in each segmentation are independent, is:

$$\mathcal{L}(O) = \left\{ \begin{array}{l} p([a])p([b])p([c])p([d]) + \\ p([a])p([b])p([cd]) + \\ p([a])p([bc])p([d]) + \\ p([a])p([bcd]) + \\ p([ab])p([c])p([d]) + \\ p([ab])p([cd]) + \\ p([abc])p([d]) \end{array} \right.$$

The decision-oriented version of the model parses O according to the most likely segmentation, thus yielding the approximation:

$$\mathcal{L}^*(O) = \max_{L \in \{L\}} \mathcal{L}(O, L)$$

To complete the definition of the model, we need a way to compute $\mathcal{L}(O, L)$ for any particular segmentation L. This amounts to giving the model a parametric expression, and providing estimation procedures for the parameters Θ. There exist many ways in which this model could be parameterized, each corresponding to different modeling of the dependencies between successive subsequences in S. In the rest of the chapter, we assume that successive subsequences are statistically independent. Although sometimes unrealistic, this simplification greatly reduces the number of parameters of the model, and achieves satisfactory results in many situations.

Other forms of dependencies can also be expressed in the multigram framework. Some have already been proposed (Deligne 1996), corresponding to more constrained models of the space of possible strings.

2.2. Estimation of Multigram Parameters

Again, assuming that the multigram subsequences are independent, the likelihood $\mathcal{L}(O, L)$ of any segmentation L of O can be expressed as:

$$\mathcal{L}(O, L) = \prod_{t=1}^{T_S} p(s_{(t)}) \tag{2}$$

and the likelihood of O is computed as the sum over all possible segmentations (see equation (1)).

An n-multigram model is consequently defined by the set of parameters Θ, consisting of the probabilities of all sequences s_i in $\mathcal{D}_S : \Theta = \{p(s_i)\}$, with $\sum_{s_i \in \mathcal{D}_S} p(s_i) = 1$.

An estimation of the parameter set Θ from a training corpus O can be stated as a maximum likelihood (ML) estimation from incomplete data (Dempster, Laird, and Rubin 1977), where the observed data are given by the string of symbols O, and the unknown data correspond to the underlying segmentation L. Thus, iterative ML estimates of Θ can be computed using an EM algorithm (Deligne 1996). Let $Q(k, k+1)$ be the following auxiliary function, computed with the likelihoods at iterations (k) and $(k+1)$:

$$Q(k, k+1) = \sum_{L \in \{L\}} \mathcal{L}^{(k)}(O, L) \log \mathcal{L}^{(k+1)}(O, L)$$

It has been shown by Dempster, Laird, and Rubin (1977) that provided $Q(k, k+1) \geq Q(k, k)$, then $\mathcal{L}^{(k+1)}(O) \geq \mathcal{L}^{(k)}(O)$. The set of parameters $\Theta^{(k+1)}$ which maximizes $Q(k, k+1)$ at iteration $(k+1)$ also leads to an increase of the corpus likelihood. Therefore, the re-estimation formula of $p^{(k+1)}(s_i)$, i.e. the probability of subsequence s_i at iteration $(k+1)$, can be derived directly by maximizing the auxiliary function $Q(k, k+1)$ over $\Theta^{(k+1)}$, under the constraint that all parameters sum up to one. Denoting by $c(s_i, L)$ the number of occurrences of the subsequence s_i in a segmentation L of the corpus, we rewrite the joint likelihood given in equation (2) so as to group together the probabilities of all identical subsequences:

$$\mathcal{L}^{(k+1)}(O, L) = \prod_i \left(p^{(k+1)}(s_i)\right)^{c(s_i, L)}$$

The auxiliary function $Q(k, k+1)$ can then be expressed as:

$$Q(k, k+1) = \sum_i \sum_{L \in \{L\}} \mathcal{L}^{(k)}(O, L) c(s_i, L) \log p^{(k+1)}(s_i)$$

which, as a function of $p^{(k+1)}(s_i)$ subject to constraints $\sum_i p^{(k+1)}(s_i) = 1$ and $p^{(k+1)}(s_i) \geq 0$, is maximized for:

$$p^{(k+1)}(s_i) = \frac{\sum_{L \in \{L\}} c(s_i, L) \times \mathcal{L}^{(k)}(O, L)}{\sum_{L \in \{L\}} c(L) \times \mathcal{L}^{(k)}(O, L)} \tag{3}$$

where $c(L) = \sum_i c(s_i, L)$ is the total number of subsequences in L. Equation (3) shows that the estimate for $p(s_i)$ is merely a weighted average of the number of occurrences of subsequence s_i within each segmentation. Since each iteration improves the model in the sense of increasing the likelihood $\mathcal{L}^{(k)}(O)$, it eventually converges to a critical point (possibly a local maximum). The re-estimation equation (3) can be implemented by means of a forward-backward algorithm (Deligne and Bimbot 1995).

A decision-oriented procedure can readily be derived from the re-estimation formula by taking into account only the most likely segmentation, $L^{*(k)}$, at iteration (k):

$$p^{(k+1)}(s_i) = \frac{c(s_i, L^{*(k)})}{c(L^{*(k)})} \tag{4}$$

The probability of each s_i is thus simply re-estimated as its relative frequency along the best segmentation at iteration (k), using a Viterbi algorithm (Viterbi 1967; Forney 1973). This procedure parses the corpus according to an ML criterion and can also be used during the estimation phase, as is done by Bimbot, Pieraccini, Levin, and Atal (1994, 1995).

The set of parameters Θ is initialized with the relative frequencies of all co-occurrences of symbols up to length n in the training corpus. The set Θ is then iteratively re-estimated until the training set likelihood does not increase significantly, or a fixed number of iterations is reached.

In practice, some pruning technique may be advantageously applied to the dictionary of subsequences, to avoid overlearning (Deligne 1996). A straightforward way to proceed consists of discarding, at each iteration, the least likely subsequences, i.e. those with a probability falling under a prespecified threshold.

2.3. Joint Multigram Model

The multigram model easily generalizes to the *joint* multigram model (Deligne, Bimbot, and Yvon 1995), to deal with the case of D observable streams of symbols, drawn from D distinct alphabets. The D strings are assumed to result from the parallel concatenation of D subsequences, of possibly different lengths. As the model allows the D matched subsequences to be of different length, it assumes a many-to-many alignment between the D strings.

Formulation of the model. In the following, we study the case of two streams $\begin{pmatrix} O = o_{(1)} \ldots o_{(T_O)} \\ \Omega = \omega_{(1)} \ldots \omega_{(T_\Omega)} \end{pmatrix}$, assumed to result from the concatenation of pairs of subsequences $\begin{pmatrix} s_{(t)} \\ \sigma_{(t)} \end{pmatrix}$. A model restricting to n the maximal length of a subsequence $s_{(t)}$ in O, and to ν the maximal length of a subsequence $\sigma_{(t)}$ in Ω, will be referred to as an (n, ν)-joint multigram model. We denote by L_O (respectively L_Ω) a segmentation of O (respectively Ω) into subsequences, and by L the corresponding joint segmentation of O and Ω into joint subsequences: $L = (L_O, L_\Omega)$. The likelihood of (O, Ω) is computed as the sum over all joint segmentations:

$$\mathcal{L}(O, \Omega) = \sum_{L \in \{L\}} \mathcal{L}(O, \Omega, L)$$

or approximated with the likelihood of the single best joint segmentation:

$$\mathcal{L}^*(O, \Omega) = \mathcal{L}(O, \Omega, L^*_O, L^*_\Omega) \tag{5}$$

where

$$L^* = (L^*_O, L^*_\Omega) = \arg \max_{L \in \{L\}} \mathcal{L}(O, \Omega, L)$$

Assuming again that adjacent joint subsequences are independent, the likelihood of a joint segmentation is the product of each pair probability:

$$\mathcal{L}(O, \Omega, L) = \prod_t p\left(\begin{matrix} s_{(t)} \\ \sigma_{(t)} \end{matrix} \right)$$

Estimation of the joint multigram parameters. The parameter estimation of a joint multigram model is based on the same principles as for the one-string multigram model. Let $\mathcal{D}_S = \{(s_i, \sigma_j)\}$ denote a dictionary that contains all the pairs of subsequences (s_i, σ_j), where s_i can be formed by combinations of $1, 2, \ldots$ up to n symbols of the vocabulary of O, and where σ_j can be formed by combinations of $1, 2, \ldots$ up to ν symbols of the vocabulary of Ω. A joint multigram model is fully defined by the set Θ of each pair probability $p(s_i, \sigma_j)$. Replacing s_i by (s_i, σ_j), and O by (O, Ω) in equation (3), we can write directly the parameter re-estimation formula at iteration $(k+1)$:

$$p^{(k+1)}(s_i, \sigma_j) = \frac{\sum_{L \in \{L\}} c(s_i, \sigma_j, L) \times \mathcal{L}^{(k)}(O, \Omega, L)}{\sum_{L \in \{L\}} c(L) \times \mathcal{L}^{(k)}(O, \Omega, L)} \tag{6}$$

where $c(s_i, \sigma_j, L)$ is the number of occurrences of the pair (s_i, σ_j) in L and $c(L)$ is the total number of matched subsequences in L. The implementation of equation (6) using the forward-backward algorithm is detailed in Deligne, Bimbot, and Yvon (1995).

The training procedure jointly parses the two strings according to a maximum likelihood criterion. It produces a dictionary of pairs of subsequences which can be used for automatic transduction purposes as explained below. As for the basic multigram model, this dictionary may be advantageously pruned to avoid over-learning. This can be done either *a posteriori* by discarding the least frequent pairs of subsequences, or *a priori* by taking into account only the pairs of subsequences which are compatible with a known (possibly approximate) pre-alignment of the symbols in the two streams.

3. Multiphone Units for Speech Synthesis

In this section, we consider application of the above model to the problem of unit selection in concatenative speech synthesis.

3.1. Some Issues in Concatenative Synthesis

Concatenative speech synthesis techniques rely on the generation of an utterance by chaining prerecorded speech segments (Harris 1953). As compared to rule-based systems, this approach is considered easier and faster to implement.

One factor that influences the quality of a concatenation system is the definition and recording of a proper set of speech units. In particular, the units have to be carefully selected to reduce the effect of the acoustic discontinuities induced by the concatenation process. For instance, the static and dynamic characteristics of a phone are highly dependent on the adjacent phones, and it has long been observed that phonemes are not appropriate speech synthesis units.

Current approaches generally use diphones. A diphone can be understood as a 'domino', composed of the speech portion located between the middle of a phone and the middle of the next phone. Using such units, all possible transitions between phonemes are stored, which greatly increases the speech quality but at the cost of a larger storage volume.

However, being relatively short units, diphones are still affected by contextual variability. For instance, Peterson – quoted by Tzoukermann (1993) – deems that, whereas English has about 1600 diphones, nearly 8000 allophonic variants need to be stored to achieve an acceptable speech quality.

Beside recording multiple allophonic variants of diphones, a solution is to use longer units. Doing so helps to decrease the number of concatenations, i.e. the number of acoustic discontinuities required to reconstruct a given utterance. It also allows us to choose during synthesis the combination of units for which contextual effects at the join ('border') will be minimal. But, as the space available for storing speech segments is usually limited, it is not realistic to use larger units systematically.

Some work has been devoted in past years to the issue of selecting appropriate units, for which concatenation is likely to be easier. In fact, diphones are not all equally affected by contextual variability. Consequently, one way of looking at the selection of an optimal set of concatenative units is to merge into longer units only those diphones which are expected to be most vulnerable to the effects of coarticulation. Various criteria have therefore been proposed to select automatically the most variable diphones. According to Olive (1990), the vulnerability of a phoneme to the effects of coarticulation is mainly related to its duration and to the strength of its articulatory gesture. A different approach is proposed by Sagisaka and Iwahashi (1995), where the expected distortion of a speech unit is related to the variety of contexts in which it may occur. Some work on this question has also been dedicated to the French language, in particular by Bimbot (1988) and Aubergé (1991).

Longer units can also be selected on the basis of their frequency. Indeed, the replacement of a highly variable diphone by a longer unit improves the quality of the synthesized speech. However, from a global point of view, the extent to which this improvement is perceived also depends on how frequently those longer units occur in utterances. Storing long sequences of diphones which are rarely encountered in the target language is a fairly uneconomical way to obtain a small increase in quality. Conversely, prioritizing the merging of those diphones which most frequently co-occur together amounts to searching for an optimal trade-off between the reduction of the expected number of concatenations and the savings in space requirements (Bimbot, Deligne, and Yvon 1995).

In the rest of this section, we show how the multigram model can be used to perform this kind of trade-off between the size of the speech segment dictionary and the expected number of concatenations needed to synthesize an utterance. Section 3.2 introduces some notation, and explains how to use the multigram model for the selection of an optimal set of speech units. Experiments run with the Malecot corpus are reported in Section 3.3.

3.2. Application of the n-Multigram Model

In the following, a unit formed by merging from 1 up to $(n-1)$ diphones is referred to as an n-multiphone unit. An n-multiphone unit is therefore represented as a sequence of 2 up to n phonemes. Its acoustic realization consists of half a phone plus $(n-2)$ phones plus an half a phone. For instance, the merging of the three diphones /as/, /sj/ and /jɔ̃/ defines a multiphone of length four: /asjɔ̃/.

A set of multiphone units for speech synthesis can be selected using the multigram model presented in Section 2. For this application, the observation stream O is a string of diphones, derived from phonemically transcribed utterances, like for instance:

il le et te ev vi id dɑ̃ ɑ̃k ki il lf fo od dʀ ʀa ai iv və ən ni iʀ

which corresponds to the diphone representation of the French sentence: *il est évident qu'il faudra y venir*. This example is extracted from the Malecot database, described immediately below.

Multiphones are derived by the training procedure described in Section 2.2. At the end of the procedure, the most likely segmentation of the training database (via equation (4)) is used to define a final set of multiphones. As an example, we list below the most frequent 5-multiphone units found after training the model on the Malecot corpus (presented in detail in the next subsection).

se	me	də	ʁsəkə	le	de	la	ʁe	al
paʁ	es	sa	lə	ilia	ʁə	ed	ʁa	ʒəsɥi
ʒə	uvule	et	sivu	avwaʁ	setad	kə	kɛlkə	ik

3.3. Experiments on the Malecot Database

The Malecot corpus is composed of phonemic transcriptions of informal conversations in French (Malécot 1972; Tubach and Boë 1985). We have simplified the transcription, using a phonemic alphabet of 35 symbols. The corpus contains about 13,000 sentences, corresponding to about 200,000 phonemes.

We have split the corpus into a training set (the first 150,000 phonemes), and a test set (the last 50,000 phonemes). In both sets, all the delimiters were discarded so that word boundaries are unknown during the learning procedure. As a result, it may happen that a multiphone sequence overlaps two successive words.

The dictionary of multiphone units is built from the training set. Then, the test set is used to assess the relevance of the units.

Evaluation protocol. The decomposition of the test sentences into multiphones is performed according to an ML criterion, which is consistent with the strategy used to derive the multiphone units during the training phase. An alternative strategy would have been to decompose the test sentences so as to minimize the number of concatenations, which, though less consistent with the training criterion, would probably have turned out to be even more favorable to the multiphone approach.

The potentialities of diphones and of n-multiphones as candidate units for concatenative speech synthesis are compared, for various values of n, using a double criterion, which takes into account both the average number of concatenations and the corresponding storage requirements. As concerns storage requirements, we adopted the following point of view. Diphone-based synthesis requires the storing of all possible diphones. For multiphone-based synthesis, the space needed corresponds to the variable-length multiphone units derived from the training procedure to which are added all missing diphones. In fact, all diphones need to be stored to ensure the possibility of synthesizing any possible utterance. The storage requirements are measured in terms of minutes of speech, on the basis of 50 ms for a border phoneme and 100 ms for each phoneme within a unit.

The number of concatenations required to reconstruct the test utterance is given as an average number per rhythmic group. Those are annotated in the Malecot corpus.

Results. Examples of decompositions obtained with models of n-multiphones with $2 \leq n \leq 6$ for several sentences in the test set are given in Table 6.1. These examples illustrate how the approach provides non-uniform segmentations into units, which are usually related to the morpheme and word structure of the sentence.

The quantitative results for sets of 2-, 3-, 4- and 5-multiphone units are displayed in Figure 6.1. For each set of units, a histogram shows the distribution of the multiphone units depending on their length. The proportion of the diphones (i.e. of the 2-multiphones) missing from each set of n-multiphones, and *a posteriori* added to it to ensure full coverage, is indicated with dashed lines. In all cases, the total number of diphones is equal to 1,225, on the basis of 35 phonemes in French.

In Fig. 6.1(a), the number of units which need to be stored is, of course, equal to the number of diphones (1,225), corresponding to about 2 minutes of

Table 6.1. Multiphone decompositions for an example set of test sentences in the Malecot corpus.

Orthographic transcription	*(Broad) phonemic transcription*
(1) c'est un peu comme ça	(1-a) s e tœ̃ pø kɔm sa
(2) je suis responsable du groupe	(2-a) ʒəsɥi ʁespɔ̃sablə dy gʁup
(3) ce qui est une chose qui n'existe pas	(3-a) sə ki e tyn ʃoz ki n ɛgzistə pa
(4) moi j'ai etudié à l'université	(4-a) mwa ʒ e etydje a l ynivɛʁsite
(5) il est évident qu'il faudra y venir	(5-a) il e tevidɑ̃ k il fodra i vəniʁ
2-multiphone (diphone) decomposition	
(1-b) se et tœ̃ œ̃p pø øk kɔɔm ms sa	
(2-b) ʒəəs sɥɥi iʁʁe es sp pɔ̃ ɔ̃s sa ab bl ləəd dy yg gʁʁu up	
(3-b) səək ki ie et ty yn nʃʃo oz zk ki in nɛɛg gz zi is st təəp pa	
(4-b) mw wa aʒʒe ee et ty yd dj je ea al ly yn ni iv vɛɛʁʁs si it te	
(5-b) il le et te ev vi id dɑ̃ ɑ̃k ki il lf fo od dʁʁa ai iv vəən ni iʁ	
3-multiphone decomposition	
(1-c) set tœ̃p pøk kɔm msa	
(2-c) ʒəəs sɥi iʁʁes spɔ̃ ɔ̃sa abl ləd dy ygʁʁup	
(3-c) sək kie ety ynʃʃoz zk kin nɛɛgz zis stəəpa	
(4-c) mwa aʒe ee et tyd dje eal lyn niv vɛʁʁsi ite	
(5-c) ile ete evi idɑ̃ ɑ̃k kil lfo od dʁʁa ai iv vən niʁ	
4-multiphone decomposition	
(1-d) setœ̃ œ̃pø øk kɔms sa	
(2-d) ʒəəsɥi iʁe es sp pɔ̃ ɔ̃sa abl ləd dy yg gʁup	
(3-d) sək kie ety ynʃʃozk kin nɛgz zi ist təəpa	
(4-d) mwaʒʒee etyd dj je eal ly yniv vɛʁs site	
(5-d) ile ete evid dɑ̃k ki ilfo od dʁʁa ai iv vəni iʁ	
5-multiphone decomposition	
(1-e) setœ̃ œ̃pøøk kɔmsa	
(2-e) ʒəsɥi iʁʁe es sp pɔ̃ ɔ̃sa abləd dy yg gʁup	
(3-e) səkie etyn nʃʃoz zk kin nɛɛgzis st təəpa	
(4-e) mwaʒe ee etyd dj je ea alyn nivɛʁʁsite	
(5-e) ile ete evi id dɑ̃k ki ilfod dʁʁa ai iv vəniʁ	
6-multiphone decomposition	
(1-f) setœ̃ œ̃pøøk kɔmsa	
(2-f) ʒəsɥi iʁe es sp pɔ̃ ɔ̃sa abləd dy yg gʁup	
(3-f) səkie et ty ynʃoz zk kin nɛɛgzist təəpa	
(4-f) mwaʒe ee etydje eal lynivɛɛʁsite	
(5-f) ile ete evi id dɑ̃k ki ilfod dʁ ʁa ai iv vəniʁ	

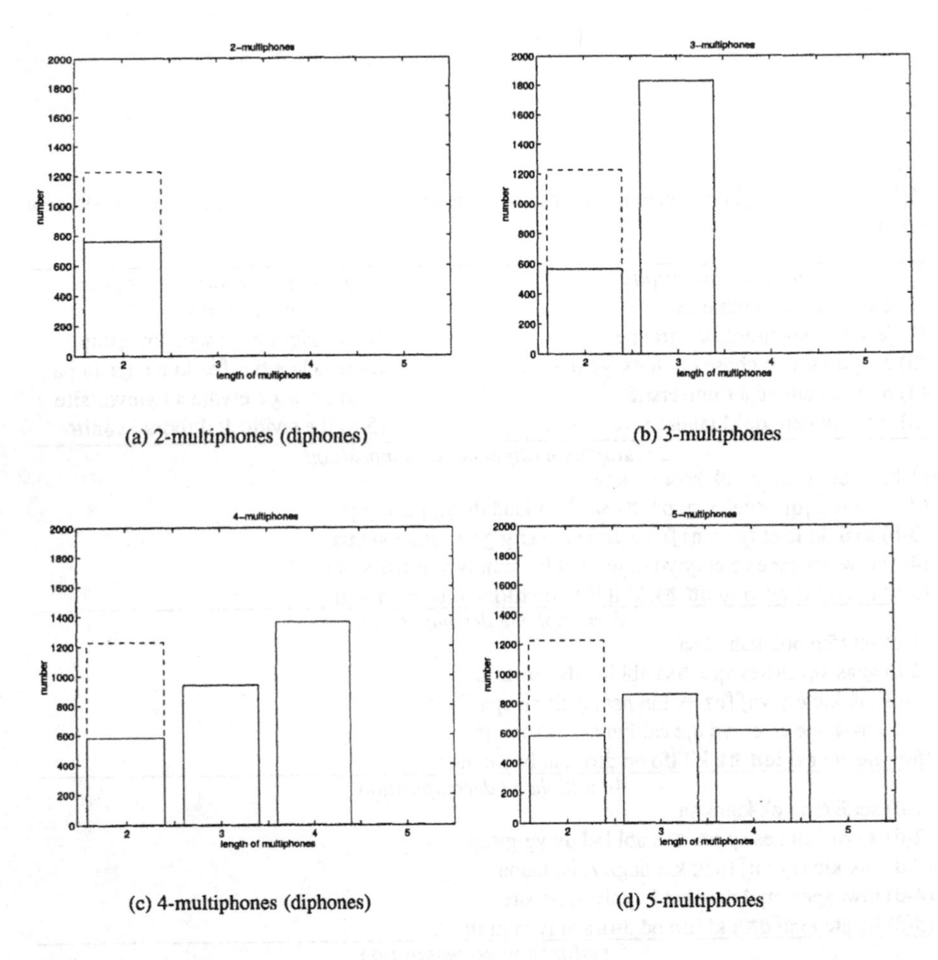

Figure 6.1. Comparative evaluation of the n-multiphones for $n = 2 \dots 5$. Histograms show the number of multiphones required to cover several sentences of the test set. Dashed lines show the number of diphones added to achieve full coverage of the test set.

speech. The expected number of concatenations is 13.5 on the test set and 13.6 on the training set. In Fig. 6.1(b), the 3,053 3-multiphone units correspond to about 8 minutes of speech. The expected number of concatenations is 7.7 on the test set and 7.5 on the training set. In Fig. 6.1(c), 3,528 units are required for 4-multiphone synthesis, corresponding to about 12 minutes of speech. The expected number of concatenations is 6.6 on the test set and 6.2 on the training set. Finally, for 5-multiphones (Fig. 6.1(d)), 3,575 units are used corresponding to some 14 minutes of speech. The expected number of concatenations is 6.3 on the test set and 5.8 on the training set.

Hence, storing 5-multiphone acoustic units multiplies by about seven the space requirements, in comparison with a diphone-based system. As a counterpart, it is expected that an utterance can be synthesized with, on average, half as many concatenations (6.3 instead of 13.5). The number of concatenations, as well as the space requirements, tend to stabilize as the maximal length of a unit increases. In other words, there seems to be some value of n beyond which any further lengthening of the units no longer proves significantly advantageous.

3.4. Comments and Perspectives

Our experiments show that the multigram model offers a viable approach for selecting an extended set of formal units for speech synthesis, resulting from a compromise between the number of such units and their utility. We expect that reducing the number of concatenations by a factor of two can only have a positive impact on the quality of the synthetic speech. However, this needs to be confirmed by experiments with actual acoustic units.

For the time being, we have considered that all concatenations would have an identical impact on the quality of the synthetic speech, which is clearly a too simplistic hypothesis. Nevertheless, the proposed approach can be readily generalized to the case when different costs are assigned to different concatenations (including an infinite cost for concatenations that must be avoided). By integrating these costs in the likelihood function (equation (2)), it is possible to derive a set of units that optimizes the compromise between the frequency of the units and their ability to be concatenated. One difficulty remains: the estimation of realistic costs.

4. Learning Letter-to-Sound Correspondences

Converting a written utterance into its phonemic counterpart is the first operation which needs to be performed by any text-to-speech system.

4.1. Background

Traditional approaches to the task of letter-to-sound conversion rely upon the application of a series of pronunciation rules, which are usually complemented by a dictionary of the most commonplace exceptions. For languages like English, or to a lesser extent French, the development of an accurate rule set by a linguist is a long and demanding task, reflecting the fact that letter-to-sound correspondences have been seriously obscured in the course of time by numerous phonological changes. Furthermore, the proliferation in real texts of loan words, technical terms, acronyms, and worst of all, proper names, which hardly respect the general pronunciation rules, makes the development of large coverage accurate rule-based systems a very challenging task.

As a consequence, a number of alternative approaches have been proposed over the past years, aiming at automatically extracting letter-to-sound correspondences from large pronunciation dictionaries. We shall not present here a detailed account of these numerous attempts, which are comprehensively reviewed elsewhere (Damper 1995; Yvon 1996a), but rather emphasize the main specifics of the multigram account of this problem.

It is a well-known fact that the pronunciation of most graphemes is influenced by the surrounding graphemes (the orthographic context). This influence is of variable length and may apply in one or the other direction. As demonstrated in the previous section, the main strength of the multigram model is that it provides a principled stochastic framework for expressing variable-length dependencies.

The general idea of the approach developed in the following sections consists of matching, on a statistical basis, variable-length subsequences of graphemes with variable-length subsequences of phonemes. Strings of graphemic and of phonetic (strictly phonemic) symbols are thus viewed as parallel concatenations of patterns, of unknown borders, where each pattern is a pair formed by a subsequence of graphemes and a subsequence of phonemes. Retrieving probabilities of co-occurrences between graphemic and phonemic subsequences is thus equivalent to estimating a probability of occurrence for a limited set of recurrent patterns. As such, it can be expressed in the joint multigram framework presented in Section 2.3. We expect this model to be capable of automatically adapting itself to the necessary context needed to disambiguate the pronunciation of a given substring.

This approach departs quite radically from most other approaches to the learning of letter-to-sound rules, as being both *unsupervised* and *holistic*. Supervision here refers to the kind of data supplied to the learning procedure. Whereas most learning procedures require a one-to-one alignment between orthographic and phonemic representations in the pronunciation dictionaries, our model is able to learn from unaligned data. In fact, the definition of a one-to-one alignment between orthography and phonology is quite arbitrary, for the proper pairing between the two representations matches in many cases more than one grapheme with more than one phoneme – which is precisely the kind of matching that our model aims at capturing.

Our model, like HMMs (Parfitt and Sharman 1991), or more recently the stochastic phonographic transducer described by Luk and Damper (1996), is also *holistic*, meaning that it selects the best pronunciation based on the entire transcription of a given string. Indeed, the best pronunciation for a given graphemic string results, in our model, from a maximum likelihood joint decoding of the orthographic and phonemic strings, optimized at the word level. As a consequence, the pronunciation assigned to a letter is indirectly conditioned by the pronunciation assigned to other letters in the word.

This makes our approach very different from techniques like multilayer perceptrons (Sejnowski and Rosenberg 1987), decision-trees (Torkolla 1993), or memory-based learning (Stanfill and Waltz 1986; van den Bosch and Daelemans 1993), which infer the pronunciation of a word by successively inferring, on the basis of its orthographic context, the best phonemic counterpart for each letter in the word. As a result, these methods provide no means to guarantee the overall consistency of a pronunciation, and are usually unable to account for so-called *long-range dependencies*.

In the rest of this section, we recast the problem of letter-to-sound conversion in the framework of the joint multigram model. We then present and discuss a series of experiments conducted with a French pronunciation dictionary.

4.2. Adaptation of the Multigram Model

Since the joint multigram model assigns a probability to any pair of subsequences, it can be used as a stochastic transducer. We study this application for the task of grapheme-to-phoneme transduction. Assume, for instance, that a joint multigram model is estimated on a training set where O and Ω are respectively a string of graphemes and a string of phonemes. The training process consists of mapping variable-length subsequences of graphemes to variable-length subsequences of phonemes. The resulting set of matched subsequences, along with their probability of co-occurrence, can be used to infer, through a decoding process, the string of phonemes Ω corresponding to a test string of graphemes O. This transduction task can be stated as a standard maximum *a posteriori* decoding problem, consisting in finding the most likely phonemic string $\widehat{\Omega}$ given the graphemic stream O:

$$\widehat{\Omega} = \arg\max_{\Omega} \mathcal{L}(\Omega|O) = \arg\max_{\Omega} \mathcal{L}(O,\ \Omega)$$

Assuming that $L^* = (L^*_O,\ L^*_\Omega)$, i.e. the most likely joint segmentation of the two strings, accounts for most of the likelihood, the inferred pronunciation results from the maximization of the approximated likelihood defined by equation (5):

$$\widehat{\Omega}^* = \arg\max_{\Omega} \mathcal{L}(O,\ \Omega,\ L^*_O,\ L^*_\Omega) \qquad (7)$$

$$= \arg\max_{\Omega} \mathcal{L}(O,\ L^*_O|\Omega,\ L^*_\Omega)\mathcal{L}(\Omega,\ L^*_\Omega) \qquad (8)$$

by application of Bayes' rule. These two equations give us two different transcription strategies. The term $\mathcal{L}(O,\ L^*_O|\Omega,\ L^*_\Omega)$ measures how well the graphemic subsequences in the segmentation L^*_O match the inferred phonemic subsequences in L^*_Ω. It is computed as $\prod p(s_{(t)}|\sigma_{(t)})$, where the conditional probabilities are deduced from the probabilities $p(s_i,\ \sigma_j)$ estimated during the training phase. The term $\mathcal{L}(\Omega, L^*_\Omega)$ measures the likelihood of the inferred pronunciation: it can be estimated as $\widetilde{\mathcal{L}}(\Omega, L^*_\Omega)$, using a language model. This decoding strategy is a way to impose syntagmatical constraints in the string Ω (here phonotactical constraints). The maximization (8) thus becomes:

$$\widetilde{\Omega}^* = \arg\max_{\Omega} \mathcal{L}(O, L^*_O | \Omega, L^*_\Omega) \widetilde{\mathcal{L}}(\Omega, L^*_\Omega) \quad (9)$$

In the experiments reported in section 4.3, the phonotactical component $\widetilde{\mathcal{L}}(\Omega, L^*_\Omega)$ is computed by modeling the succession of the phonemic subsequences with a bigram model. The conditional probabilities attached to those successions can be estimated on a parsed version of the phonemic training string, after the last iteration of the estimation algorithm.

4.3. Experiments on the BDLEX Database

The database used for our experiments is BDLEX (Pérennou, de Calmès, Ferrane, and Pécatte 1992), a French lexicon containing about 23,000 words and compounds. In addition to providing an accurate phonological description of each form (including pronunciation variants) and correspondences between orthographical and phonological representations, this lexicon also contains a morpho-syntactic description for each entry. For each word, we extracted one single phonemic transcription, and discarded duplicate lines (graphemic forms for which one single transcription was matched with more than one syntactic tag). The resulting list of 22,553 words (strings O) and their transcriptions (strings Ω) was randomly divided into a non-overlapping training set (20,298 words) and test set (2,255 words).

Evaluation protocol. The experimental protocol proceeds in three stages:

1 **Initialization:** the dictionary of joint subsequences is built by computing the relative frequency of all possible many-to-many alignments in the training corpus.

2 **Learning:** 10 iterations of Viterbi learning (see Section 2.2) are carried out.

3 **Decoding:** Two alternative decoding strategies have been explored, according either to equation (7) or (9). Transcription accuracy is computed using the unseen test set, on a per word basis, by counting the number of full agreements between the model transcriptions and the correct ones. Per phoneme scores are also evaluated, using the following formula:

$$P_C = 100 \times \frac{N_P - N_I - N_D - N_S}{N_P}$$

Table 6.2. Percentage word and phoneme accuracies of transcriptions with varying size parameters.

			Word accuracy (%)			
n	$\nu = 1$	$\nu = 2$	$\nu = 3$	$\nu = 4$	$\nu = 5$	$\nu = 6$
2	25.59	34.01	34.59	35.12	35.12	35.17
3	27.05	43.95	47.63	37.78	47.10	47.41
4	26.65	47.01	54.46	44.92	54.72	53.44
5	26.65	47.18	57.43	44.39	56.14	49.98
6	26.65	39.02	57.29	53.66	–	–

			Phoneme accuracy (%)			
n	$\nu = 1$	$\nu = 2$	$\nu = 3$	$\nu = 4$	$\nu = 5$	$\nu = 6$
2	76.51	80.80	81.10	81.24	81.28	81.26
3	77.59	85.91	87.51	84.59	87.02	87.02
4	77.36	86.57	89.48	87.64	89.40	88.52
5	77.36	86.53	89.93	87.55	90.14	87.76
6	77.36	83.67	89.57	89.00	–	–

where N_P is the total number of phonemes in the test set, and N_I, N_D and N_S are the total numbers of insertions, deletions and substitutions, respectively, necessary to transform a phonemic string produced by our transcription system into the corresponding reference string.

In the next section, we mainly focus on a comparative evaluation of various learning and decoding strategies. The aims are to find how to limit over-training and to assess the benefit of an additional bigram model based on equation (9).

Results. Our first experiment aims at evaluating the influence of the two size parameters n and ν, which specify respectively the maximum number of graphemes and phonemes in a joint subsequence. Table 6.2 reports scores obtained for n varying between 2 and 6, and ν between 1 and 6.

Performance improves with the maximum size of the joint subsequences, as long as both parameters remain close. This latter point reflects the fact that, in French, the graphemic and phonemic representations of words roughly have the same length. Additional experiments with larger size parameters have, however, led us to believe that further increase of these parameters was not profitable in terms of improved generalization capability.

During initialization, all possible joint segmentations of a given word are explored, leading to a very large dictionary containing a substantial number of very rare joint subsequences. These infrequent pairs are likely to reflect idiosyncratic properties of the training lexicon rather than well-attested matching pairs. The consequence is a degradation of the performance when decoding a test set.

Table 6.3. Percentage word and phoneme accuracies of transcriptions for the (5, 5) model with various frequency thresholds.

Thresholds (Θ_1, Θ_2)	*Word accuracy (%)*	*Phoneme accuracy (%)*
(3, 0)	64.52	92.31
(1, 0)	63.77	92.05
(5, 0)	63.95	92.00
(5, 1)	63.55	91.77

A straightforward way to reduce this effect of over-learning consists of discarding during training all joint subsequences whose frequency falls below a predefined threshold. To evaluate this strategy, we have reproduced the previous experiment with a (5, 5) model, using frequency thresholds (θ_1 at initialization and θ_2 during iteration) taking values in the set $\{0, 1, 3, 5, 10\}$. The four best results are reproduced in Table 6.3.

These experiments, reproduced for several other models, reveal that in any case, thresholding has to be more selective during the initialization than during iterations. The second observation is that thresholding, even moderate, is a very effective way of significantly increasing the performance: the best strategy resulted in an 8 percentage point improvement of the per word correctness. However, the exhaustive storing of all paired combinations of symbols at initialization turns out to be very time and memory consuming.

In view of this, a heuristic was tested, which consists of discarding *a priori* all mappings which differ too much from a given initial joint segmentation of the two strings. (In our experiments, two subsequences can be paired only if they have their first symbol in the same pair in the initial rough alignment.) This initial joint segmentation need not be an accurate alignment: here, it was obtained by first applying a low-order joint multigram mapping, limiting to two the size of the paired subsequences.

Constraining the initial pairs to comply roughly with an approximate pre-mapping based on small subsequences enabled us to evaluate the performance of models allowing subsequences of up to seven symbols. The scores, obtained by preliminary application of a (2, 2) multigram model, are shown in the left part of Table 6.4. They are better than the results shown in Table 6.3, which are based on a standard initialization, where all possible pairings are considered. This confirms the importance of reducing the set of all possible mappings to enhance the generalization capability of the model.

Using a reference pre-alignment that is only approximate does not seem to affect transcription accuracy. Indeed, limiting the set of initial pairs to those which are compatible with the exact phonemic alignment provided with the BDLEX database yields only marginal improvements. The scores relative to those experiments are given in the right part of Table 6.4.

Table 6.4. Percentage word and phoneme accuracies for various models using pre-aligned transcriptions.

	Initialization			
	(2,2) Joint Model		*Manual Alignment*	
Model	*Word (%)*	*Phoneme (%)*	*Word (%)*	*Phoneme (%)*
(3, 3)	48.87	87.72	50.11	88.04
(4, 3)	59.33	90.59	60.13	90.80
(5, 3)	60.93	90.85	62.04	91.06
(5, 5)	61.73	91.52	62.79	91.73
(6, 5)	59.60	90.66	60.93	91.03
(6, 6)	58.40	90.39	59.91	90.68
(7, 5)	56.41	89.74	58.40	90.26
(7, 6)	52.99	88.84	53.92	89.06
(7, 7)	51.35	88.39	52.20	88.54

Our best results are obtained with a (5, 5) model; a further increase of the maximal length of a subsequence degrades the performance. The deterioration of the scores indicates an over-training problem when subsequences of more than five symbols are used. This suggests that the heuristics used to discriminate between reliable and spurious subsequences during the initialization stage prove insufficient for long subsequences.

This behavior severely affects the performance, as many incorrect mappings precisely arise from the limitations placed on the size of the subsequences. For instance, the only decomposition of the French word *mangeur* which allows for unambiguous pronunciation of each subsequence is *(m)(angeur)*, which involves a subsequence of six graphemes. Any segmentation obtained with a (5, 5) model is thus likely to result in an erroneous transcription. At the same time, increasing the maximal size of a subsequence reduces the reliability of the model. For lack of a heuristic good enough to enhance the generalization capability of high-order models, we explore the possibility of capturing long-distance dependencies through the use of a syntagmatic model during the decoding process.

Decoding with a syntagmatic model. All results reported so far were obtained using the decoding procedure of equation (7). As indicated by equation (9), decoding may also benefit from the computation of a bigram model of phonemic sequences.

For each of the joint multigram models trained in the previous experiments, a bigram model was learned using the segmentation produced by the last training iteration. This model, which captures the most salient phonotactic constraints, was then used during decoding according to equation (9). It turns out that, for every parameter setting, this enhanced decoding strategy allows us to improve

Table 6.5. percentage word and phoneme accuracies of transcriptions using a bigram model with various size parameters.

Model	*Word accuracy (%)*	*Phoneme accuracy (%)*
(2, 4)	62.57	91.66
(2, 3)	62.35	91.54
(2, 2)	61.24	91.30
(3, 3)	61.82	91.24

significantly the transcription accuracy. The four best results for this series of experiments are given in Table 6.5.

The improvement is especially significant for low-order models: the per phoneme accuracy for a (3, 3) model raises from 87.51% to 91.24%. The best performance attained so far with a thresholding heuristic, 92.31% with a (5, 5) model, is outperformed using a (3, 2) model, which achieves a 93.72% accuracy. Higher-order models, however, do not seem to benefit from this enhanced decoding strategy. This result can be explained as follows. For low-order models, contextual dependencies – the length of which exceed the authorized size of a subsequence – are captured by the syntagmatic model. However, when larger and arguably less ambiguous subsequences come into play, the bigram estimates become less reliable, resulting in an overall degradation of performance.

These experiments nonetheless demonstrate the advantage of using an auxiliary bigram model. This allows us to obtain a substantial improvement in terms of performance, while limiting the computational burden of the parameter estimation procedure.

4.4. Discussion

Considering the fact that our approach is completely unsupervised, these first results are rather encouraging. However, they seem to reach a ceiling around 94% phoneme accuracy, which is some way below the performance obtained using alternative self-learning algorithms. For instance, experiments carried out on the same data using dynamically-expanding context (Torkolla 1993), a decision-tree learning algorithm, resulted in a 98% phoneme accuracy.

Careful examination of example joint segmentations (see Table 6.6) produced during the learning stage can help improve our understanding of the errors made by the model. The first series of words (i.e. above the horizontal line in the table) illustrates the capacity of the model to extract accurately many-to-many pairings, and also to retrieve morphologically relevant segmentations. Consider for instance the segmentation of *verbal*, an adjectival derivation based on the lemma *verb*, or the verbal form *heurter*, built upon the noun

Table 6.6. Segmentation examples using a (5, 5) model.

Graphemic	*Phonemic*
(verb)(al)	(vɛRb)(al)
(heurt)(er)	(œRt)(e)
(légal)(ité)	(legal)(ite)
(en)(flamm)(er)	(ɑ̃)(flam)(e)
(civil)(is)(ation)	(sivil)(iz)(asjɔ̃)
(spoli)(er)	(spɔlj)(e)
(désec)(houer)	(dezeʃ)(ue)
(ébauc)(her)	(eboʃ)(e)
(torre)(ntiel)	(tɔR)(ɑ̃sjɛl)
(débra)(ncher)	(debr)(ɑ̃ʃe)

heurt, etc. Even more significantly, in the verb *enflammer*, the two affixes are correctly identified.

The fact that most learned mappings agree with the pronunciation rules, and are relevant from a morphological point of view, is not enough to warrant a nearly zero error rate during a decoding test. To ensure perfectly accurate transcriptions, an additional requirement would be that the graphemic subsequence involved in any mapping can be pronounced without ambiguity.

Our model extracts those pairings that co-occur most frequently, but there is no obvious reason why these should also correspond to unambiguous subsequences of graphemes. This fact is illustrated by the second series of words (below the horizontal line). In the first example, the tendency of the model to favor decompositions into morpheme-like units even jeopardizes the accuracy of other transcriptions. The infinitive verbal affix *er* is indeed correctly identified, but the resulting segmentation dangerously isolates the final *i* of *spoli*, the pronunciation of which as /j/ depends on the adjacent vowel /e/. Used in another context, where a consonant would follow the *i*, this subsequence is likely to be mispronounced. The other examples illustrate cases where a digraph is broken by the segmentation: *ch* in the first two words, and a vowel-plus-nasal-consonant pair in the last two words. Consequently, resulting joint subsequences are highly unreliable, especially the first one of our last example where the vowel *a* at the end of the subsequence *débra* has lost its phonemic counterpart /ɑ̃/, which is mapped with the second subsequence.

In our experiments, the expected benefit of modeling longer – thus less ambiguous – subsequences turns out to be limited. It evidences the difficulty for our current training procedure to provide reliable estimates for an increased number of parameters. In other words, the transcription accuracy achieved with higher-order models reflects a trade-off between the disambiguation of the subsequences, and the deterioration of the reliability of the estimates. For lack of a learning procedure allowing us to discard all ambiguous subsequences, we studied the effectiveness of solving possible pronunciation ambiguities by placing phonotactic constraints on the decoding process. The use of a syntagmatic model at the decoding stage proves fairly efficient in solving many of the ambiguities inherent in the low-order models. However, because of the over-training problem, it does not yield significant improvements with the highest-order models.

Various theoretical extensions of the multigram model have been investigated to alleviate these problems and are described by Deligne (1996), which in particular demonstrates the possibility of relaxing the assumption of independence. However, as the use of more complex statistical models requires the estimation of an increased number of parameters, some additional constraints should still be placed on the training procedure. A possible solution to the parameter estimation problem is suggested by the related work of Luk and Damper (1996): using a similar theoretical framework, the authors take advantage of various learning biases to control very cautiously the growth of the number of distinct joint subsequences, thus limiting the number of parameters to be estimated. As a result their model achieves a level of accuracy which suggests that the introduction of a learning bias is a very effective way to improve parameter estimation.

5. General Discussion and Perspectives

This chapter has introduced a general statistical model which allows variable-length dependencies within streams of symbols to be captured. We have also presented two applications of this model in the domain of speech synthesis.

The first application is related to the optimal selection of variable-length formal units for concatenative speech synthesis systems. We have shown that our model allows a trade-off between the size of the set of speech units and the expected number of concatenations needed to generate an utterance. Our approach provides a principled framework for directly deriving appropriate speech synthesis units from a corpus. Further extensions of the model, where not only the frequency but also the relative contextual sensitivity of the various units would be taken into account, offer promising perspectives.

The second application is related to automatic text-to-phoneme conversion. We have shown that the one-dimensional multigram model can be extended to represent parallel streams – in that case a string of graphemes and a string of phonemes – giving rise to the joint multigram framework. Based on this extension, we have proposed a stochastic grapheme-to-phoneme transducer which is able to learn pronunciation rules without supervision, and to infer the correct transcription of many unknown words. However, the modeling of more than a single stream considerably enlarges the parameter space. As a result, the straightforward generalization of the estimation procedure formerly used in the one-dimensional case proved insufficiently reliable, providing estimates highly vulnerable to the effects of over-training. Additional constraints, chosen so as to reflect some structural properties known to be inherent to the described strings, should advantageously be placed on the training procedure, or incorporated in the formulation of the model. For the description of grapheme-to-phoneme correspondences, these structural properties could take the form of, for instance, overlapping constraints between adjacent subsequences, as suggested by Deligne (1996).

Potential areas of practical applications for the multigrams exceed by far the two case studies reported here. The description of sentences, for instance, as concatenations of variable-length subsequences of words, based on the multigram assumptions, leads to the definition of a statistical language model (Deligne and Bimbot 1995), with multiple possible applications in natural language processing (part-of-speech tagging, statistical translation, automatic spell-checking, word-sense disambiguation, etc.). Other work using the multigram model in the domain of speech recognition (Deligne 1996; Deligne and Bimbot 1997) has already been reported. It is based on an extension of the original model enabling description of strings of continuous-valued data such as spectral vectors. This extension is used as an automatic inference procedure to derive an inventory of non-uniform acoustic units usable as speech recognition units. Generally speaking, the multigram model is a good candidate for any task which requires the description of data subject to sequential dependencies. It must be understood as a step towards a more data-driven approach to pattern processing, where sequential patterns are directly derived from the data. Ultimately, we wish to eliminate the need for deciding *a priori*, and possibly arbitrarily, the nature of the basic pattern units.

Chapter 7

TREETALK: MEMORY-BASED WORD PHONEMISATION

Walter Daelemans
Center for Dutch Language and Speech, University of Antwerp

Antal van den Bosch
ILK/Computational Linguistics, Tilburg University

Abstract We propose a memory-based (similarity-based) approach to learning the mapping of words into phonetic representations for use in speech synthesis systems. The main advantage of memory-based data-driven techniques is their high accuracy; the main disadvantage is processing speed. We introduce a hybrid between memory-based and decision-tree-based learning (TRIBL) which optimises the trade-off between efficiency and accuracy. TRIBL was used in TREETALK, a methodology for fast engineering of word-to-pronunciation conversion systems. We also show that, for English, a single TRIBL classifier trained on predicting phonetic transcription and word stress at the same time performs better than a 'modular' approach in which different classifiers corresponding to linguistically relevant representations such as morphological and syllable structure are separately trained and integrated.

1. Introduction

A central component in a typical text-to-speech system is a mapping between the orthography of words and their associated pronunciation in some phonetic or phonemic alphabet with diacritics (denoting e.g. syllable boundaries and word stress). We call this mapping *phonemisation*. Table 7.1 provides examples in a few languages of input and output of a phonemisation module.

In a traditional knowledge engineering approach, several linguistic processes and knowledge sources are presupposed to be essential in achieving this task. The MITtalk system (Allen, Hunnicutt, and Klatt 1987) is a good example of this approach. The architecture of their phonemisation module

R. I. Damper (ed.), Data-Driven Techniques in Speech Synthesis, 149-174.

Table 7.1. Example words with their phonetic transcription (including stress markers).

Word (input)	*Language*	*Phonetic transcription (output)*
belofte	Dutch	/bə'lɔftə/
biochemistry	British English	/ˌbaɪəʊ'kɛmɪstrɪ/
hésiter	French	/ezi'te/
cabar	Scottish Gaelic	/'kapər/

includes explicitly-implemented, linguistically-motivated abstractions such as morphological analysis, rules for spelling changes at morphological boundaries, phonetic rules, etc. Each of these modules and their interaction have to be handcrafted, and redone for each new language, language variant, or type of phonetic transcription. Also, lists of exceptions not covered by the developed rules have to be collected and listed explicitly.

The obvious problems with such an approach include high knowledge engineering costs for development of a single phonemisation module, limited robustness and adaptability of modules once they are developed, and lack of reusability for additional languages or dialects. Data-oriented techniques using statistical or machine learning techniques have therefore become increasingly popular (starting with Stanfill and Waltz 1986 and Sejnowski and Rosenberg 1987; see Damper 1995 for a more recent overview). Advantages of these systems include fast development times when training material is available, high accuracy, robustness, and applicability to all languages for which data in the form of machine-readable pronunciation dictionaries are available.

In previous research (van den Bosch and Daelemans 1993; Daelemans and van den Bosch 1993, 1997; van den Bosch 1997), we applied memory-based techniques (Stanfill and Waltz 1986; Aha, Kibler, and Albert 1991) to the phonemisation task in British and American English, Dutch, and French with good results. Memory-based learning is also referred to as *lazy learning* (see Aha 1997 for an overview of theory and applications of lazy learning techniques and applications) because all training examples are stored in memory without any further effort invested in processing at learning time. No abstractions are built at learning time from the examples as is the case in *eager learning* techniques such as rule induction, top-down induction of decision trees, connectionism, and statistical methods such as hidden Markov models. At processing time, a new input is compared to all examples in memory, and the output is extrapolated from those memory items that are most similar to the input according to some similarity function. For the phonemisation task, when using a well-designed similarity function, memory-based techniques turned out to have a higher generalisation accuracy (i.e. accuracy on words the system was not trained on) than a number of eager approaches (van den Bosch 1997), a result which generalises to a large number of natural language processing tasks. Empirically, it can be shown that in learning natural language processing tasks, abstraction is harmful to generalisation accuracy (see Daelemans 1996 and van den Bosch 1997 for tentative explanations). Furthermore, resulting trained systems are highly competitive when compared to knowledge-based handcrafted systems (van den Bosch and Daelemans 1993).

Unfortunately, memory-based techniques are computationally expensive, as (i) all examples have to be stored in memory, and (ii) every input pattern has to be compared to all memory items in order to find the most similar items. Existing indexing and compression techniques either only work well for numeric feature values, or they involve some form of abstraction (e.g. when using induced decision trees as an index to the case base). We cannot use the former because the phonemisation task will require symbolic (nominal) feature values only, and we have established empirically that the latter are harmful to generalisation accuracy. This chapter addresses this issue and proposes a solution in which an optimal tradeoff can be found between memory and processing efficiency on the one hand, and processing accuracy on the other hand.

In Section 2, we relate our own previous memory-based phonemisation approach to other research using analogical reasoning for phonemisation, and provide empirical results comparing it to various other learning methodologies. In Section 3, we introduce the TRIBL learning algorithm which allows dynamic selection of a tradeoff between efficiency and accuracy, and report empirical results for phonemisation in English using TRIBL in the TREETALK system. We continue (Section 4) with a discussion of the interaction of linguistic modules in a memory-based learning approach to phonemisation (drawing on van den Bosch 1997 and van den Bosch, Weijters, van den Herik, and Daelemans 1997), including a comparison of some differently structured modular systems, before concluding (Section 5).

2. Memory-Based Phonemisation

To determine the unknown phonemisation of a word given its spelling, it is worthwhile looking at the words with a similar spelling for which the phonemisation is known. Words with similar spellings have similar phonemisations. The novel word *veat* is pronounced /vit/ because it resembles among others the known words *heat*, *feat*, *eat*, *veal*, and *seat* in all of which the substrings *(v)ea(t)* are pronounced /(v)i(t)/. This type of *memory-based reasoning* has been acknowledged by many researchers as a useful approach in building phonemisation modules. It pays, however, to keep in memory the phonemisations of all known words, so as to prevent a word like *great* being pronounced /grit/ instead of /greɪt/. Together, these insights suggest an approach in which (i) a lexicon of words with their associated phonemisations is kept in memory, and (ii) a similarity-based method is used to compute the phonemisations of words not in the lexicon.

MBRtalk (Stanfill and Waltz 1986; Stanfill 1987, 1988) was the first application of this approach to phonemisation. Their memory-based reasoning is a paradigm which places recall from memory at the foundation of intelligence. Every letter in every word is represented as a record in the database. Each record contains a fixed number of fields (features). In this case, features are the letter symbol to be transcribed, the transcription, the four letters to the left and to the right (a window of nine), and a stress marker. A novel test word is entered as a record with some fields filled (the predictors – the context in this case) and others empty (the 'goals' – transcription and stress marker). This target is compared to each frame in the database, using a computational measure of (dis)similarity, and the most similar records are used to extrapolate the value for the goal field. The approach is therefore an adaptation for symbolic (non-numeric) features of the k-NN algorithm (Cover and Hart 1967; Hart 1968; Gates 1972; Dasarathy 1980) from statistical pattern recognition.

The most basic distance metric for patterns with symbolic features is the *overlap metric* given in equations (1) and (2); where $\Delta(X, Y)$ is the distance between patterns X and Y, represented by n features, w_i is a weight for feature i, and $\delta()$ is the distance per feature:

$$\Delta(X, Y) = \sum_{i=1}^{n} w_i \delta(x_i, y_i) \tag{1}$$

where:

$$\delta(x_i, y_i) = \begin{cases} 0 & \text{if } x_i = y_i \\ 1 & \text{otherwise} \end{cases} \tag{2}$$

This metric, which we will call IB1 in this chapter, is obviously deficient when different features have a different relevance for solving the task, as is the case in phonemisation. The metric that Stanfill and Waltz (1986) propose is therefore more sophisticated: It combines feature relevance and value difference metrics in a single value.

Feature relevance metrics assign weights to features according to their relevance in solving the task. For example, in phonemisation, the target letter to be transcribed is obviously the most relevant feature, and should therefore be assigned more weight than the context features. Value difference metrics assign different distances or similarities to different pairs of values of the same feature. Again in phonemisation, graphemes *f* and *v* are more similar than *f* and *a*, and when matching the feature values of a new record with those of the database, this should be taken into account. The specific value difference metric used by Stanfill and Waltz was simplified and made symmetric later by Cost and Salzberg (1993).

Stanfill and Waltz reported superior generalisation accuracy to NETtalk (Sejnowski and Rosenberg 1987), the famous neural network (back-propagation) learning approach to phonemisation, on the same data. Weijters (1991) compared NETtalk to a more simple memory-based reasoning procedure using a weighted overlap metric. Feature relevance weights were hand-set, descending in strength from the target grapheme to be described towards context features further away. This approach also outperformed NETtalk. Wolpert (1990) performed a similar memory-based learning experiment with the same results.

Van den Bosch and Daelemans (1993) combined the overlap metric with *information gain*, an information-theoretic feature relevance weighting method, inspired by the work of Quinlan (1986, 1993) on top-down induction of decision trees. Information gain was also used by Lucassen and Mercer (1984) to guide a decision tree building process for learning phonemisation. Information gain (IG) weighting looks at each feature in isolation, and measures how much information it contributes to our knowledge of the correct class label. The information gain of feature f is measured by computing the difference in uncertainty (i.e. entropy) between the situations without and with knowledge of the value of that feature:

$$w_f = H(C) - \sum_{v \in V_f} p(v) H(C|v) \tag{3}$$

where C is the set of class labels, V_f is the set of values for feature f, and $H(C) = -\sum_{c \in C} p(c) \log_2 p(c)$ is the entropy of the class labels. The probabilities are estimated from relative frequencies in the training set. The resulting IG values can then be used as weights in equation (1). The k-NN algorithm with this metric is called IB1-IG (Daelemans and van den Bosch 1992).

While our approach focuses on classification of single graphemes in context with respect to their phonemic mapping, other approaches have been proposed that map *variable-width chunks* of graphemes to chunks of phonemes by analogy (Sullivan and Damper 1992, 1993; Pirrelli and Federici 1994; Yvon 1996b; Damper and Eastmond 1997) most of which build on Glushko's (1979) psycholinguistically-oriented single-route model of reading aloud, and Dedina and Nusbaum's (1991) PRONOUNCE model for chunk-based text-to-sound conversion.

2.1. Phonemisation with IGTREE

The most important problem with memory-based approaches is their time and space complexity. Storing and using all patterns extracted from all words in the lexicon with their associated phoneme is prohibitively inefficient unless implemented on massively parallel machines (Stanfill and Waltz 1986). Previously (Daelemans and van den Bosch 1997), we presented an optimised approximation of memory-based learning using so-called IGTrees. (In this chapter, we refer to the approach to automatic phonemisation as IGTREE, and to the data structures which the approach uses as IGTrees.). In this section, we will describe the design of this phonemisation approach, its general philosophy, and the results of empirical comparisons. Subsequently, in Section 3, we show how the IGTREE approach can be seen as one end of a continuum and 'normal' lazy learning (IB1-IG) as the other, and how an optimal efficiency-versus-accuracy tradeoff can be found between these extremes.

The approach we take for phonemisation could be described as *lexicon-based generalisation*. We need a data structure from which we can retrieve known phonemisations of words, and at the same time extrapolate to new, previously-unseen words by their similarity to the words in the lexicon. Our solution is to base generalisation on the capacity of our algorithm to learn automatically to find those parts of words on which similarity matching can safely be performed. The system stores single grapheme-phoneme correspondences with a minimal context sufficient to be certain that the mapping is unambiguous (in the given lexicon). These induced contexts range from very small (corresponding to general rules) to very large (corresponding to lexical structures), and everything in between. For example, for English, *v* followed by *o* is transcribed as /v/ (a general rule), and *i* preceded by *es* and followed by *den* is transcribed /ɪ/, a very specific context necessary to make the distinction between the pronunciation of grapheme *i* in *president* and in *preside*.

Daelemans and van den Bosch (1997) proposed the following general architecture for the TREETALK word phonemisation system to implement this approach. Figure 7.1 illustrates the architecture diagrammatically.

Automatic alignment. The spelling and the phonetic transcription of a word often differ in length (e.g. *rookie* → /ruki/). Our phonemisation approach demands, however, that the two representations be of equal length, so that each individual graphemic symbol can be mapped to a single phonetic symbol (e.g. *r oo k ie* → /r u k i/). We try, therefore, to align the two representations by adding null phonemes in such a way that graphemes or strings of graphemes are consistently associated with the same phonetic symbols (e.g. *rookie* → /ru–ki–/, where '–' depicts a phonemic null). Our alignment algorithm works in a purely probabilistic way by capturing all possible phoneme-grapheme mappings with their frequency, and selecting the most probable mapping at word-pronunciation pair level on the basis of the multiplied individual letter-phoneme association probabilities. Empirical tests indicate that there is no significant loss in generalisation accuracy when using our automatic alignment instead of a handcrafted one.

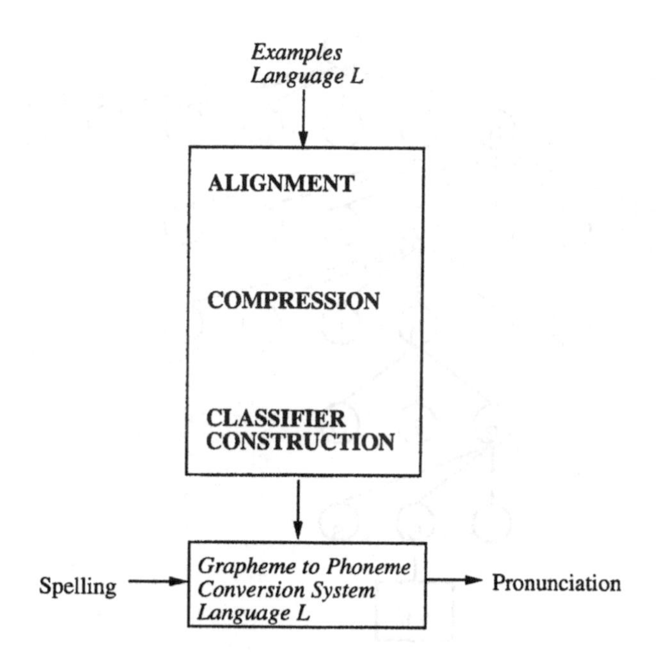

Figure 7.1. General architecture of the TREETALK word phonemisation system.

Alternative approaches to alignment are possible. For example, Van Coile (1990) uses a hidden Markov phoneme model in conjunction with the Viterbi algorithm (Viterbi 1967; Forney 1973); Luk and Damper (1996, 1998) combine stochastic grammar induction with Viterbi decoding using a maximum likelihood criterion.

Lexicon compression. In our approach, all lexical data (all available aligned spelling-pronunciation pairs) have to be kept in memory. But since (i) we only need for each letter in each word the minimal context necessary to find a unique corresponding phoneme, and (ii) the same grapheme-context-phoneme combinations may reappear in many words, a considerable amount of compression is possible. To achieve this compression, we use IGTrees (Daelemans, van den Bosch, and Weijters 1997). An IGTree is a decision tree in which the root node represents the feature with the highest relevance for solving the task, the nodes connected to the root represent the feature with the second highest relevance, etc. The arcs connecting a node to the nodes at the level below represent the different values attested for that feature in the lexicon. In the phonemisation application, paths in the IGTree represent grapheme-context-phoneme combinations. Connections between the root node and the next level represent the graphemes to be classified, connections between

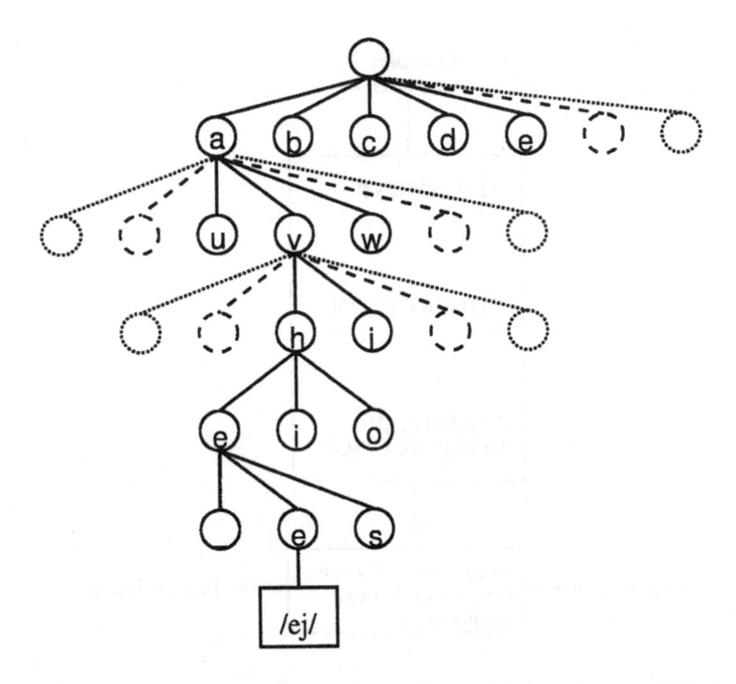

Figure 7.2. Retrieval of the pronunciation of *a* in the word *behave* from an IGTree. The path represents the minimally disambiguating context *ehave*.

other levels of the tree represent context graphemes in order of relevance, and leaf nodes represent the phonemes associated with the grapheme-context paths. Figure 7.2 shows part of an IGTree for English phonemisation.

To determine the relevance order of graphemes, we used information gain as in equation (3), which we adopted from other decision tree learning methods (e.g. Quinlan 1993). However, the use of mutual information in phonemisation has been proposed as early as 1984 by Lucassen and Mercer. Not surprisingly, the information gain (and hence relevance) of the grapheme to be transcribed is highest, followed by the nearest context graphemes to the left and the right, with context letters further away becoming decreasingly important. The depth of a path in an IGTree reflects in a certain sense the ambiguity of the mapping it represents. Leaves near the top of the tree typically belong to highly regular pronunciations. The result of this compression using IGTrees is that all transcriptions in the lexicon can be retrieved from this data structure very quickly, and the lexicon is compressed into a fraction of its original size.

Generalising from the compressed lexicon. To retrieve the phonemisation of a word that was not in the learning material, each letter of the new word is taken as a starting point of tree search. The search then traverses the tree, up to the point where either (i) it successfully meets a leaf node, or (ii) search fails as the specific context of the new word was not encountered in the learning material and, consequently, was not stored as a path in the tree. In case (i), the phonemic label of the leaf node is simply taken as the phonemic mapping of the new word's letter. In case (ii), the exact matching strategy is taken over by a *best guess* strategy.

In present versions of the system, the best guess strategy is implemented in a straightforward way. When building a path in the tree, the construction algorithm constantly has to check whether an unambiguous phonemic mapping has been reached. At each node, the algorithm searches in the learning material for all phonemic mappings of the path at that point of extension. In cases when there is more than one possible phonemic mapping, the algorithm computes the most probable mapping at that point. Computation is based on occurrences: The most frequent mapping in the learning material is preferred. (In case of ties, a random choice is made.) This extra information is stored with each non-ending node. When a search fails, the system returns the most probable phonemic mapping stored in the node at which the search fails. Finally, the phonemisation of a new word is assembled by concatenating IGTREE's phonemic classifications of each of its graphemes in context.

Decision trees for phonemisation were introduced by Lucassen and Mercer (1984), who also used an information-theoretic metric as a guiding principle in constructing the tree. Their context features are preceding and following letters and preceding phonemes, coded as binary features, which necessitates a recursive search for the most informative binary features using mutual information. The application of machine learning algorithms such as ID3 and C4.5 (Quinlan 1986, 1993) to phonemisation is analogous to this approach (see e.g. Dietterich, Hild, and Bakiri 1995). The IGTREE approach differs in that it keeps all information in the training data necessary for classification in the tree, unlike the previous approaches which prune or restrict the tree in order to improve generalisation. In IGTREE, generalisation is achieved by the way the word-level problem is cut up into letter-level parts, and by the defaults computed on the nodes of the tree.

The IGTREE approach, although developed independently with a completely different motivation, also turns out to be equivalent to Kohonen's (1986) dynamically expanding context approach (DEC), applied to phonemisation by Torkolla (1993). DEC extracts rules from the data according to a predefined specificity hierarchy (in this case, specificity of context and the intuition that context further away from the focus position is decreasingly relevant). Starting with rules transforming a single grapheme into a phoneme, context is added until, given a particular training set, the rules extracted are unambiguous in the training set, i.e. the context and grapheme combination determine a unique phoneme. As the rules are stored as a tree structure, the resulting method is almost identical to the van den Bosch and Daelemans (1993) version of

Table 7.2. Examples of instances generated for task GS from the word *booking*.

Instance number	*Left context*			*Focus letter*	*Right context*			*Classification*
1	_	_	_	*b*	*o*	*o*	*k*	/b/1
2	_	_	*b*	*o*	*o*	*k*	*i*	/u/0
3	_	*b*	*o*	*o*	*k*	*i*	*n*	/–/0
4	*b*	*o*	*o*	*k*	*i*	*n*	*g*	/k/0
5	*o*	*o*	*k*	*i*	*n*	*g*	_	/ɪ/0
6	*o*	*k*	*i*	*n*	*g*	_	_	/ŋ/0
7	*k*	*i*	*n*	*g*	_	_	_	/–/0

IGTREE. The only difference is that here information gain is used to decide automatically upon the 'specificity hierarchy' instead of handcrafting it. Finally, an approach to phonemisation similar to van den Bosch and Daelemans (1993) and Torkolla (1993) has been used in the Onomastica project for the transcription of proper names (Andersen and Dalsgaard 1994).

2.2. Experiments with IGTREE

We base our experiments on a corpus of English word pronunciations extracted from the CELEX lexical databases (van der Wouden 1990; Burnage 1990) maintained at the Centre for Lexical Research of the University of Nijmegen. The corpus contains 77,565 unique word pronunciations. Each listed word pronunciation is accompanied by additional information on morphological analysis, syllabification, and stress assignment. We define the phonemic mapping of a grapheme in context as its associated phoneme plus a stress marker indicating whether the phoneme is an initial phoneme of a syllable receiving primary stress (marker 1), secondary stress (marker 2), or no stress (marker 0). This formulation thus integrates grapheme-phoneme conversion and stress assignment in one task; we refer to it henceforth as the GS task.

Table 7.2 displays example instances and their combined phoneme/stress marker classifications generated on the basis of the example word *booking*. The phonemes with stress markers are denoted by composite labels. For example, the first instance in Table 7.2, _ _ _ *book*, maps to class label /b/1, denoting a /b/ which is the first phoneme of the syllable /bu/ receiving primary stress.

On the basis of the CELEX corpus of 77,565 words with their corresponding phonemic transcription with stress markers, a database containing 675,745 instances was constructed. The number of actually-occurring combinations of phonemes and stress markers is 159, which is fewer than the (Cartesian) product of the number of occurring subclasses ($62 \times 3 = 186$); some phoneme/stress marker combinations do not occur in the data.

To allow for a comparison between memory-based (lazy) learning algorithms and alternative learning algorithms, we selected the lazy-learning algorithms IB1 (Aha, Kibler, and Albert 1991; Daelemans, van den Bosch, and Weijters 1997) and IB1-IG (Daelemans and van den Bosch 1992; Daelemans, van den Bosch, and Weijters 1997), and IGTREE (Daelemans, van den Bosch, and Weijters 1997), the optimised decision-tree approximation of IB1-IG, all described above. Moreover, we selected two additional eager learning algorithms. First, we selected C5.0, an updated implementation of C4.5 (Quinlan 1993). The latter is a decision-tree algorithm similar to IGTREE, but which recomputes information gain at each added non-ending node, and which is supplied with various additional features. From these features, we selected subsets of values, and default pruning (for details, see Quinlan 1993). Second, we selected the connectionist back-propagation learning algorithm (Rumelhart, Hinton, and Williams 1986), henceforth BP – a good example of eager learning since it spends a considerable amount of effort in recoding the training material as real-valued weights of connections between simple processing units. BP is run on a single hidden-layer perceptron, using local codings of input ($7 \times 42 = 294$ units) and output (159 units), and 200 hidden units (for details, see van den Bosch 1997). Finally, to provide a baseline accuracy, the simple DC algorithm (van den Bosch 1997) was also selected for application to the GS task: DC classifies new instances either by copying the classification of duplicate instances in memory when found (i.e. exploiting the overlap between training material and new material), or, when the latter fails, by guessing the most frequent classification in the whole memory (i.e. exploiting the bias in the training material to the best uninformed guess).

Figure 7.3 displays all generalisation accuracies in terms of incorrectly classified test instances, a results summary, and the overlap and bias errors. A test instance is classified incorrectly when the phoneme part is misclassified, or the stress-marker part, or both. Performance is measured in terms of generalisation accuracy, i.e. the precision with which test instances are classified. In Figure 7.3, accuracies are displayed as generalisation errors, i.e. percentages of misclassified test instances.

The results displayed in Figure 7.3 indicate that IB1-IG performs best on test instances. The generalisation accuracy of IB1-IG is slightly but significantly better than that of IGTREE (a one-tailed t-test yields $t(19) = 1.94$, $p < 0.05$). All other differences are significant with $p < 0.001$, the smallest difference being between IGTREE and IB1 ($t(19) = 11.48$, $p < 0.001$). Thus, memory-based learning yields adequate performance when it is augmented with an information-gain weighted similarity metric in IB1-IG. IGTREE, the approximate optimisation of IB1-IG, generates a comparable, yet significantly higher number of errors than IB1-IG. While eager learning in IGTREE is competitive with IB1-IG, eager learning in C5.0 and BP results in considerably

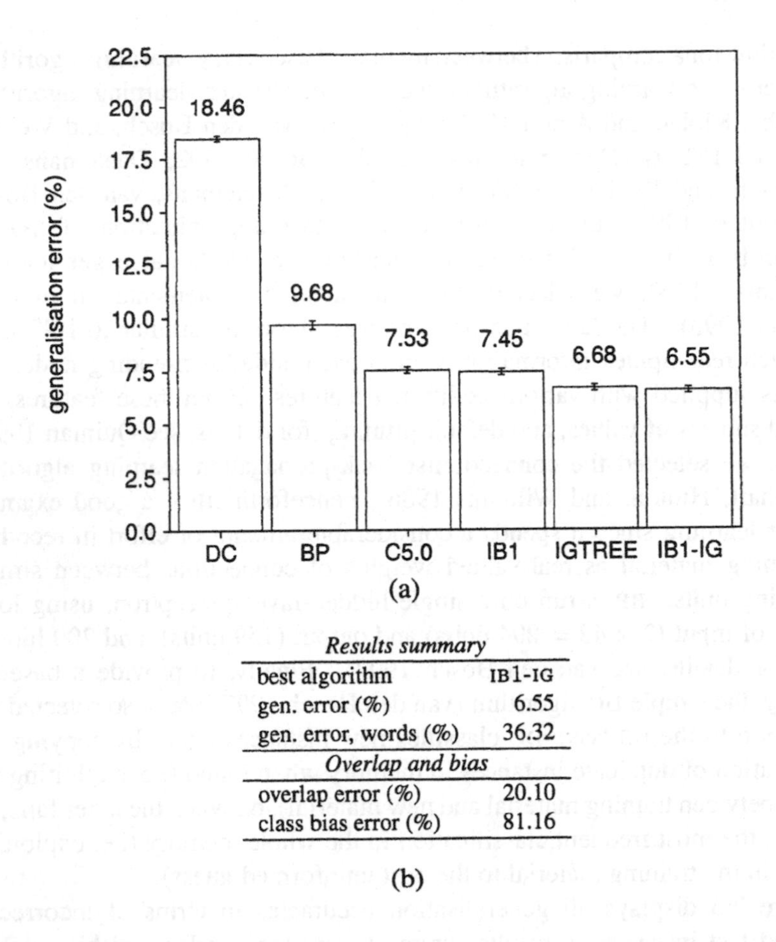

(a)

Results summary	
best algorithm	IB1-IG
gen. error (%)	6.55
gen. error, words (%)	36.32
Overlap and bias	
overlap error (%)	20.10
class bias error (%)	81.16

(b)

Figure 7.3. Generalisation accuracy of IB1, IB1-IG, IGTREE, C5.0, BP, and DC on the GS task: (a) generalisation errors in terms of the percentage incorrectly classified test instances of five algorithms; (b) results summary, bias and overlap errors.

lower accuracies than for IB1-IG. Both algorithms abstract from their learning material more eagerly than IGTREE does, either by explicit forgetting of feature-value information through pruning (in C5.0) or by compressed encoding (in BP). Abstraction as performed by C5.0 and BP appears harmful for generalisation performance for this task.

In sum, when optimal accuracy is desired for the GS task, the results point to IB1-IG as the most suitable algorithm. However, when computational efficiency is also desired in the target application, there appears to be a trade-off between IB1-IG and IGTREE. Table 7.3 displays the memory storage (in terms of kilobytes needed for storing all feature-value and class information) and processing costs (in terms of seconds needed to classify one

Table 7.3. Average memory storage (in kilobytes) and processing time per test instance (in seconds) obtained with IB1-IG and IGTREE applied to the GS dataset.

Algorithm	*Storage (KB)*	*Time per instance (s)*
IB1-IG	4,751	0.4826
IGTREE	760	0.0052

test instance, on average) of both algorithms. (Figures are for a PC equipped with a Pentium 75 MHz processor running Linux.) The slight advantage in generalisation accuracy of IB1-IG over IGTREE is at the cost of considerably more memory storage; IGTREE's memory compression compared to that of IB1-IG is 84.0%. Moreover, IGTREE is about 93 times faster than IB1-IG.

3. TRIBL and TREETALK

The results discussed in the previous section show that, for the phonemisation task, there is a slight (yet significant) performance loss when using IGTREE compared to pure memory-based learning as used in IB1-IG. For some other tasks, in which the relevance of the predictive features cannot be ordered in a straightforward way, IB1-IG or even IB1 may perform significantly better than IGTREE; e.g. when applied to morphological segmentation and stress assignment of English words (van den Bosch, Daelemans, and Weijters 1996; van den Bosch 1997). Clearly, what is needed is an algorithm that allows us to find an optimal position (from the point of view of the user of the resulting system) between the extremes of pure memory-based learning with feature weighting (IB1-IG) which is most accurate but least efficient, and IGTREE which is most efficient, but may result in lower generalisation accuracy.

3.1. The TRIBL Learning Method

The TRIBL method is a hybrid generalisation of IGTREE and memory-based learning in IB1-IG. TRIBL searches for an optimal trade-off between (i) minimisation of storage requirements by building an IGTREE (and consequently also optimisation of search time) and (ii) maximal generalisation accuracy. To achieve this, a parameter is set determining the switch from IGTREE to memory-based learning in IB1-IG in learning as well as in classification. The parameter value denotes the last level at which IGTREE compresses homogeneous subsets of instances into leaves or non-terminal nodes. Below this level, all remaining uninspected feature values of instances in non-homogeneous subsets (i.e. instances that are not yet disambiguated) are stored uncompressed in *instance-base nodes*. During search, the normal IGTREE search algorithm is used for the levels that have been constructed

Table 7.4. Average per-instance generalisation performance (with standard deviation) of IGTREE, IB1, IB1-IG and TRIBL, with memory storage and processing time per test instance, applied to the GS dataset.

Algorithm	*Generalisation accuracy (%)*	*Storage (KB)*	*Time per instance (s)*
IB1	92.54 ±0.16	4,751	0.4386
IB1-IG	93.45 ±0.15	4,751	0.4826
TRIBL-1	93.45 ±0.15	4,157	0.0075
TRIBL-2	93.45 ±0.15	3,551	0.0054
TRIBL-3	93.43 ±0.15	2,683	0.0053
TRIBL-4	93.34 ±0.15	1,601	0.0052
TRIBL-5	93.32 ±0.15	912	0.0052
IGTREE	93.32 ±0.15	760	0.0052

by IGTREE; when IGTREE search has not ended at the level marked as the switch point between IGTREE and IB1-IG, one of the instance-base nodes is accessed, and IB1-IG is employed on the sub-instance-base stored in this instance-base node.

3.2. Experiments with TREETALK

The architecture of the phonemisation system TREETALK described in Section 2.1 was modified to use TRIBL instead of IGTREE, trained on the same aligned data. (See Daelemans, van den Bosch, and Zavrel (1997) for a discussion of results of TRIBL on several non-linguistic benchmark datasets.) The empirical results are shown in Table 7.4. These results were obtained by means of 10-fold cross validation (CV) experiments (see Weiss and Kulikowski 1991 and p. 54 of van den Bosch 1997 for details) on the full GS data. The table repeats the results obtained with the pure memory-based learning methods IB1 and IB1-IG, the results with IGTREE, and lists in between the results obtained with different values for the various TRIBL switch points from IGTREE to memory-based learning. We show both accuracy and efficiency in terms of storage needed and processing time.

The results listed in Table 7.4 show that TRIBL indeed offers a trade-off between memory storage costs and processing time on the one hand, and generalisation accuracy on the other hand. As regards memory storage costs, the table shows a gradual decrease of the memory needed; the memory costs of IB1-IG can be reduced to about 50% (in the case of TRIBL-3) without significant loss of performance. With TRIBL-3, processing time is reduced considerably, and is hardly different from IGTREE's processing time. Even TRIBL-1 obtains a marked reduction in processing time, without any loss in performance; generating nodes representing the values of the most important feature (i.e. the middle grapheme) partitions the full instance base into 42 considerably smaller subsets (one for each letter of the alphabet). The alphabet in the English CELEX corpus contains 26 letters, 12 with occasional diacritics, and the 3 non-letter characters ., ', and -. With each search, only one out of 42 subsets needs to be searched to find the best classifications.

Here, we have tested all possible switch points between IGTREE and IB1-IG. We have also investigated two heuristics for finding an optimal switch point automatically. The results in Table 7.4 suggest that TRIBL-3 maintains IB1-IG's accuracy as well as IGTREE's speed.

First, we assert that the *branching factors* per level in IGTREE-constructed trees (i.e. the fan-out of feature-value tests per node per level) offer both an explanation for the performance of the various TRIBL versions as well as a potentially adequate switch point heuristic. The higher the branching factor of a node at a certain level in the tree, the smaller the subsets represented by the node's child nodes, i.e. the less consumptive the search. Alternatively, when a node has a low branching factor, it may not pay off to create daughter nodes, as compared to leaving the subset uncompressed, to be searched with a memory-based strategy. In the trees generated by IGTREE on the GS data, the first two levels have high average branching factors: 42 (i.e. the maximal branching factor) and 18, respectively. The lower five levels display average branching factors ranging from 4 to 1; the first two levels are clearly outliers. TRIBL-3 optimally reflects this branching factor difference: It builds a two-level tree on the basis of the 2 most important features, and leaves the remaining 5 feature values uncompressed. Since TRIBL-3 indeed matches the accuracy of IB1-IG and the processing speed of IGTREE, it can be said to be an optimal form of TRIBL for the GS task.

Second, we have experimented with a heuristic for estimating an optimal switch point between IGTREE and IB1-IG based on *average feature information gain*. When the information gain of a feature exceeds the sum of the average information gain of all features plus one standard deviation of the average, then the feature is used for constructing a decision-tree level (Daelemans, van den Bosch, and Zavrel 1997). This argues for TRIBL-1 as optimal, since only the information gain of the middle grapheme is markedly higher than the threshold value, while all other features fall below this value. This heuristic should yield a safer switch point than the branching factor heuristic. More systematic experiments are needed to determine the relative efficacy of the two.

With IGTREE and TRIBL, processing efficiency is sufficiently high to allow experimenting with architectures containing several separately-trained modules, mimicking the architecture of traditional knowledge-based systems which included modules for morphological analysis, syllabification, etc. In the next section, we investigate the benefits of such modularisation for obtaining optimal generalisation accuracy.

4. Modularity and Linguistic Representations

Mainstream phonological and morphological theories, influenced by Chomskian linguistic theory across the board since the publication of Chomsky and Halle (1968) have generally adopted the idea of abstraction levels in various guises (e.g. levels, tapes, tiers, grids) (Goldsmith 1976; Liberman and Prince 1977; Koskenniemi 1984; Mohanan 1986) as essential abstractions in modelling morpho-phonological processing tasks, such as word phonemisation.

According to these leading morpho-phonological theories, systems that (learn to) convert spelled words to phonemic words in one pass, i.e. without making use of abstraction levels, such as the learning algorithms trained on the GS task, are assumed to be unable to generalise to new cases. Going through the relevant abstraction levels is deemed essential to yield correct conversions. This assumption implies that if one wants to build a system that converts text to speech, one should implement explicitly all relevant levels of abstraction. As indicated earlier, such explicit implementations of abstraction levels can indeed be witnessed in many speech synthesisers, implemented as (sequential) modules (Allen, Hunnicutt, and Klatt 1987; Daelemans 1988).

This strong claim prompts the question: could the generalisation accuracy of IB1-IG, IGTREE, or TRIBL on the GS task be surpassed by systems which employ abstraction levels explicitly? We performed systematic experiments to investigate this question (see van den Bosch 1997 for a full description). We trained and tested a word-phonemisation system reflecting linguistic assumptions on abstraction levels quite closely: The model is composed of 5 sequentially-coupled modules. Second, we trained and tested a model in which the number of modules was reduced to 3, integrating 2 pairs of levels of abstraction. Third, we compared the generalisation accuracies of both modular systems with the results on the GS task reported in Section 3.

The resource of word-phonemisation instances used for these experiments is again the CELEX lexical database of English (Burnage 1990). We added to our corpus of 77,565 words with their pronunciations with stress markers all information related to these words' morphological segmentation, syllabification, and stress assignments. Each abstraction level is computed in the newly-tested systems by a specific module, each one performing a different morpho-phonological subtask. Analogous to our procedure for constructing instances for the GS task, an instance base is constructed for each subtask containing instances produced by windowing and attaching to each instance the classification appropriate for the (sub)task under investigation. Table 7.5 displays example instances derived from the example word *booking*. For each (sub)task, an instance base of 675,745 instances is built, as before.

In the table, six classification fields are shown, one of which is a composite field; each field refers to one of the (sub)tasks investigated here. M stands for

Table 7.5. Examples of instances generated from the word *booking*, with classifications for all of the subtasks investigated, viz. M, A, G, Y, S, and GS.

(a) Letter-window instances

Instance number	*Left context*			*Focus*	*Right context*			M	A	G	S	GS
1	-	-	-	*b*	*o*	*o*	*k*	1	1	/b/	1	/b/1
2	-	-	*b*	*o*	*o*	*k*	*i*	0	1	/u/	0	/u/0
3	-	*b*	*o*	*o*	*k*	*i*	*n*	0	0	/–/	0	/–/0
4	*b*	*o*	*o*	*k*	*i*	*n*	*g*	0	1	/k/	0	/k/0
5	*o*	*o*	*k*	*i*	*n*	*g*	-	1	1	/ɪ/	0	/ɪ/0
6	*o*	*k*	*i*	*n*	*g*	-	-	0	1	/ŋ/	0	/ŋ/0
7	*k*	*i*	*n*	*g*	-	-	-	0	0	/–/	0	/–/0

(b) Phoneme-window instances

Instance number	*Left context*			*Focus*	*Right context*			Y	S
1	-	-	-	/b/	/u/	/–/	/k/	1	1
2	-	-	/b/	/u/	/–/	/k/	/ɪ/	0	0
3	-	/b/	/u/	/–/	/k/	/ɪ/	/ŋ/	0	0
4	/b/	/u/	/–/	/k/	/ɪ/	/ŋ/	/–/	1	0
5	/u/	/–/	/k/	/ɪ/	/ŋ/	/–/	-	0	0
6	/–/	/k/	/ɪ/	/ŋ/	/–/	-	-	0	0
7	/k/	/ɪ/	/ŋ/	/–/	-	-	-	0	0

morphological decomposition; A is graphemic parsing; G is grapheme-phoneme conversion; Y is syllabification; S is stress assignment, and GS is integrated grapheme-phoneme conversion and stress assignment. (Graphemic parsing is not represented in the CELEX data. We used the automatic alignment algorithm mentioned earlier (cf. Daelemans and van den Bosch 1997) to determine which letters are the first or only letters of a grapheme (class 1) or not (class 0.) The example instances in Table 7.5 show that each (sub)task is phrased as a classification task on the basis of windows of letters or phonemes (the stress assignment task S is investigated with both letters and phonemes as input). As with the GS task, each window represents a snapshot of a part of a word or phonemic transcription, and is labelled by the classification associated with the middle letter of the window. For example, the first letter-window instance ---*book* is linked with label 1 for the morphological segmentation task (M), since the middle letter *b* is the first letter of the morpheme *book*; the other instance labelled with morphological-segmentation class 1 is the instance with *i* in the middle, since *i* is the first letter of the (inflectional) morpheme *ing*. Classifications may either be binary (1 or 0) for the segmentation tasks (M, A and Y), or have more values, such as 62 possible phonemes (G) or three stress markers (primary, secondary, or no stress, S), or a combination of these classes (159 combined phonemes and stress markers, GS).

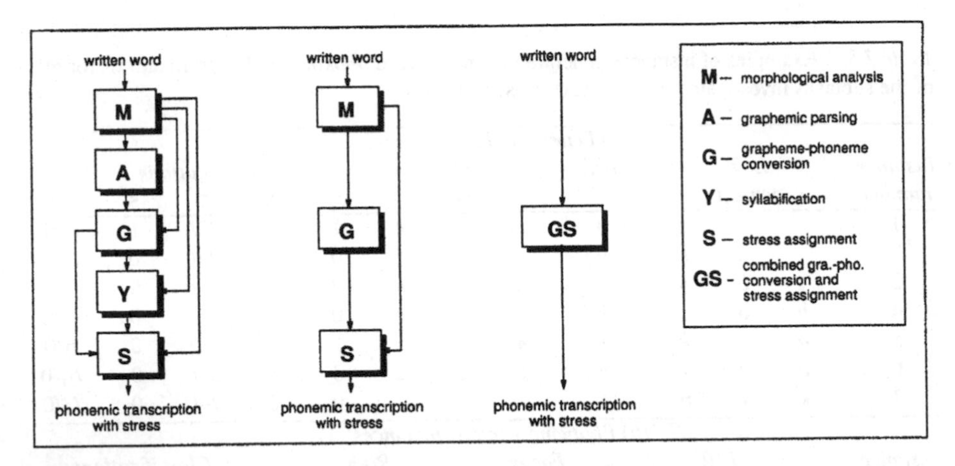

Figure 7.4. Architectures of the three investigated word-phonemisation systems. Left: M-A-G-Y-S; middle: M-G-S; right: GS. Rectangular boxes represent modules; the letter in the box corresponds to the subtask as listed in the legend (far right). Arrows depict data flows from the raw input or a module, to a module or the output.

4.1. Experiments

Our experiments are grouped into three series, each involving the application of IGTREE to a particular word-phonemisation system. IGTREE is selected rather than IB1-IG or TRIBL, since it is more efficient for the construction of a multi-module system. It is not feasible to hold 5 instance bases in memory at the same time, while keeping 5 decision trees in memory simultaneously is possible on a moderate-specification personal computer. The architectures of the two modular systems are displayed in Figure 7.4, which also shows the simple architecture of the system performing the GS task. We briefly outline both modular systems and report on the experiments needed for constructing them.

M-A-G-Y-S. The architecture of the M-A-G-Y-S system is inspired by SOUND1 (Hunnicutt 1976, 1980), the word phonemisation subsystem of the MITtalk text-to-speech system (Allen, Hunnicutt, and Klatt 1987). When MITtalk is faced with an unknown word, SOUND1 produces on the basis of that word a phonemic transcription with stress markers. This word phonemisation process is divided into the following five processing components:

1 *morphological segmentation*, which we implement as the module M;

2 *graphemic parsing*, module A;

3 *grapheme-phoneme conversion*, module G;

4 *syllabification*, module Y;

5 *stress assignment*, module S.

The architecture of the M-A-G-Y-S system is depicted to the left of Figure 7.4. The representations utilised include direct output from previous modules, as well as representations from earlier modules. For example, the S module takes as input the syllable boundaries generated by the Y module, but also the phoneme string generated by the G module, and the morpheme boundaries generated by the M module.

M-A-G-Y-S is put to the test by applying IGTREE in 10-fold CV experiments to the five subtasks, connecting the modules after training, and measuring the combined score on correctly classified phonemes and stress markers, which is the desired output of the word-phonemisation system. An individual module can be trained on data from CELEX directly as input, but this method ignores the fact that modules in a working modular system can be expected to generate some amount of error. When one module generates an error, the subsequent module receives this error as input, assumes it is correct, and may generate another error.

In a five-module system, this type of cascading error may seriously hamper generalisation accuracy. To counteract this potential disadvantage, modules can also be trained on the output of previous modules. We cannot expect to learn to repair completely random, irregular errors, but whenever a previous module makes consistent errors on a specific input, this may be recognised by the subsequent module. Having detected a consistent error, the subsequent module is then able to repair the error and continue with successful processing. Earlier experiments performed on the tasks investigated in this chapter have shown that classification errors on test instances are indeed consistently and significantly decreased when modules are trained on the output of previous modules rather than on data extracted directly from CELEX (van den Bosch 1997). Therefore, we train the M-A-G-Y-S system with IGTREE in the variant in which modules are trained on the output of other modules. We henceforth refer to this type of training as *adaptive*, referring to the adaptation of a module to the errors of a previous module.

Figure 7.5 displays the results obtained with IGTREE under the adaptive variant of M-A-G-Y-S. The figure shows all percentages (displayed above the bars; error bars on top of the main bars indicate standard deviations) of incorrectly classified instances for each of the five subtasks, and a joint error on incorrectly classified phonemes with stress markers, which is the desired output of the system. The latter classification error, labelled PS in Figure 7.5, regards classification of an instance as incorrect if either or both of the phoneme and stress marker is incorrect. The figure shows that the joint error on phonemes and stress markers is 10.59% of test instances, on average. Computed in terms of transcribed words, only 35.89% of all test words are converted to stressed phonemic transcriptions flawlessly. The joint error is lower than the sum of the

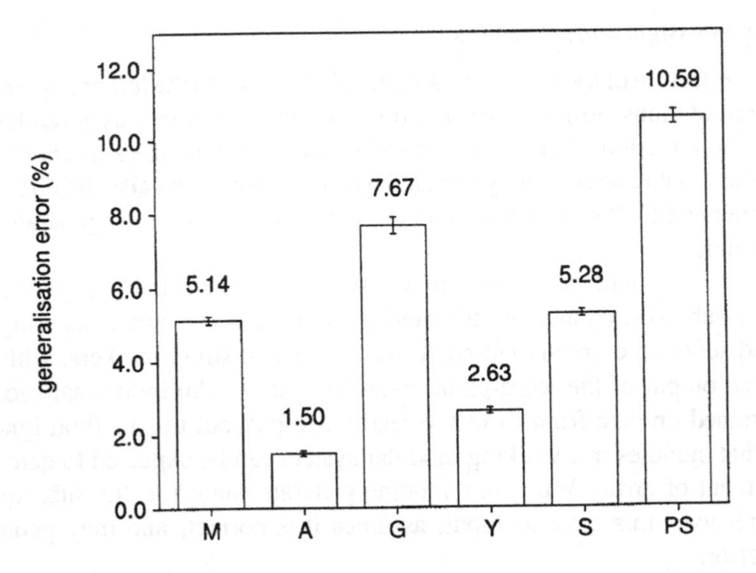

Figure 7.5. Generalisation errors on the M-A-G-Y-S system in terms of the percentage of incorrectly classified test instances by IGTREE on the five subtasks M, A, G, Y and S, and on phonemes and stress markers jointly (PS).

errors on the G subtask and the S subtask, viz. 12.95%, suggesting that about 20% of the incorrectly classified test instances involve an incorrect classification of both the phoneme and the stress marker.

M-G-S. The subtasks of graphemic parsing (A) and grapheme-phoneme conversion (G) are clearly related. While A attempts to parse a letter string into graphemes, G converts graphemes to phonemes. Although they are performed independently in M-A-G-Y-S, they can be integrated easily when the class-1-instances of the A task are mapped to their associated phoneme rather than 1, and the class-0-instances are mapped to a phonemic null, /–/, rather than 0 (cf. Table 7.5). This task integration is also used in the NETtalk model (Sejnowski and Rosenberg 1987). A similar argument can be made for integrating the syllabification and stress assignment modules into a single stress-assignment module. Stress markers, in our definition of the stress-assignment subtask, are placed solely on the positions which are also marked as syllable boundaries (i.e. on syllable-initial phonemes). Removing the syllabification subtask makes finding those syllable boundaries which are relevant for stress assignment an integrated part of stress assignment. Syllabification (Y) and stress assignment (S) can thus be integrated in a single stress-assignment module S.

When both pairs of modules are reduced to single modules, the three-module system M-G-S is obtained. Figure 7.4 displays the architecture of the M-G-S system in the middle. Experiments on this system were performed

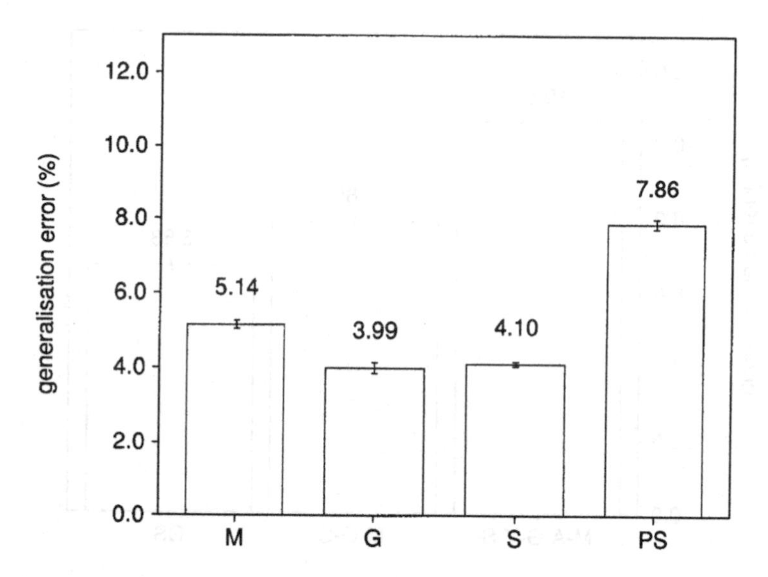

Figure 7.6. Generalisation errors on the M-G-S system in terms of the percentage of incorrectly classified test instances by IGTREE on the three subtasks M, G, and S, and on phonemes and stress markers jointly (PS).

analogous to those with the M-A-G-Y-S system. Figure 7.6 displays the average percentages of generalisation errors generated by IGTREE on the three subtasks and phonemes and stress markers jointly (the error bar labelled PS).

Removing graphemic parsing (A) and syllabification (Y) as explicit in-between modules yields better accuracies on the grapheme-phoneme conversion (G) and stress assignment (S) subtasks than in the M-A-G-Y-S system. Both differences are significant; for G, $t(19) = 43.70$, $p < 0.001$ and for S, $t(19) = 32.00$, $p < 0.001$. The joint accuracy on phonemes and stress markers is also significantly better in the M-G-S system than in the M-A-G-Y-S system, $t(19) = 37.50$, $p < 0.001$. Unlike M-A-G-Y-S, the sum of the errors on phonemes and stress markers, 8.09%, is barely more than the joint error on PS instances of 7.86%: there is hardly any overlap in instances with incorrectly classified phonemes and with incorrectly placed stress markers. The percentage of flawlessly processed test words is 44.89%, which is markedly better than the 35.89% of M-A-G-Y-S.

4.2. Comparing the Modular Systems

Figure 7.7 repeats the results obtained from applying IGTREE to each of the three systems, in terms of the percentages of incorrectly classified PS instances. IGTREE yields significantly better generalisation accuracy on phonemes and stress markers, both jointly and independently, trained on GS, as compared

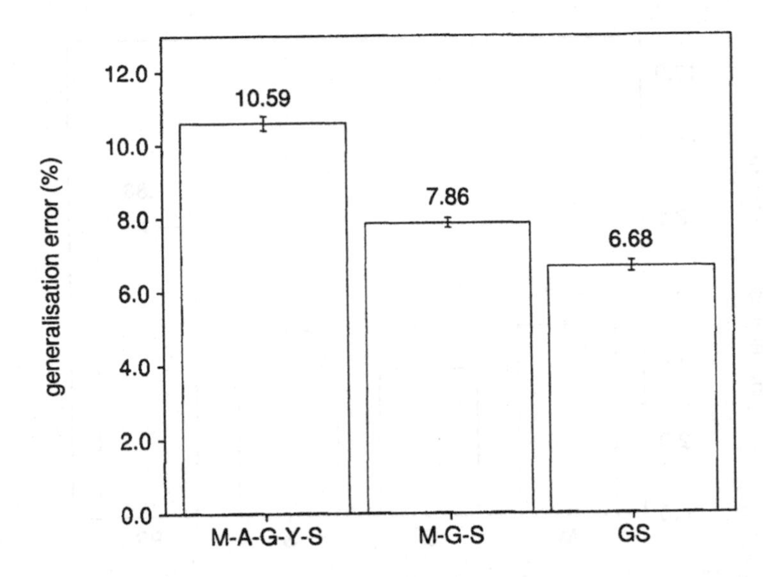

Figure 7.7. Generalisation performances of IGTREE in terms of the percentage of incorrectly processed PS instances, applied under the adaptive variant to the two modular systems M-A-G-Y-S and M-G-S, and applied to the single-module system GS.

to M-A-G-Y-S and M-G-S. In terms of PS instances, the accuracy on GS is significantly better than that of M-G-S with $t(19) = 40.48$, $p < 0.001$, and that of M-A-G-Y-S with $t(19) = 6.90$, $p < 0.001$. Thus, under our experimental conditions and using IGTREE as learning algorithm, optimal generalisation accuracy on word phonemisation is obtained with GS, the system that does not incorporate any explicit decomposition of the word-phonemisation task.

As an additional comparison between the three systems, we analysed the positive and negative effects of learning the subtasks in their specific systems' context. The particular sequence of the five modules as in the M-A-G-Y-S system reflects a number of assumptions on the *utility* of using output from one subtask as input to another. Morphological knowledge is useful as input to grapheme-phoneme conversion (e.g. to avoid pronouncing *ph* in *loophole* as /f/, or *red* in *barred* as /rɛd/); graphemic parsing is similarly useful as input to grapheme-phoneme conversion (e.g. to avoid a non-null pronunciation of *gh* in *through*); etc. Thus, feeding the output of a module *A* into a subsequent module *B* implies that one expects to perform better on module *B* with *A*'s input than without. The results obtained with the M-A-G-Y-S, M-G-S, and GS systems can serve as tests for their respective underlying utility assumptions, when they are compared to the accuracies obtained with their subtasks learned in isolation.

To measure the utility of including the outputs of modules as inputs to other modules, we performed the following experiments:

Table 7.6. Utilities of learning subtasks (M, A, G, Y, and S) as modules or partial tasks in the M-A-G-Y-S, M-G-S, and GS systems. For each module in each system, the utility of training the module with ideal data (middle) and actual, modular data under the adaptive variant (right) is compared against the accuracy obtained with learning the subtasks in isolation (left). Accuracies are given in percentage of incorrectly classified test instances.

Subtask	*Isolated error (%)*	*Ideal error (%)*	*Utility*	*Actual error (%)*	*Utility*
		M-A-G-Y-S			
M	5.14	5.14	0.00	5.14	0.00
A	1.39	1.66	−0.27	1.50	−0.11
G	3.72	3.68	+0.04	7.67	−3.95
Y	0.45	0.75	−0.30	2.63	−2.16
S	7.96	2.67	+5.29	5.28	+2.68
		M-G-S			
M	5.14	5.14	0.00	5.14	0.00
G	3.72	3.66	+0.06	3.99	−0.27
S	7.96	3.97	+3.99	4.10	+3.86
		GS			
G	3.72	–	–	3.79	−0.07
S	4.71	–	–	3.76	+0.95

1 IGTREE was applied in 10-fold CV experiments to each of the five subtasks M, A, G, Y and S, only using letters (with the M, A, G and S subtasks) or phonemes (with the Y and S subtasks) as input, and their respective classification as output (cf. Table 7.5). The input is directly extracted from CELEX. These experiments provide the baseline score for each subtask, and are referred to as the *isolated* experiments.

2 We applied IGTREE in 10-fold CV experiments to all subtasks of the M-A-G-Y-S, M-G-S, and GS systems, training and testing on input extracted directly from CELEX. The results from these experiments reflect the putative accuracy of the modular systems if each module performed perfectly flawless. We refer to these experiments as *ideal*.

From the results of these experiments we measure, for each subtask in each of the three systems, the utility of including the input of preceding modules, for the ideal case (with input straight from CELEX) as well as for the actual case (with input from preceding modules). The utility measure is the difference between IGTREE's generalisation error on the subtask in modular context (either ideal or actual) and its accuracy on the same subtask in isolation. Table 7.6 lists all computed utilities.

For the M-A-G-Y-S system, it can be seen that the only large utility, even in the ideal case, could be obtained with the stress-assignment subtask. In the isolated case, the input consists of phonemes; in the M-A-G-Y-S system, the input contains morpheme boundaries, phonemes, and syllable boundaries. The ideal positive effect on the S module of 5.29% fewer errors turns out to be a positive effect of 2.68% in the actual system. The latter positive effect is outweighed by a rather large negative utility on the grapheme-phoneme conversion task of −3.95%. Neither the A nor Y subtasks profit from morphological boundaries as input, even in the ideal case; in the actual M-A-G-Y-S system, the utility of including morphological boundaries from M and phonemes from G in the syllabification module Y is markedly negative at −2.16%.

In the M-G-S system, the utilities are generally less negative than in the M-A-G-Y-S system. There is a small utility in the ideal case with morphological boundaries included as input to grapheme-phoneme conversion; in the actual M-G-S system, the utility is negative (−0.27%). The stress-assignment module benefits from including morphological boundaries and phonemes in its input, both in the ideal case and in the actual M-G-S system.

The GS system does not contain separate modules, but it is possible to compare the errors made on phonemes and stress assignments separately to the results obtained on the subtasks learned in isolation. Grapheme-phoneme conversion is learned with almost the same accuracy in isolation or when learned as a partial task of the GS task. Learning the grapheme-phoneme task, IGTREE is neither helped nor hampered significantly by learning stress assignment simultaneously. There is a positive utility from learning stress assignment, however. When stress assignment is learned in isolation with letters as input, IGTREE classifies 4.71% of test instances incorrectly on average. (This is a lower error than obtained with learning stress assignment on the basis of phonemes, indicating that stress assignment should take letters as input rather than phonemes.) When the stress-assignment task is learned along with grapheme-phoneme conversion in the GS system, a marked improvement is obtained: 0.95% fewer classification errors are made.

Summarising, comparing the accuracies on modular subtasks to the accuracies on their isolated counterpart tasks shows only a few positive utility gains in the actual system, all obtained with stress assignment. The largest utility is found on the stress-assignment subtask of M-G-S. However, this positive utility does not lead to optimal accuracy on the S subtask; in the GS system, stress assignment is performed with letters as input, yielding the best accuracy on stress assignment in our investigations, viz. 3.76% incorrectly classified test instances.

5. Conclusion

We have presented a memory-based approach to learning word phonemisation. The approach is data-oriented and language-independent. Its only prerequisite is a reasonably large word-pronunciation corpus, which is available for many languages. Furthermore, we have argued that a memory-based approach to learning word phonemisation should be favoured over other generic machine-learning approaches aimed at abstraction of learning material. We have demonstrated that retaining full memory of all training instances consistently yields optimal generalisation accuracy. The more a learning algorithm attempts to abstract from individual learning instances, the more its generalisation accuracy is harmed. This has been demonstrated on a corpus of 77,565 English word-pronunciation pairs; IB1-IG, a memory-based learning algorithm with an information-gain weighted similarity metric, outperformed other algorithms and obtained a generalisation accuracy of 93.5% correctly classified phonemes (including stress markers, 63.7% flawlessly converted test words).

A disadvantage of memory-based learning is its large computational demands in terms of memory and slow processing; we therefore developed IGTREE, a decision-tree learning algorithm which is based on IB1-IG. IGTREE uses a fraction of IB1-IG's memory requirements and is much faster; however, it performs slightly worse in terms of accuracy. Subsequently, we presented and tested TRIBL, a hybrid between IB1-IG and IGTREE, and demonstrated that this hybrid could indeed maintain IB1-IG's performance while processing at the speed of IGTREE. Memory-based learning can be optimised, albeit carefully, by finding an appropriate hybrid of decision-tree learning of the most relevant features of the data, and memory-based learning of the remaining features.

Finally, we demonstrated that the explicit modelling of abstraction levels as sequentially-trained memory-based modules, as suggested by mainstream morpho-phonological theories, does not improve performance on word phonemisation. The best performance is obtained by training a learning algorithm on the direct conversion of grapheme strings to a combined class representing phonemic mapping and stress assignment simultaneously. Although positive utility gains of sequenced modules are found, they are overshadowed by errors propagated through the modules; the more modules in the system, the more the overall output of the system suffers in this way. Within the scope of our experimental settings, we conclude that abstraction levels are best left implicit: our results strongly suggest that the most relevant information for phonetic mapping and stress assignment is stored implicitly in the written words themselves, in relatively local areas.

Further research is needed to measure the effects of widening the window used for generating instances. We have assumed that a 3-1-3 window (three left-context graphemes, a focus grapheme, and three right-context graphemes) is sufficient and adequate but this is clearly rather dangerous, since a considerable number of word-pronunciation instances remain ambiguous with this context. Other areas of relevant future investigation include using transcribed text as training material, using previous classifications as input features, learning meta-word phenomena such as intonation contours and sentence accents, and incorporating trained GS-modules (e.g. optimised versions of TRIBL) in text-to-speech synthesis applications.

Acknowledgements

This research was carried out in the context of the "Induction of Linguistic Knowledge" research programme, partially supported by the Foundation for Language Speech and Logic (TSL), which is funded by the Netherlands Organisation for Scientific Research (NWO). Part of the second author's work was performed at the Department of Computer Science of the Universiteit Maastricht. We thank Jakub Zavrel, Bertjan Busser, the other members of the Tilburg ILK group, Ton Weijters and Eric Postma for fruitful discussions and feedback.

Chapter 8

LEARNABLE PHONETIC REPRESENTATIONS IN A CONNECTIONIST TTS SYSTEM – I: TEXT TO PHONETICS

Andrew D. Cohen
Independent Consultant

Abstract Results from connectionist experiments in text-to-speech conversion suggest that non-symbolic intermediate ('phonetic') representations may have a useful part to play in the design of a synthesis system. A similar strategy suggests itself in the subsequent stage when speech is produced from the intermediate representation, which makes it possible to bypass a symbolic, phonemic stage in the overall system, once trained. (This second stage is dealt with in a later chapter.) Error can still be calculated in terms of phonemes correct, but this is not necessarily a good measure of the naturalness and acceptability of the output speech. In contrast to other trainable text-to-speech systems, emphasis is laid here on the fundamental importance of phonetic and phonological sources of variability, and their separation from the underlying physical and temporal events. As far as possible, this phonetic/phonological capability should be built into the system prior to training on the main task at hand, as this corresponds more closely to the way these skills are acquired in human beings.

1. Introduction

Text-to-speech (TTS) systems generally break the problem down into two stages: (1) conversion of text to some intermediate representation of pronunciation and (2) conversion of this representation into synthetic speech. The choice of intermediate representation is bound to have an important impact on what information can and cannot be easily captured and, therefore, on the quality and nature of the speech output. By tradition, this intermediate representation is symbolic. Almost invariably, it is *phonemic*: many specific examples of this will be found elsewhere in this book. In this chapter, we

R. I. Damper (ed.), Data-Driven Techniques in Speech Synthesis, 175-197.

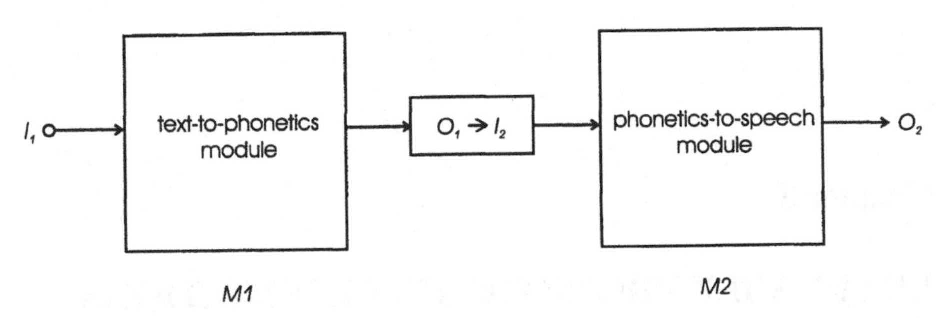

Figure 8.1. The complete SOMtalk architecture consists of two modules, $M1$ and $M2$. Module $M1$ takes text as input I_1 and produces as output O_1 coordinates in the D_phon phonetic space. This is then converted into coordinates in the D_dip diphone space, which act as input I_2 to module $M2$. This module produces synthetic speech as output O_2. This chapter is concerned with the text-to-phonetic module M_1 only. The phonetics-to-speech module is the subject of Chapter 12. Note that the 'intermediate' representations, O_1 and I_2, by their reference to coordinates in phonetic and diphone spaces respectively, are non-symbolic.

argue for a different, non-symbolic representation that – for want of a better description – we will call *phonetic*.

In any speech synthesis system, it is clearly worthwhile to include phonetic and phonological knowledge. In fact, given fundamental limitations on schemes for blank-slate learning (cf. Wolpert and Macready 1995, 1997), it is probably essential to incorporate a bias towards phonological sources of variability in the representations used. Many phoneticians have argued that coarticulation cannot account for all the variability found in natural speech, and that a dynamic non-segmental phonetics is needed to reflect the true nature of spoken language (Coleman and Local 1992; Local 1994). A major issue, therefore, is how this type of phonetics can be rendered into a form that is usable with self-learning schemes such as neural networks. Compositional structures such as trees, rewrite rules, or complex layered hierarchies are not very compatible with the vector/matrix operations performed by most neural networks (Chalmers 1990; Adamson and Damper 1999). Many phenomena (notably assimilation) can be handled by a monostratal non-segmental phonology (Local 1992), in which there are no phonological processes. We propose to replace traditional, segmental phonetic notations with a non-symbolic, learnable representation called D_phon (Cohen 1995) based on a trajectory through a two-dimensional (2D) phonetic space derived from the self-organising map or SOM (Kohonen 1982, 1990), and to represent phonological factors in terms of attractor basins developed through auto-associative (AA) networks. Accordingly we call the complete system SOMtalk.

Figure 8.1 shows the overall structure of SOMtalk. It consists of two modules, $M1$ and $M2$. Module $M1$ takes text as input I_1 and produces as output O_1 coordinates in the D_phon phonetic space. A simple transformation is then made from D_phon coordinates to coordinates in a diphone space, called D_dip (Cohen 1995). Module $M2$ takes this representation as input I_2 and produces synthetic speech as output O_2. This chapter is concerned with the text-to-phonetics module M_1 only. The phonetics-to-speech module is the subject of Chapter 12. These spaces are built empirically from each dataset.

Clearly, human beings who learn to convert text to speech have prior knowledge of the phonetics and phonology of their language (not all phonetic strings are possible and not all possible phonetic strings are words) as they are speakers of a particular language before they learn to read it. In speech recognition, efforts are being made to incorporate phonetic knowledge both within neural network approaches (Djezzar and Haton 1995) and with statistical methods (Deng and Sun 1994; Deng 1997). In text-to-speech (TTS) conversion, a phonological approach has been taken in automatic generation of triphone units (Yoshida, Nakajima, Hakoda, and Hirokawa 1996). Use of neural networks is now widespread in speech synthesis, mainly as a result of the advantages of trainability and parsimony they offer. Some examples are Reidi (1995), Karaali, Corrigan, and Gerson (1996) and Lagana, Lavagetto, and Storace (1996). Other data-driven methods, such as hidden Markov models (HMMs) (Parfitt and Sharman 1991; Luk and Damper 1993a, 1993b; Donovan 1996; Donovan and Woodland 1999), are being used, in most cases based on a segmental phonemic notation (hence the emphasis on problems such as determining durations of phones, and modelling coarticulatory effects). However, the approaches based on technologies primarily developed for recognition (especially HMMs) represent modifications of highly complex systems in which the underlying assumptions are unchanged. These assumptions are motivated by the need to keep training and decoding computationally tractable and this makes it difficult to reflect in the implementation recent trends in phonetic and phonological theory. Specifically, to avoid the segmental approach (and its attendant difficulties), new ways of representing systems of phonetic and phonological contrasts are required which emphasise the continuous, dynamic nature of speech. For example, we may represent phonological contrasts in terms of transitions rather than phones. The two-stage approach adopted here (Fig. 8.1) lays fundamental emphasis on phonetic and phonological factors, and their separation from the underlying physical and temporal events.

We explore ways of incorporating prior knowledge about a phonological problem (here conversion of text to 'phonetics') into neural networks, which operate in a 'black box' manner relatively opaque to interpretation, and which start from a *tabula rasa* and the dataset alone. As many authors in the neural network field have noted, it is better to focus on issues such as representations and prior knowledge rather than the learning dynamics, since there are no general-purpose learning algorithms which are good for an arbitrary prior (Wolpert and Macready 1995, 1997). Hence, there is a need to introduce domain-specific (here phonetic/phonological) knowledge in modules that can influence the development of other modules, or modify their output.

The remainder of this chapter is structured as follows. Section 2 gives some preliminaries and background to the problem. Section 3 details the dictionary used and the form of the inputs and outputs, I_1 and O_1 respectively, to the text-to-phonetics module M_1. Subsequently, Section 4 describes how phonetic knowledge may be built into the learning procedures using a form of error correction. This has been carried out with auto-associator ('corrector') networks – a rudimentary form of knowledge transfer – and through an ambiguity measure derived from using multiple neural networks. In Section 5, the idea of model selection is introduced, and we note that none of the regularisation schemes developed so far offers a complete solution. This suggests there will be no general means of incorporating any prior constraints thought to be important, and therefore specific codings must be developed for each domain. In Section 5.5, the issue of how much correction to apply is discussed, and some basic schemes are described. In Section 6, we give results for single networks with correction, and develop a more sophisticated correction scheme using ambiguity in combination with other procedures, which shows much improved performance over a single network. Some introductory material on the bias/variance dilemma, neural networks and combining multiple networks is included, but no derivations are given for the formulas used. The latter can be found in the sources cited. However, this work does not fall neatly within the machine-learning tradition because it attempts to incorporate prior phonetic knowledge from outside the dataset. Finally, Section 7 concludes with some pointers to future work.

2. Problem Background

The text-to-phoneme (or grapheme-phoneme) conversion problem has been tackled using a variety of schemes from the neural network and machine-learning communities, including the work of Lucassen and Mercer (1984), Sejnowski and Rosenberg (1987), McCulloch, Bedworth, and Bridle (1987), Wolpert (1992), van den Bosch and Daelemans (1993), Dietterich, Hild, and Bakiri (1995) and Adamson and Damper (1996). (See Damper 1995 for a detailed review which also includes treatment of some of the rule-based methods.) Many of these use a letter-by-letter approach, which makes no attempt to exploit linguistic information, such as a human reader's tacit knowledge of phonetics and phonology. While analogy-based approaches (Dedina and Nusbaum 1991; Sullivan and Damper 1993; Damper and Eastmond 1997) naturally use larger-size units than the letter, there is a need to go further in considering wider linguistic/phonetic phenomena than those based on the grapheme, but it remains a problem to see how this might be done in terms of segmental phonetics. Many such wider phonetic phenomena, not just stress, exceed both letter- and grapheme-based units in their scope. This knowledge need not necessarily be introduced through rewrite rules, rule orderings, finite-state automata, feature matrices, auto-segmental representations or other formal devices which have no phonetic or phonological content in themselves. Rather, it is possible to continue a trend towards richer representations: there are implicit, non-symbolic ways of introducing phonetic structure which use a complete phonetic space, rather than attempt to find the correct atoms or features into which to break it up. In this work, we use a topographic, ordered phonetic space D_phon (and a simple extension D_dip to diphone space – see Chapter 12) to introduce a distance measure based on a 2D feature space.

Learning a grapheme-phoneme mapping has typically been viewed as a multiclass classification problem, probably because each phoneme (class of sound) has been viewed as a discrete entity. The boolean nature of most phonetic features (and the use of phonemes themselves) leads very naturally to the task being viewed as learning a complex, discrete mapping between sets of binary strings. Sejnowski and Rosenberg (S&R) in their pioneering NETtalk introduced a distributed output encoding consisting of a string of binary phonetic features. The actual output was the 'closest' of the 324 valid binary strings, according to some measure such as the angle between the two output strings considered as vectors. In this way, the nature of the discrete mapping onto a limited set of classes is maintained, although the output space would contain 2^{26} strings if all the possible (but including non-permitted) combinations of phonetic features were allowed.

One way to generate this kind of distributed encoding is through error-correcting codes, which have long been used in the machine-learning literature. The use of these codes requires considerable insight into the problem at hand. For example, the process of achieving a good separation in Hamming space of each codeword for a specific problem has by no means been automated, and manual correction may be necessary. As noted by Dietterich and Bakiri (1991), performance may improve as the length of the code increases, but the transparency of the results is correspondingly diminished. A detailed comparison between ID3 and back-propagation has been made by Dietterich, Hild, and Bakiri (1995), and with error-correcting codes by Dietterich and Bakiri (1995). However, these codes are best suited to multiclass classification problems, and cannot be used when function approximation is sought, requiring that outputs not in the training set be acceptable as correct answers. The latter requires interpolation, something decision tree-based classifiers such as ID3 and CART (Quinlan 1979, 1982, 1986; Breiman, Friedman, Olshen, and Stone 1984) are not naturally equipped to do. In NETtalk, the output was the nearest correct (i.e. phonetically permitted) pattern, where the distance measure was either Euclidean or based on the angle between the vectors (both gave similar results). If a suitable metric space can be learned from features known to be significant in the domain, then many of these difficulties can be avoided, and the metric may be adapted to the task.

Additionally, if the two domains (text and phonemes) are viewed as essentially symbolic, a question arises as to whether a sub-symbolic, data-driven procedure is appropriate for this task. An alternative is to view the phonetic domain as a continuous phoneme space, in which points lying between phonemes represent intermediate sounds (although not impossible combinations of binary phonetic features). In this way, the task and the view of what counts as successful generalisation can be recast in terms of function approximation, where interpolation may generate entirely new allowable outputs not seen in the original training set. Prior knowledge of likely or 'good' outputs can be incorporated through techniques such as attractor basins and use of a phonetic or diphone metric (D_phon or D_dip) in the output pattern space which build structure into the network, rather than pruning or massaging of the dataset. It is hypothesised that success in this task, as well as in the more general problem of modelling variability, is a matter of using good representations rather than complex computations. The methodology employed here does not rely on knowing a good representation for a task in advance, but on learning in one or more modules which then transfer their capabilities to other modules in the system, or which improve the performance of the latter in some way.

3. Data Inputs and Outputs to Module $M1$

In this section, we describe the dictionary used to produce training datasets, and define the input/output encodings employed.

3.1. Dictionary

The dataset is the (approximately) 20,000 words used by Sejnowski and Rosenberg (1987)), converted to a Southern British English pronunciation using the Medical Research Council (MRC) psycholinguistic database (Coltheart 1981). The S&R database has the useful property of having been (manually) aligned so that the number of symbols in the input and target are equal, with 'silent' letters mapped onto the null phoneme. This same alignment scheme has been adopted, although the MRC notation contains many two-character symbols, and so it has been transcribed into a notation which uses only one-character symbols.

The adapted MRC notation is then converted to one derived from a phoneme set based on a set of 10 pseudo-articulatory features which gave 43 phoneme symbols. These features themselves are not used, except indirectly through the phonetic SOM derived from them (see Section 4 later). Actual target values are generated from the coordinates of a (12 × 8) SOM. The lexical stress information in the S&R database is discarded, as so many of the stress markings would be subject to modification by prosodic factors in a practical TTS system. Whether *subject*, for example, is a noun or a verb is clearly a sentence-level matter. Prosody – the specification of an intonation contour and durations – is considered to be a separate matter best handled at a higher level in the TTS process. It has been suggested that durations be treated primarily at a syllable level (Campbell 1992), although clearly one could go further in this direction (cf. Coleman and Local 1992). Simplifying the problem in this way also allows successful convergence of network training with many fewer hidden units than used in previous studies (Sejnowski and Rosenberg 1987; Dietterich, Hild, and Bakiri 1990). Convergence can be achieved with as few as 40 hidden units on a training set of 4000 words.

3.2. Encoding of Inputs and Outputs

A moving window of 10 input letters is used with 22 bits to encode a letter, giving 220 inputs to the neural network(s) to be described in Section 4 below. In NETtalk, 26 bits per letter were used in a 1-out-of-*n* encoding scheme, which is here modified by using 21 bits plus 1 bit for the five vowel letters (*a*, *e*, *i*, *o*, *u*), giving 22 bits per letter and, hence, 220 inputs for a 10-letter input vector. For each vowel, the vowel bit is set plus one other; for each consonant one bit is set in a 1-out-of-*n* scheme. A space is encoded as all zeros. In the 'letter space' thus created, vowels and consonants are separately clustered and within these two clusters each point is at an equal Hamming distance. A pretraining set of 4000 words gave 26,560 patterns, the full 19,988-word training set gave 166,160 patterns. (Since two phonemes are translated at a time – see immediately below – it might be possible to reduce this number by a half with a window shifted by two letters/phonemic symbols per step. This, however, has not been tried here.)

The choice of output encoding is of particular importance. Output is the two phonemes for the central two letters of the window. This generates two sets of coordinates on the (12 × 8) phonetic SOM. Hence, 20 units for each (x, y) coordinate pair in a 1-out-of-*n* encoding scheme gave 40 units in the output layer. Various other schemes have been tried for encoding coordinates, but little or no improvement in speed of learning or test set performance was obtained.

The pattern set is generated by taking each word and its phonemic transcription, and advancing them through the 10 character input window by 1 character at a time. So, for example, the word *government* (phonetic transcription `g^v@nm@nt` in my notation, which reflects the representation developed by the SOM) would produce nine patterns as shown in Table 8.1. In

Table 8.1. Sequence of input and output windows for the word *government*.

Input window	*Output window*	*Coordinates in phoneme space*
. . . . go v e r n	g ^	(6,8) (4,1)
. . . g ov e r n m	^ v	(4,1) (11,1)
. . g o ve r n m e	v @	(11,1) (2,2)
. g o v er n m e n	@ -	(2,2) (1,3)
g o v e rn m e n t	- n	(1,3) (7,1)
o v e r nm e n t .	n m	(7,1) (8,8)
v e r n me n t . .	m @	(8,8) (1,3)
e r n m en t . . .	@ n	(1,3) (7,6)
r n m e nt	n t	(7,6) (5,4)

this table, the symbol '–' represents the null phonetic character, used for letters which do not map directly to a sound, following the alignment method of S&R.

S&R used a distributed representation for the output patterns, which is well known to have many useful properties in neural network training (e.g. Hinton, McClelland, and Rumelhart 1986). This used a 26-bit code based on 21 articulatory phonetic features (position of tongue, whether fricative, nasal, liquid, glide and so on) and a further 5 bits were used to encode stress information. In this way, similar phonemes have similar target vectors, but there is no attempt to regulate exact distances between phonemes (e.g. the Hamming distance is meaningless in their encoding, as is the magnitude of each vector), and the overall complexity of the encoding is high. Non-phonetic features such as pause, full-stop and silent are introduced to deal with features which are essentially prosodic, and are really on a different level to the purely phonetic articulatory features. This information is left out of the problem as treated here: no attempt is made to learn prosodic features such as intonation, stress and so on, which are probably better handled with a symbolic procedure and a specialised mark-up language on the text in question.

4. Detailed Architecture of the Text-to-Phonetics Module

The detailed architecture of the txt-to-phonetics module M_1 is depicted in Figure 8.2. It consists of a self-organising map (SOM) which takes inputs from MLP_1 (a multilayer perceptron) and AA_1 (an autoassociative network). The SOM produces output coordinates in D_phon phonetic space. Both MLP_1 and AA_1 are trained on a variant of error back-propagation (Rumelhart, Hinton, and Williams 1986).

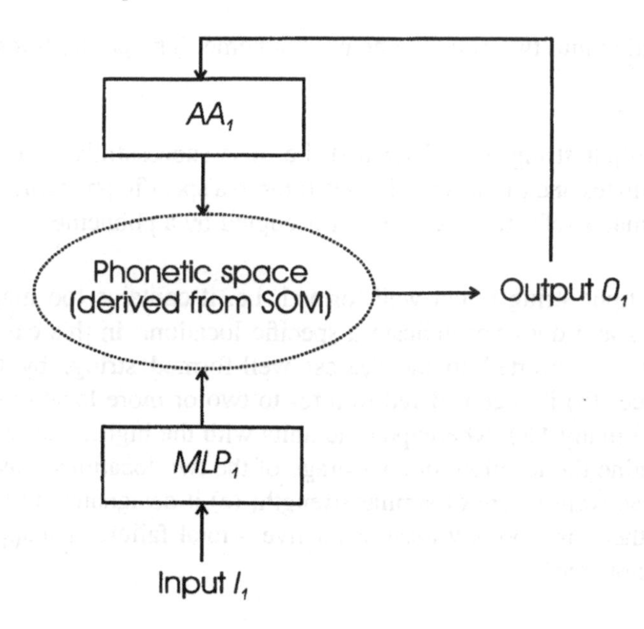

Figure 8.2. Architecture of the text-to-phonetics module $M1$. See text for details.

4.1. SOM Network

At the heart of SOMtalk is a pretrained self-organising map which takes phonetic features as input and produces coordinates in D_phon space as outputs. Full details appear in Cohen (1997).

4.2. MLP Network

MLP_1 takes text as input encoded as described in Section 3.2. The output is D_phon coordinates, i.e. location of the winner neuron in the SOM corresponding to the target phonetic features.

4.3. AA Network

The output from MLP_1 in Figure 8.2 may not correspond to a legal output. Hence, we use an auto-associator (AA_1 in Fig. 8.2) in an attempt to 'correct' it. The auto-associator is trained to reproduce legal coordinates in D_phon space as its output. In neural network terms, it embodies a set of attractors. In operation, AA_1 updates recursively (as shown in the figure) for a set number of 'corrector' cycles (see Section 5.5).

Final output from the text-to-phonetics module M_1 consists of a binary string which is to be interpreted as the coordinates of two consecutive phonemes. This string is divided into two halves: one per phoneme. The possibilities for each string are:

1 The output string is well-formed, i.e. it is successfully converted into coordinates and (a) the coordinates refer to a specific phoneme, or (b) the coordinates refer to a location not occupied by a phoneme.

2 The output string is not well-formed, i.e. it contains too many or too few 1's and does not indicate a specific location. In this case: (a) the output is converted to the nearest well-formed string, by Euclidean distance; (b) it is considered to refer to two or more locations (if there are too many 1's), whereupon the units with the highest activation may determine the location, or an average of the two locations may be taken if the activations are of similar strength; (c) it designates no location at all if there are too few locations active – total failure is mapped to the null phoneme.

Each case can be handled with a variety of different error-correction methods, based on the particular distance metric selected (in the phoneme space or output activations space). In fact, case 2 was found to occur only rarely. Case 1(a) accounted for the majority of the outputs. Case 1(b) can be handled either by choosing the nearest phoneme coordinate (by Euclidean distance) which does label a phoneme, or by choosing the phoneme which most strongly excites the SOM at that coordinate. Because the two-symbol window advances by one symbol at each step, there are two candidates for each letter – apart from the initial letter of the word. (An issue which arises is how to determine the order of the two symbols. For the sake of brevity, we sidestep the issue here. For full details, see Cohen 1997.) In the event of a conflict, solution 2(b) is adopted.

5. Model Selection

In this section, we outline several important issues relating to the implementation of the various connectionist models incorporated in SOMtalk.

5.1. Bias and Variance

Assuming the output representations and distance metric have been determined, it is then possible to define the bias and variance of a model which may fit the data. In statistical terms, the precise meaning of bias (it also has a generic meaning somewhat similar to the machine-learning usage) is the expected value of $(T - \theta)$, where T is an estimator of the parameter θ.

The expected value of a function of a continuous random variable x, having probability density function $p(x)$, over the range $[a, b]$ is defined as:

$$E[f(x)] = \int_a^b f(x)p(x)dx$$

If the probability density is rectangular, then $E[f(x)]$ is equal to the mean and the variance of the distribution is $E[(f(x) - E[f(x)])^2]$. There is usually held to be a trade-off between the bias and variance of a model (Geman, Bienenstock, and Doursat 1992; Bishop 1995). Geman et al. have decomposed mean-squared error of an estimator (here a feed-forward network) on a dataset D into bias and variance terms as follows:

$$\begin{aligned} E_D\left[(f(I;D) - E[T|I])^2\right] = \left(E_D[(f(I;D)] - E[T|I]\right)^2 + \\ E_D\left[\left(f(I;D) - E_D[f(I;D)]\right)^2\right] \end{aligned} \quad (1)$$

where the input I and target T are treated here as scalars for simplicity. E_D refers to the expectation with respect to the dataset D. The regression $E[T|I]$ is the best fit in the mean squared error sense, so the left-hand side is a measure of the effectiveness of $f()$ as a predictor of T. The first term on the right-hand side (bias squared) is a measure of how well the estimator has fit the data on the average, while the second term (variance) measures how much change there is in the estimator from sample to sample (smoothness of the fit). Many other decompositions are possible, particularly in classifiers (Breiman 1996b), and this kind of breakdown has proved most fruitful – especially in the theory of multiple networks as seen below.

In these experiments, early stopping via a held-out set (of the same size as the training set) has been used, despite its lack of a theoretical basis, for a number of reasons:

1. Use of k-fold cross-validation (Weiss and Kulikowski 1991), particularly for k equal to the dataset size, is too computationally intensive for large datasets, as is Bayesian estimation.

2. There is no data shortage in this domain as far as a computationally feasible system is concerned – any particular dataset will always be incomplete in the sense that not all possible contexts for each phonetic item will occur. Larger datasets will allow a more complex model leading to improved performance, up to some asymptotic point not yet found in these experiments.

3. Error reduction on the training set is not particularly important for this study, since 100% correct performance can easily be obtained on all

19,988 words treated as training data, and compression is not the main goal here.

4 We are principally interested in the total error reduction that can be achieved (on the unseen data) using the methods described here, and the nature of the remaining error, not in some theoretically optimal level of error on a particular training or test set. For these reasons, the performance obtained will not be optimal.

5 We also note that over-fitting is not such an important issue with multiple networks – a certain amount can be beneficial (Sollich and Krogh 1996).

5.2. Generalisation and Error Measure

The error function used here is the mean squared error (MSE), $E()$, for a set of q input-output pairs, D. (This should not be confused with the expectation, $E[]$, of the previous section.) The input-output pairs are given as:

$$D = \{I_i^q \in \{0,1\}^i, T_k^q \in \{0,1\}^k\}$$

where input and output target, I and T, are of dimension i and k respectively. This error can be formulated in terms of a two-layer network as:

$$E = \frac{1}{2}\sum_q \sum_k \left[T_k^q - f\left(\sum_j u_{kj} f\left(\sum_i w_{ji} I_i^q\right)\right)\right]^2$$

where w_{ji} and u_{kj} are weights from input-to-hidden and hidden-to-output layers respectively, and $f()$ is a continuous differentiable function, here the logistic. Bias terms on each unit are omitted for conciseness. The delta-bar-delta algorithm of Jacobs (1988) is used for weight updates, which are done in the on-line, stochastic mode.

In the case of text-to-speech conversion, we have no prior knowledge of the underlying functional form of the distribution of the data which could be directly applied to a surface-fitter such as a neural network, although a great deal of prior knowledge of this task exists at a symbolic level. Attempts to encode some of this knowledge were described in Cohen (1995). In the absence of information on the underlying function to be fitted, there is no way to determine the optimal number of weights except by experiment – although this is found not to be a critical issue in this study, as also found by Dietterich and Bakiri (1995). Similarly, smoothing remains a basically unsolved problem, since the model complexity can always be increased relative to the amount of data available.

5.3. Multiple Networks

The very idea of model selection implies that a best single model must be selected from a competing set. In fact, combinations of models may work better in certain problems by making network training less sensitive to the partition of the available data into test and train subsets. There is now a large literature on this topic: for approaches which work from the data alone, see Breiman (1992, 1996b) and references cited therein. Another family of methods is based on *mixtures of experts* (Jacobs, Jordan, Nowlan, and Hinton 1991; Jordan and Jacobs 1994) which use some prior knowledge or heuristic to determine architecture prior to training.

We can also make use of the combined approach in determining an optimal number of passes through the AA ('corrector') networks described in Section 4.3. Krogh and Vedelsby (1995) have shown that the generalisation error for multiple networks can be expressed as:

$$E = \overline{E} - \overline{A} \qquad (2)$$

where $\overline{E}$ is a weighted average of error from each network over all patterns, and $\overline{A}$ is a similar weighted sum of ambiguities, where the weights w_n are positive and sum to one. Equation (2) derives from similar reasoning to that underlying the bias/variance trade-off given in equation (1), except that the averaging is done over network outputs, rather than possible training sets. In the following, as above, outputs and targets are assumed for simplicity to be scalars; the generalisation to vectors for our networks is straightforward. Bars are used to denote weighted averages over all networks. The quadratic error on a pattern x for a network n is given in the usual way by $\varepsilon_n(x) = [T - f_n(x)]^2$. For the multiple networks:

$$e(x) = \left[T - \overline{f}(x)\right]^2$$

where $\overline{f}(x) = \sum_n w_n f_n(x)$. The *ambiguity*, or disagreement among networks, on input x for a single network n is defined as:

$$a_n(x) = \left[f_n(x) - \overline{f}(x)\right]^2 \qquad (3)$$

The ambiguity for multiple networks is given by:

$$\overline{a}(x) = \sum_n w_n a_n(x) = \sum_n w_n \left[f_n(x) - \overline{f}(x)\right]^2$$

this being the variance of the weighted multiple networks around the weighted mean. Some algebraic manipulation, and using the fact that the weights w_n sum to one, yields:

$$\begin{aligned} \overline{a}(x) &= \sum_n w_n \varepsilon_n(x) - e(x) \\ &= \overline{\varepsilon}(x) - e(x) \end{aligned}$$

by the definition of $\overline{\varepsilon}(x)$. Hence:

$$e(x) = \overline{\varepsilon}(x) - \overline{a}(x) \tag{4}$$

Using capital letters to stand for averages over the input distribution, we have the ensemble generalisation error equal to:

$$E = \sum_n w_n E_n - \sum_n w_n A_n$$

which can be written in shortform as equation (2) above, from which it is clear that $E < \overline{E}$. Where there is no reason to believe more strongly in any of N individual networks (weights equal to $1/N$):

$$E \leq \frac{1}{N} \sum_n E_n \quad \text{where} \quad E_n = \int \varepsilon_n(x) P(x) dx$$

Similarly:

$$A_n = \int a_n(x) P(x) dx \quad \text{and} \quad E = \int e(x) P(x) dx$$

A key point is that $\overline{A}$ can be estimated using unlabelled data – we do not require the target value T for a given input x. Krogh and Vedelsby use these ideas in conjunction with cross validation, using non-overlapping test sets to determine an optimal N (number of networks) and K (size of test set). They also suggest using the method in an active learning scheme to determine which patterns contain maximal information about the function to be fitted. From equation (4), it follows that:

$$\overline{\varepsilon}(x) \geq \overline{a}(x)$$

Therefore, a pattern producing a large average ambiguity will have a large average error. These ideas can also be used with error-correcting AA networks to determine the optimal number of corrector cycles (since obviously we cannot make use of the target values to do this). Each network feeds to an identical auto-associator, so that following one cycle, each of the outputs will be individually corrected to a new output. Networks with similar outputs are likely to move to similar regions of activation space following the correction, but where there are significant differences these are likely to be exaggerated – unless the outputs are in the close neighbourhood of examples in the AA training set. The network ambiguity provides a self-adjusting parameter for controlling the amount of correction to be applied.

It can be seen that the combination of multiple models will be most useful when they disagree most. In the context of a phonetic task, a pattern with a high level of ambiguity is likely to mean an irregular spelling, since the relevant contexts will not appear uniformly across each training set. (In this study, each network is trained on non-overlapping datasets.) If we assume that the networks have not overfitted their data, we may take the training errors themselves as estimates for actual network error in testing. This assumption is not likely to be totally correct, but some of the overfitting due to variance can be averaged out in combining networks (see e.g. Perrone and Cooper 1994, and references therein).

5.4. Training the AA Network

As for the MLPs, AA networks were trained on the delta-bar-delta variant of back-propagation, with updating in on-line mode. During the forward pass, the units of AA_1 in Figure 8.2 are updated as follows:

$$AA_j(t+1) = \alpha AA_j(t) + O_j(t)$$

where $AA_j(t)$ and $O_j(t)$ are the AA weights and output activations respectively at time t, and α is the amount of decay ($\alpha < 1$). The closer α gets to unity, the further back in time the weighted moving average is extended – obviously at the expense of more detailed recall of the immediately preceding time steps. Here, the average word consists of 8 phonemes, but in general the immediately preceding 2 or 3 phonemes are most significant in forming attractor basins. Orthographic dependencies may spread further forward or back, but these are taken care of by the 10-letter input window (see Section 3.2) since we are not dealing with stress or any prosodic effects. There is a recursive cycle in which the encoder network is fed information about its past state (i.e. from its own outputs at one time step previously) so that attractor spaces will develop given enough training.

5.5. Optimal Number of Correction Cycles

It is a problem to determine the optimal number of correction cycles to produce the best output. Obviously, we cannot (except for comparison purposes) make use of the actual target values (implying 'divine guidance') or the procedure becomes no more than an error accounting scheme, useless for data without target values. In the divine-guidance scheme, actual MSE is used to determine the optimal number of cycles. A maximum of 10 cycles is allowed (although, in fact, no more than 7 are ever used). The following schemes have been tried:

1 Use a fixed number of cycles. Values from 1 to 5 have been tried, but this offers the worst performance improvement in terms of overall error (about 1–3 percentage points.)

2 Cycle until the output is stable (e.g. no change for three cycles), using the maximum-permitted 10 cycles (as with all the methods which vary the number of cycles on a per-pattern basis). This can be done at the level of output activation or output diphone. Problems with this method are that it does not always help the error (and it is difficult to detect these cases in advance), and the pathological cases in which the output cycles into an illegal phoneme location and gets stuck. As a result, performance is not good with this procedure (improvements of 5–8 percentage points only).

3 Cycle until the distance to the nearest legal phoneme position (for each of the two output phonemes) is at a minimum. Compared to using the MSE, more cycles are used and performance is not as good.

In the case of multiple networks we have additional options:

4 Maximise ambiguity, regardless of error. This involves recalculating the ambiguity for all networks at each cycle on a particular pattern (see equation (3) above).

5 Maximise ambiguity, while trying to hold error low through one of the above stabilisation procedures.

6 Minimise ambiguity instead of maximising it.

7 Apply any of procedures 1–3 on the final output, after combining the multiple networks.

In practice, it was found that there was little or no advantage to be gained from using the corrector network on the combined output. There is also no guarantee that the stabilisation procedures will actually hold error low, although they prevent the pathological cases referred to above. Used alone, they allow too many corrector cycles, which can actually increase overall error on some individual networks, although the error is always reduced when the networks are combined. Using two or more options connected with logical AND will always restrict the number of cycles. We wish to produce the maximum disagreement between the networks, but not allow illegal phoneme strings (which could have a high ambiguity since they are not in the neighbourhood of corrector network examples) to distort the final output.

6. Results

Initially, the methodology has been applied to the grapheme-phoneme conversion problem to measure the viability of embedding prior knowledge. The important point is not so much the total error (efforts have been made to avoid optimising on a particular test set by changing test sets on each run) but the amount of correction offered by the AA network and other correction schemes. An on-line form of back-propagation has been used, with the words presented in a randomised order (after each pattern has been seen once). The training set error was typically 5–10 percentage points less than test set error at the time of peak performance when training was stopped. There was no problem achieving 100% correct on the training set by continuing to train past this point. The AA corrector has been used on individual networks (i.e. there is a single MLP) and multiple MLP networks, as now described.

6.1. Results on the Individual Networks

Two sets of 4000 words were drawn at random (without replacement) from the original NETtalk dataset, which made up the training set and test set for each run on each architecture. To avoid optimising on a particular test set, further (non-intersecting) test sets were used each time a training parameter changed. Apart from this, the standard cross-validation procedure was used to prevent over-fitting to the training set. Only the number of MLP hidden units was varied, as using two hidden layers was not found to bring any increase in performance. A third set of the 4000 most frequently-occurring words (according to Kučera and Francis 1967) was used to train the auto-associator. None of these words appears in the test sets.

The procedure for determining the output was to cycle through the AA network (method 2 above) until no changes in output phoneme occurred for three iterations. All the results given in Table 8.2 are for the test set (unseen data). A typical test set of 4000 words gave 29,551 pattern pairs (10-letter input window to 2 phoneme output window) which equals 59,102 individual output instances to learn. Hence, an improvement of 5 percentage points amounts to an extra 2955 instances correct in this context.

In Table 8.2, bit correction refers to the procedure 2(a). These results represent typical runs of about 15–25 presentations of each pattern, stopping when a five-point moving average of error on a cross-validation set (of size equal to the training set) begins to rise. No substantial effort has been made to get the uncorrected error as low as it might be, given these architectures. The error represents the percentage of incorrect phonemes output on the unseen test data.

Table 8.3 shows the pattern of errors when random target coordinates are used. Comparison with Table 8.2 shows the importance of the D_phon

Table 8.2. Typical results for single networks using structured SOM representation to obtain target coordinates: Percentage of test set incorrect (by phoneme).

Hidden units	*No correction*	*Bit correction*	*AA network active*
65	29.49	28.81	25.10
70	42.58	41.70	36.90
80	38.13	37.77	32.83
120	34.18	34.04	29.37
140	29.40	28.87	25.96
160	31.04	30.47	25.66
180	36.67	36.53	30.90

Table 8.3. Effect of using random target coordinates on composition of error (values in percentages of total error) – 120 hidden units in both cases. ±AA denotes AA units used/not used.

±AA	*%V as V*	*%V as C*	*%C as C*	*%C as V*	*Total error (%)*
+AA	3.18	5.72	46.27	12.83	68.00
−AA	3.07	5.57	46.54	13.05	68.24

representation, as base error is far higher and the reduction achieved with the AA network much smaller in the latter (with random target coordinates). In Table 8.2, the basic principle is shown to work in that correction of the output takes places but the effect is quite small at this stage. Further work is needed to refine the exact order of each stage in calculating the output, and the influence of each stage upon the next. Breakdown of the error shows the effect more clearly in that learning something about the CV and VC structure of the data makes a vowel less likely to appear as a consonant and *vice versa* as depicted in Figures 8.3 and 8.4. In the other categories – vowels misidentified as a different vowel and consonants misidentified as a different consonant – the error reduction is smaller. Error is quoted in percentage terms in these figures to facilitate comparison. Absolute error figures for performance comparison are given below.

6.2. Results with Multiple Networks

Each of the N networks is trained on a non-overlapping partition D_n of the training set D, such that the final output will be given by:

$$f(I; D) = \sum_{n=1}^{N} w_\alpha f_n(I; D_n))$$

where $\sum_n w_\alpha = 1$.

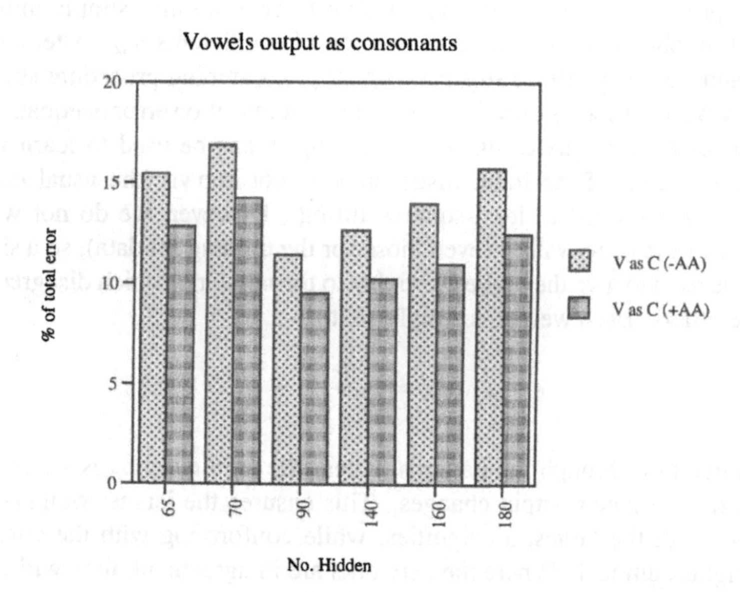

Figure 8.3. Effect of AA units on composition of error for different size networks: vowels misidentified as consonants. ±AA denotes AA units used/not used.

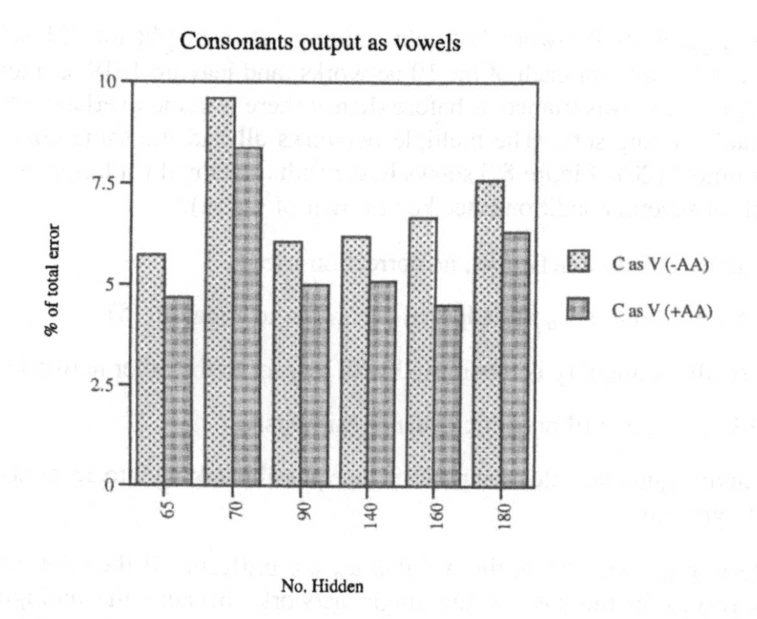

Figure 8.4. Effect of AA units on composition of error for different size networks: consonants misidentified as vowels.

If the probability density function of D is known, we may simply minimise the MSE to obtain the optimal combination of coefficients w_α. Alternatively, Krogh and Vedelsby (1995) suggest a linear programming procedure such that each network contributes exactly $w_\alpha E$ to the generalisation error of equation (2). If estimates of E are used, unlabelled examples can be used to learn w_α by gradient descent. If the input distribution is not known (the usual case), a possibility is to resort to least-squares fitting. However, we do not wish to make use of the target values (even those for the training set data), so a simpler method is used to give the highest weights to the networks which disagree most with the others. Each weight is simply set to:

$$w_n = \frac{a_n}{\sum_n a_n} \tag{5}$$

at each iteration through the patterns, where the ambiguity a_n is altered each time as the weighted output changes. This ensures the largest weights go to networks with the largest ambiguities, while conforming with the constraint that weights sum to 1. Where the networks are in agreement, they will tend to be weighted the same. For the opposite effect:

$$w_n = 1 - \frac{a_n}{\sum_n a_n} \tag{6}$$

The original 19,988-word NETtalk training set was split into 11 subsets, giving 1817 words for each of the 10 networks, and leaving 1819 as a test set. The AA network was trained as before (hence there is some overlap with each individual training set). The multiple networks all had the same number of hidden units (120). Figure 8.5 shows best results obtained (in terms of MSE) for each of several conditions (see key to right of figure):

no adjust: standard networks, no correction used.

adjust we: weights w_α are adjusted according to equation (5).

Max Amb: ambiguity is maximised with respect to the other networks.

Stabilise: cycle until phonetic distance minimised.

DG: divine guidance, the best performance possible taking into account actual target values.

Unless otherwise stated, the weights w_α are uniform. If they are not, this affects results in the case of the single networks because the ambiguity is recalculated based on the modified output (produced from the weighted average) on a per-pattern basis. In general, we have observed a good correlation between ambiguity and error.

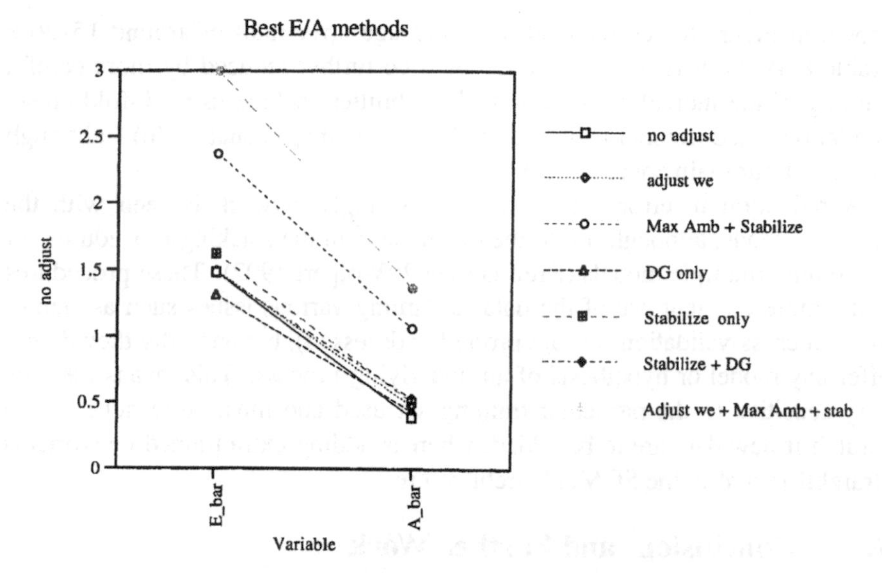

Figure 8.5. Methods for cycling corrector network which produced lowest mean square error using $\overline{E}$ and $\overline{A}$ to control error as in equation (2).

Table 8.4. Test set errors for SOMtalk with and without error correction versus benchmark systems (i.e. identical systems but without a corrector network). Percent incorrect figures are phoneme scores. The figure for 'divine guidance' was 15.0%.

Study	*Network type*	*Training set size*	*Test set size*	*Test set incorrect (%)*
Benchmark	Single (140)	4000	4000	29.4
Benchmark	Single (140) + error correction	4000	4000	25.7
SOMtalk	Multiple	18169	1819	19.6
SOMtalk	Multiple + error correction	18169	1819	16.5

Low ambiguity indicates a high bias in the overall combined networks, since they even agree on patterns outside their training set. In this case, we may expect the final combined generalisation error to be very similar to the weighted average of the individual errors. High ambiguity indicates a high variance, and in this case we can expect the combined generalisation error to be less than the weighted average of the individual network errors.

Overall actual error levels are in line with the predicted multiple generalisation error levels of Figure 8.5. The flatter the slopes and the lower the error, the better the predicted performance ($\overline{E} - \overline{A}$ reduces). MSE does not always correspond exactly with error at the phoneme/diphone level, but it is not possible for them to drift too far apart. The best performing methods as shown in Figure 8.5 correspond to percentage error rates of around 15–20% (Table 8.4), but it is believed these could be further reduced by more careful training of the individual networks. Possibilities include using k-fold cross-validation or a different method to set the w_α – perhaps equation (6) – although this is not our main concern here.

A reduction in error compared with a single network is seen with the above methods, although not to the extent seen in the stacking procedures for combining multiple models (Breiman 1992; Wolpert 1992). These procedures make more efficient use of the data (assuming various issues such as size of internal cross-validation sets are properly addressed), but naturally they do not offer any model or hypothesis of an underlying process. This means they are very sensitive to the particular training set used and must be retrained from scratch if new data are to be added, whereas adding extra trained networks is straightforward in the SOMtalk architecture.

7. Conclusions and Further Work

In these experiments, the aim has not been to achieve the best possible performance: systems which have access to a large stored dictionary will always produce better performance. Instead, a small well-defined problem has been explored with techniques which may well represent overkill for a small database. However, in the case of converting a phonetic representation to speech (as in Chapter 12), where we wish to train on spectral coefficients, the amounts of data involved would be far larger and it is (given current computational resources) infeasible to train a single large network.

We have been able to investigate the use of a set of frequently-occurring words and the effect of various interpretable error-correction procedures. In addition, by using the equations of Krogh and Vedelsby, we have an explicit way of balancing individual network error against network disagreements (ambiguity) which allows us to explore the error-correcting properties of a particular representation. The AA network provides a performance improvement with both single and multiple networks. The use of multiple networks allows extra possibilities in error correction and provides more information about an input than a single network would (e.g. about its regularity/irregularity). Hence, several new parameters can be introduced by the corrector network method which can be used to control the amount of correction applied, and which are more easily related to the linguistic level than many statistical procedures (e.g. those used in stacking). However, the main problem addressed is how to incorporate our implicit phonetic and phonological knowledge into a learning scheme for text-to-phoneme conversion.

Future work could extend these methods either (1) to improve the performance from an engineering perspective or, as a longer term goal, (2) to model more accurately human performance. Many parameters could be adjusted in pursuit of (1) to control depth of attractor basins, number of feedback weights, number of individual networks, procedures for maximising ambiguity and so on. Also, within this framework, it is possible to formulate hypotheses about regularity of particular spellings versus the amount of correction required. However, major improvements will probably only come about by incorporating constraints from what we know of phonetics and phonology, which amounts to the pursuit of (2). We also claim that this kind of approach has something to offer in the second stage of synthesis (i.e. the actual speech production from the intermediate 'linguistic' representation – see Chapter 12), where a major problem is to incorporate appropriate variability so as to avoid monotony. This variability is properly seen as arising from phonetic and phonological sources. Hence, there is a need for learning schemes which can be applied in non-segmental phonological modelling.

Chapter 9

USING THE TILT INTONATION MODEL: A DATA-DRIVEN APPROACH

Alan W. Black

Language Technologies Institute, Carnegie Mellon University

Kurt E. Dusterhoff

Phonetic Systems UK Ltd, Bishops Cleeve

Paul A. Taylor

Centre for Speech Technology Research, University of Edinburgh

Abstract This chapter describes the use of the Tilt intonation model as a data-driven approach to building computational models of intonation for fundamental frequency (f_0) generation in a speech synthesis system. The chapter presents the theoretical issues behind the design of the model as well as a comprehensive description of the theory underlying it. Automatic labelling issues and synthesis functions are discussed. A series of three substantial experiments is presented showing the training of an f_0 generation algorithm using the Tilt model. The f_0 generation algorithm is trained from a database of news stories labelled with Tilt events. The resulting model can generate natural sounding contours from the information available at f_0 generation time in a speech synthesiser (e.g. syllable position, segmental information, position in phrase, position in utterance, lexical stress). The accuracy of the generated contours was tested against unseen data and compares favourably with similar experiments done on the same database using different intonation theories. A detailed discussion of the features and results is given, including comparison with related work.

1. Background

Within the fields of speech processing, intonation has played an important role. Although many claim it to be central to both the recognition and generation

R. I. Damper (ed.), Data-Driven Techniques in Speech Synthesis, 199-214.

of speech, in actual fact it has often been left on the sidelines. Either it is actively removed, as in many speech recognisers, or only very simplistic models are used, as in many synthesisers. As the quality of speech recognition and synthesis improves, greater demands are being made on the contribution of intonation to those fields. If synthesis quality is to improve, intonation modules in synthesisers must move beyond the simple declarative intonation that most systems currently produce.

However, with a greater demand comes greater complexity. New speaking styles, new voices, even new languages, must be catered for quickly, easily and reliably. In the past, many intonation theories constructed rules relating labelling systems to fundamental frequency (f_0) contour generation. Moving to a new domain required significant work changing many low-level hand-selected parameters.

In this chapter, we present an algorithm which uses the *Tilt intonation model* to generate f_0 contours from high-level linguistic information. We first describe the Tilt model, discussing its formal properties, how a Tilt labelling may be automatically derived from an utterance, and then how an f_0 contour may be generated from a Tilt labelling.

To show that this representation is a useful abstraction of the f_0 contour we present a series of three experiments showing how models can be trained from Tilt labelled data that predict Tilt labelling from the features that are available in a speech synthesis system at f_0 generation time. The resulting models are used to generate f_0 contours which sound natural and, when measured against original contours, produce results as good as other similar f_0 generation experiments that use different intonation theories.

2. Tilt Intonation Model

In this section, we present an overview of the Tilt intonation model.

2.1. Intonational Events

The Tilt model is based on the concept of *intonational events*. Unlike traditional phonetic descriptions, in which speech is described by a linear sequence of contiguous phones, intonation is best described as a series of events, which happen every so often and which are not necessarily contiguous. There are two basic types of intonational event, the *pitch accents* and the *boundary tones*. In contrast to many other theories of intonation, there is no further categorisation into types of pitch accents or boundary tones. Classification of events is described solely by continuous parameters.

The intonational events of an utterance are related to that utterance's syllables by a series of *links*. All pitch accents are linked to exactly one syllable. Boundary tones are linked to phrase-final syllables. A syllable can be linked to zero, one or two events (in which case it is linked to one pitch accent and one boundary tone).

2.2. Event Classification

This subsection describes how an f_0 contour can be given a complete representation using the Tilt model. The process of analysing f_0 contours to produce Tilt representations involves the following steps:

1 Locate events by determining approximate start and end points.

2 Calculate the rise/fall/connection (RFC) parameters (see below) on each event separately.

3 Calculate the Tilt parameters from the RFC parameters.

4 Complete the analysis by determining values for non-event sections of the contour.

Locating events. The first step involves making an initial identification of accents and boundaries in an utterance. This can be done by hand or automatically. The way in which the identification is performed does not affect subsequent processing.

RFC analysis. The rise/fall/connection (RFC) model (Taylor 1994) describes events by using a rise element, a fall element or a rise followed by a fall element. The shape of fall elements is described by:

$$f_0(t) = \begin{cases} A - 2A(t/D)^2 & 0 < t < D/2 \\ 2A(1 - t/D)^2 & D/2 < t < D \end{cases} \tag{1}$$

where A is element amplitude and D is element duration. This function is plotted in Figure 9.1. Rise elements are simply the inverse of this function. Each rise and fall has an amplitude and duration parameter which can be used to scale the curve in the frequency and time dimensions respectively.

Given the initial event identification, the following algorithm can determine the rise and fall parameters automatically. An f_0 contour search region is defined around each event. The f_0 contour within this region is smoothed and a peak picking algorithm is used to determine whether the contour here should be characterised by a rise-fall (a peak is present) or only a rise or only a fall (no peak). The rises and falls are parameterised by choosing values of A and D which best fit the f_0 contour region.

In this way, each event can be characterised by four values: rise amplitude, rise duration, fall amplitude, and fall duration. In a rise-fall, all four values are filled. In a rise-only or fall-only event, the missing values are set to zero.

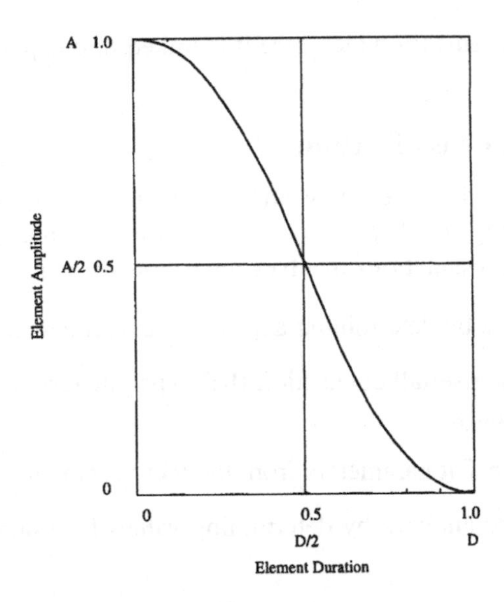

Figure 9.1. A fall element.

The RFC model effectively codes intonational events with a small number of values. The values are easily interpretable with respect to the f_0 contour. However, these values are not ideal as far as a linguistic analysis is concerned. The main problem is that the values are not independent. For instance, there is a high correlation between the size of a rise and the size of a fall. Furthermore, the shape or pitch pattern of an event, independent of amplitude, is seen to be the strongest indicator of its linguistic function (c.f. Hyman 1978; Bolinger 1986; Pierrehumbert and Hirschberg 1990), which is somewhat hidden as all four variables affect the event shape.

Tilt analysis. Tilt analysis is a further stage of processing which converts the four RFC values into three values: *duration*, *amplitude* and *tilt* known collectively as *Tilt* parameters.

Duration is the sum of the rise and fall duration:

$$D_{\text{event}} = D_{\text{rise}} + D_{\text{fall}}$$

Amplitude is the sum of the absolute amplitudes of the rise and fall elements:

$$A_{\text{event}} = |A_{\text{rise}}| + |A_{\text{fall}}|$$

The tilt parameter itself is a measure of the overall shape of the event, independent of amplitude or duration. The tilt parameter as calculated from amplitudes is:

$$\text{tilt}_A = \frac{|A_{\text{rise}}| - |A_{\text{fall}}|}{|A_{\text{rise}}| + |A_{\text{fall}}|}$$

The tilt value is the difference divided by the sum of the absolute values of the rise and fall amplitudes. As such, it varies between -1 (indicating a zero rise) and +1 (which indicates a zero fall). A value of 0 indicates that the rise and fall amplitudes are equal. An equivalent tilt value also exists for duration:

$$\text{tilt}_D = \frac{D_{\text{rise}} - D_{\text{fall}}}{D_{\text{rise}} + D_{\text{fall}}}$$

Using these equations directly, RFC events can be converted to Tilt events. By the reverse process of solving the equations, Tilt events can be converted back to RFC descriptions.

The tilt parameter effectively expresses the relative sizes of the rise and fall amplitudes or durations. Empirical evidence has shown that the amplitude and duration values are strongly correlated and if a single composite value is used for both amplitude and duration, there is very little information loss. Information loss is calculated by resynthesising the f_0 contour of the event and comparing it to the f_0 contour synthesised from the RFC representation. The single value is made from averaging the amplitude and duration values.

In this way, the four RFC parameters are reduced to three Tilt parameters with an insignificant information loss. Furthermore, the three Tilt parameters exhibit a much higher degree of independence. It is important to understand exactly what the use of a composite value implies. A single value implies that the ratio of the rise amplitude to the fall amplitude is the same as the ratio of the rise duration to the fall duration. It does not impose any restriction on what the ratio of the rise amplitude to the rise duration or the ratio of the fall amplitude to the fall duration can be. That is, the rise and fall gradients are assumed to be the same absolute value and of opposite sign.

2.3. Intonational Events

Although the intonational event representations as described above encode significant information in the f_0 contour, this information alone is not sufficient to encode and decode fully (i.e. synthesise) a complete f_0 contour. Specifically, information must be provided regarding the absolute positions in time and amplitude of the events. This information can be represented in several ways depending on the use to which the representations are put.

Full event formalism. In this representation, two additional values are required for each event (regardless of whether it is represented in RFC or Tilt parameters). The first, `start_f0`, specifies the f_0 value of the start of the event. The second value represents the position of the event in time. A useful measure we have found for this is the distance between the peak of the event (that is the juncture between the rise and fall) and the start of the vowel of the syllable that the event is linked to. This measure is attractive because the alignment of accents with regard to their associated syllables has been shown to have an important influence on the linguistic perception of the intonation (Ladd 1996). In the absence of links or a syllable description, the absolute position of the event in time can also be used. This is less helpful because it carries no real linguistic function. However, it still serves to allow unambiguous synthesis of f_0 contours from event descriptions.

In the event formalism, then, an utterance is represented by a list of events, each with the three basic parameters (amplitude, duration and tilt), and the two positional parameters (`start_f0` and position).

Segmental formalism. In the segmental formalism, utterances are represented by a contiguous sequence of intonational elements. Because events are often not contiguous, a dummy element called a *connection* is used to fill the gaps between events. A *silence* element is also used for silent portions of an utterance. Each connection and silence element has a duration and amplitude, where the amplitude represents the difference in f_0 from the beginning to the end of the event. These extra elements provide enough information for the events to be located in time and f_0 without additional information on the events themselves.

These two representations – event and segmental – are formally equivalent and can be interchanged depending on the requirements of the implementation.

2.4. Generation of f_0 Contours

Given a Tilt representation of an utterance, it is an easy matter to convert this into an f_0 contour. First, the Tilt parameters are converted into RFC descriptions using the following equations:

$$
\begin{aligned}
A_{\text{rise}} &= \frac{A_{\text{sum}}(1 + \text{tilt}_A)}{2} \\
A_{\text{fall}} &= \frac{A_{\text{sum}}(1 - \text{tilt}_A)}{2} \\
D_{\text{rise}} &= \frac{D_{\text{sum}}(1 + \text{tilt}_D)}{2} \\
D_{\text{fall}} &= \frac{D_{\text{sum}}(1 - \text{tilt}_D)}{2}
\end{aligned}
$$

An RFC representation can be converted into an f_0 contour by using equation (1) to synthesise rise and fall elements and straight lines to synthesise connections.

2.5. Key Features of Tilt

In the Tilt model, differences between events are described entirely in terms of the Tilt parameters. No categorical classification is performed. Differences between accents and boundaries are encoded in their continuous parameters.

3. Training Tilt Models

To use the Tilt model for speech synthesis from text, we need to be able to predict the Tilt elements (accent, boundary, connection, silence) and their parameters from information that can be found in the text. Thus, the features we use in our models are available at f_0 generation time in a synthesis system – including phrasing, accented syllables, lexical stress and segmental information.

For these experiments we are only building models for prediction of the Tilt parameters. We are assuming Tilt events (accents and boundaries) are identified. The prediction of which syllables are related to accents and boundaries is not covered here. The features are derived from a labelled speech database, to avoid confusing the training by introducing errors from other modules in the synthesis system. The features are only those which are available at f_0 generation time in a synthesis system.

3.1. Speech Database

The speech database used for creating and testing the models is the Boston University FM Radio corpus, speaker `f2b` (Ostendorf, Price, and Shattuck-Hufnagel 1995). This database consists of approximately 45 minutes (114 utterances) of single-speaker female American English news-reader speech. The database is labelled with segment, syllable and word boundaries and includes lexical stress markings. The database is additionally labelled with ToBI intonation labels (Silverman, Beckman, Pitrelli, Ostendorf, Wightman, Price, Pierrehumbert, and Hirschberg 1992).

In addition to the basic labelling, we further labelled the database with Tilt elements (accents, boundaries, connections, silence) in two ways. One set of Tilt element labels was derived from the existing ToBI accent and boundary labels. The second set was hand-labelled. For each set of Tilt element labels, their parameters were automatically derived using the techniques described above. Thus, we had two different sets of full Tilt labellings for the database.

The data were then split into training and test sets. The training set consisted of 86 utterances and the test set consisted of 28 utterances.

3.2. Extracting Training Data

For each syllable in the database linked with a Tilt element label, a set of roughly 40 context features was extracted. Features were selected from similar studies in intonation prediction, relevant literature and preliminary experimentation. Our starting point was the feature set used in Black and Hunt (1996), but this was significantly augmented.

The final feature set can be classed into five groups. Many of the feature values were extracted not only from the syllable associated with an element, but from the two preceding and two succeeding syllables as well.

The first class consists of lexical stress of the current syllable and the two before and two following. The second class concerns phrasal positioning. These features are:

- The number of syllables from the previous event (i.e. accent or boundary) and to the next.
- The number of syllables from the previous major phrase break and to the next.
- The number of stressed syllables from the previous major phrase break and to the next.
- The number of accented syllables from the previous major phrase break and to the next.
- The phrase break index (0-4) of the syllable in a window of two before and two after.

The third class of features contains information about composition of the current syllable. The basic syllable features are the percentage of the syllable which is unvoiced, the syllable duration and the position of the syllable within its word. In addition, onset and coda type (following van Santen and Hirschberg 1994) and onset and rhyme length (Silverman and Pierrehumbert 1990; van Santen and Hirschberg 1994); are also included in this feature class. There are three onset and coda types, which are determined by the least sonorant constituent of the onset or coda. These are −V (voiceless consonants), +V−S (voiced obstruents) and +S (sonorants). Glottal stops which precede null-onset syllables are classed as +V−S. Empty codas are classed as +S. The onset and coda features were selected from relevant literature specifically for use in the prediction of the peak position parameter. They proved very useful in the peak prediction model as well as others, as detailed below.

The fourth feature class concerns information on the presence of accents and boundaries. When using Tilt labels, as in Experiments 2 and 3 (see Section 4.1 below) we use two features: `tilt_accented` and `tilt_boundary` which return 1 if the syllable is related to an accent or a boundary respectively (and 0 if it is not). In Experiment 2 (see Section 4.1), where we use ToBI labels, we use a feature `accented` which is 1 if the syllable is related to any ToBI label (accent, phrase accent or tone), `tobi_accent` which returns the name of any related accent and `tobi_tone` which returns the name of any related tone (end tone or pitch accent). In all three experiments, we use these features on the current syllable and a window of two syllables before and two after.

The final feature class concerns suprasegmental element structure. Rather than being tied to syllables, this class looks backward and forward two *intonation elements* from the element associated with the current syllable. This view of intonation is necessary as multiple elements may occur on a single syllable. Conversely, syllables often have no intonation element; hence, a fixed syllable window may not include any elements at all.

3.3. CART Models

The mechanism used to build models was regression trees using standard CART techniques (Breiman, Friedman, Olshen, and Stone 1984). In this technique, a decision tree of questions about features may be automatically derived to optimise (partially) the prediction of one particular parameter. Each node in the decision tree consists of a question and two sub-trees, one for `yes` answers and one for `no`. The leaves of trees contain the means and standard deviations of example data points that are classified by the answers required to reach that node.

For our experiments, we use the CART technique to predict each of the Tilt parameters for each element type. For example, in the model which predicts accent peak position, for every syllable associated with an accent we collect the values for all of the features described above. We then grow a decision tree based on the subset of those features which best predicts the peak position parameter.

The trees are grown by finding the question that partitions the data such that the standard deviations of the two partitions is minimal. This partitioning continues until some stop limit is reached (e.g. minimal number of data points in partition), creating a `yes/no` question at each partitioning. This is a greedy algorithm in that it finds the best question at a particular time, rather than looking for a globally optimised set of questions, which would be computationally prohibitive. Questions may be about continuous or discrete features and are created automatically, such as: "Is it more than three syllables to the next major phrase boundary" or "Is the next syllable accented?"

Although CART building techniques are robust to a certain degree of noise in the data and interdependence between features, we found that we got better results by excluding some features for some models. A certain amount of hand-optimising of the features used for each model was carried out.

3.4. Measure of Accuracy

Measuring the accuracy of a generated f_0 contour is not easy; small changes in the contour are perceptually important at some stages while similar changes elsewhere may be irrelevant. To have some measure of accuracy, however, we follow others (Ross and Ostendorf 1994; Black and Hunt 1996) and use the root mean squared error (RMSE) between the generated contour and the original (smoothed) contour. We also use the correlation between the generated contour and original. The RMSE magnitude is dependent on the f_0 range of the speaker (larger for females than for males) as well as the actual error, while the correlation is more independent. Note that for these examples, the segment durations are the same in the generated examples as in the originals. Therefore, voiced sections and unvoiced sections of the signal will always align. RMSE and correlation are only calculated during the voiced sections.

The RMSE given by such experiments may seem large when compared to an RMSE found when resynthesising a contour derived from lower level parameters, e.g. directly from a Tilt labelling derived from an utterance (Taylor and Black 1994). If the generated contours did not sound reasonable, this would be more worrying. The RMSE in these experiments can be heavily affected by an accent that is partly offset, even by a few milliseconds which greatly increases the error at some points in the contour. We claim that RMSE is a good measure for f_0 comparison between similar studies on a given dataset.

In addition to RMSE, we provide correlation as an additional measure of the relation between the contours. If the f_0 contour were a straight line at the average pitch for the speaker, the RMSE would be 36.2 Hz for our test set but the correlation would be below 0.0. Thus, optimising both RMSE and correlation is necessary.

In addition to the overall comparison, we also recorded the accuracy of each of the individual models built on our test data. This helped us to concentrate on particular areas for improvement (e.g. peak position).

4. Experiments and Results

The basic method for the experiments was to build CART models for each of the parameters of each Tilt element type: that is, five each for accents and boundaries, and one each for connections and silences. These were trained from our training set. For testing, we used the decision trees to predict parameters for each syllable in our test set that was associated with a Tilt element. Then using the procedure described above, we generated an f_0 contour from from the Tilt description. The resulting f_0 contour could then be compared with the original extracted from the test sentence.

Table 9.1. RMSE and correlation of individual models.

	accent parameter models				
	`start_f0`	*amplitude*	*duration*	*tilt*	*peak*
RMSE	32.45	48.20	0.054	0.422	0.053
Correlation	0.546	0.447	0.561	0.558	0.472
	boundary parameter models				
	`start_f0`	*amplitude*	*duration*	*tilt*	*peak*
RMSE	28.74	40.92	0.066	0.517	0.079
Correlation	0.530	0.408	0.778	0.768	0.695
	`start_f0` *parameter for other models*				
		connections		*silence*	
RMSE		31.54		28.17	
Correlation		0.441		0.810	

4.1. Three Experiments

Three different experiments were carried out:

1 The first Tilt labelling was used (where the Tilt elements were derived from the ToBI accents and tones). No reference to these ToBI labels was made in the features used to build the CART models.

2 As in the previous experiment, the first Tilt element labelling was used, but this time the ToBI accent and tones labels were used in building the CART models.

3 The second Tilt labelling was used (where the Tilt elements were hand-labelled). No ToBI labels were used in the features to build the CART models.

Experiment 1. This experiment consisted of generating the parameter prediction models from the set of Tilt labels automatically derived from the ToBI labels provided with the speech database. No ToBI labels were used in the generation of these trees. Table 9.1 shows the RMSE and correlation of each of the 12 models.

For testing, we used these models on the element-aligned syllables from our 28 test utterances and then generated an f_0 contour from the predicted Tilt parameters. The resulting f_0 contours look and sound similar to the originals when used to resynthesise. We then measured the generated f_0 contours against the original ones. The RMSE averaged over the 28 test utterances is 32.5 Hz, and the average correlation is 0.60.

Experiment 2. The second experiment groups data according to the same Tilt element labels as were used in Experiment 1. The difference is that rather

Table 9.2. Comparison of results from the three experiments.

	Labels		
	1 (automatic)	*2 (with ToBI)*	*3 (hand)*
RMSE	32.5	34.0	33.9
Correlation	0.60	0.55	0.57

than include the Tilt element labels in the context feature set, ToBI labels were included. That is, we were trying to determine whether the extra differentiation in ToBI accents and tones would improve the prediction of the Tilt parameters.

The individual model results are similar to those in the first experiment. Six of the parameter models score slightly higher than in the first test, while six score lower. As with the first experiment, the generated f_0 contours are generally similar to the smoothed original. Against the same test set, the average RMSE and correlation show slightly lower accuracy at 34.0 Hz and 0.55, respectively. The poorer result is possibly due to the fragmenting of the data space introduced by the ToBI labels and the lack of data to compensate for the extra information.

Experiment 3. The third experiment uses hand-labelled Tilt elements both in grouping the data and in the extracted feature set. As with the second experiment, half of the individual models show improvement over Experiment 1, while half are lower. The average accuracy falls in between the first two experiments, with RMSE of 33.9 Hz and correlation of 0.57.

4.2. Comparison and Discussion of Results

A summary of the results for three experiments is given in Table 9.2. These results compare favourably with other models which generate f_0 contours directly from intonation abstractions. For example, on the same data as we used here, Black and Hunt (1996) achieve a RMSE of 34.8 Hz and a correlation of 0.62 while predicting from ToBI labels. Ross and Ostendorf (1994) achieve a RMSE of 33 Hz on their own intonation model, again on the same database. (Although those experiments were carried out on the same database, they may have had different training and test sets.)

The results show that these models are at least as good as, or better than, other intonation experiments done on the same database. Our informal listening tests confirm that the quality of the generated f_0 is acceptable.

The contours pictured in Figures 9.2 and 9.3 show a general similarity between the smoothed original f_0 (above in both figures) and the generated f_0 (below). While some differences are noticeable, most do not affect either the level of intelligibility or the general nature of the speech. Figure 9.2 shows a generated contour which matches its original very closely, and contains

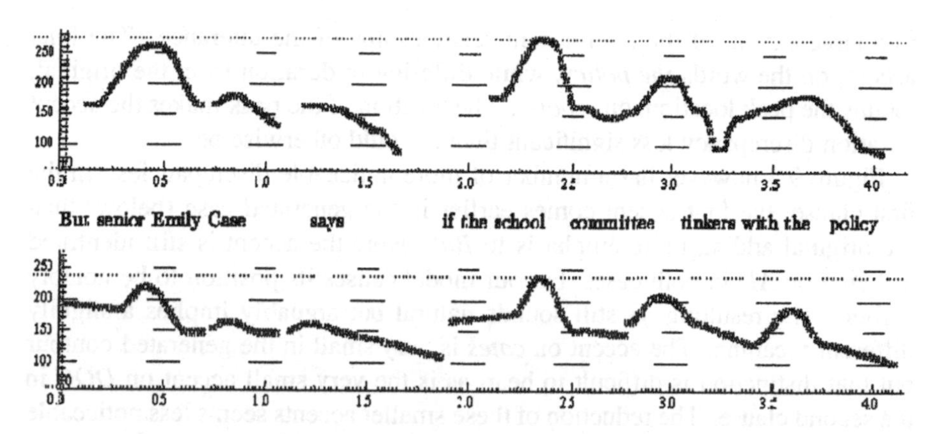

Figure 9.2. Original smoothed f_0 contour (above) and contour generated from predicted Tilt parameters (below).

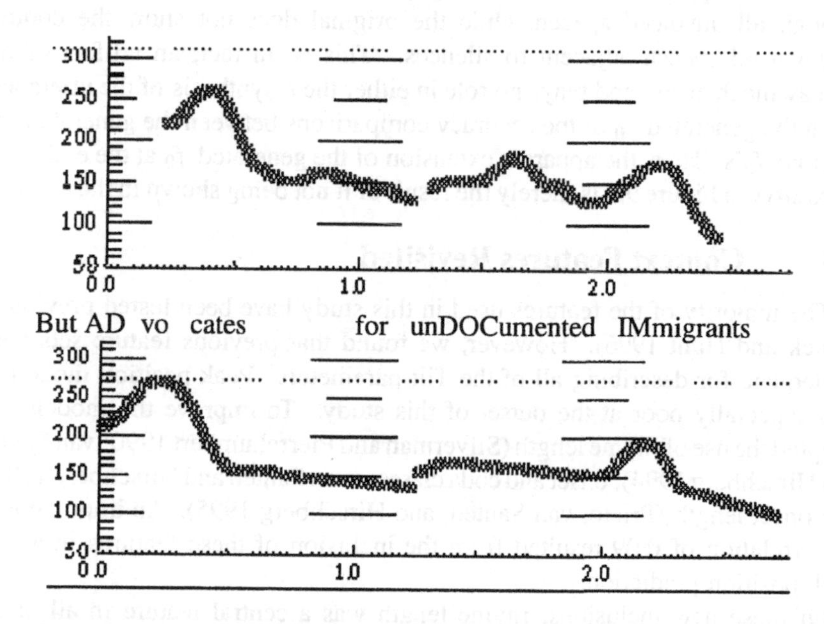

Figure 9.3. Original smoothed f_0 contour (above) and contour generated from predicted Tilt parameters (below).

no great discrepancies. Figure 9.3 shows a generated contour which contains examples of a variety of discrepancies which are noticeable in resynthesis. These examples were both generated from Experiment 1.

Certain points deserve further comment. Figure 9.2 shows six basic accents, two falling boundaries, and a pitch reset. Each of the six accent peaks is located in close proximity to the corresponding peak in the original contour and the generated pitch reset is close in f_0 to the original. This shows the retention, to a large extent, of the inter-accent relationships of the utterance. The final accent, on the words *the policy*, while differing in duration from the original, retains the peak location, on *–pol–*. The location of the peak makes the accent duration discrepancy less significant than it would otherwise be.

Figure 9.3 however has a number of more noticeable discrepancies. In the first clause, the first accent comes earlier in the generated case (below) than the original adding more emphasis to *But*. Note the accent is still identified as being on *AD* in both cases, but our model causes its position to be notably earlier. The resulting f_0 still sounds natural but arguably implies a slightly different meaning. The accent on *cates* is very small in the generated contour but that distinction is difficult to hear, as is the very small accent on *DOC* in the second clause. The reduction of these smaller accents seems less noticeable than the movement of the first accent in the example.

A further apparent error is that the generated contours show the full contour through all unvoiced speech while the original does not show the contour in unvoiced speech adjacent to silences. This is, in fact, an artifact of our display mechanism, and plays no role in either the resynthesis of the utterances using the generated f_0 or the accuracy comparisons between the generated and original f_0's. Thus, the apparent extension of the generated f_0 at the end of the word *says* in Figure 9.2 is merely the result of it not being shown in the original.

4.3. Context Features Revisited

The majority of the features used in this study have been tested previously (Black and Hunt 1996). However, we found that previous feature sets were inadequate for describing all of the Tilt parameters. Peak position modelling was especially poor at the outset of this study. To improve the models, we adopted the use of rhyme length (Silverman and Pierrehumbert 1990; van Santen and Hirschberg 1994), onset and coda classes (van Santen and Hirschberg 1994), and onset length (Prieto, van Santen, and Hirschberg 1995). An improvement in correlation of 0.09 resulted from the inclusion of these features in accent peak position prediction.

Of these new inclusions, rhyme length was a central feature in all of the peak position predictions (accents, boundaries, with and without ToBI). Coda classification was useful in non-ToBI boundary peak alignment, as well as non-peak predictions (accent amplitude and boundary durations, without ToBI). Onset type was among the optimised features in accent peak prediction, but not in boundary peak prediction. Onset length was only among the features used in boundary peak prediction (with ToBI), but appeared in both ToBI and non-ToBI sets for accent duration and accent starting f_0. Thus, although not all of the features intended to aid the prediction of peak position were necessary in all of the optimised feature sets, they did improve the prediction of that parameter, as well as contributing to other models.

Of the features previously tested, several played very noticeable roles. The lexical stress features comprised 20% (3 out of 14) of the optimised feature set for boundary duration (Experiment 1), and almost 30% (2 out of 7) of the optimised set for boundary `start_f0` (Experiment 2). The phrasal features, which make up almost 25% of the total feature set, contributed to all of the models. In the experiment which used ToBI tone features, the accent and tone features contributed heavily, but were also notably absent from the peak position models.

The suprasegmental intonation class played an interesting role in the optimised feature sets. In the Tilt-only experiments, the five features (Tilt element type with a window of two ahead and two behind) contributed at least two features to the subsets in 9 of 12 trees (only one in boundary peak and starting f_0 prediction and none in accent peak prediction). Each of the context features was included in at least one of the parameter prediction models.

5. Conclusion

This chapter has shown how the Tilt intonation model may be used within a speech synthesis system to build models that predict reasonable f_0 contours. The contours generated seem to capture not just speaker characteristics, but the news-reader style of the database as well.

There are advantages of using the Tilt intonation model rather than other popular intonation models. Events are easy to label as there are only two types. This makes hand-labelling more accurate and auto-labelling easier. No decisions are required about the type of the accent or boundary, as the variation in them is automatically derived from the simple event label and the f_0 contour. Although auto-labelling is not discussed detail here, other work has described some experiments using our event auto-labelling system in speech recognition (Taylor, King, Isard, Wright, and Kowtko 1997), and we are improving our auto-labeller so that extracting a Tilt labelling becomes a fully automatic process. A fuller description of Tilt and of auto-labelling is given in Taylor (2000).

The mapping from Tilt labels to f_0 and the mapping from f_0 to Tilt labels is fully defined as part of the model. The work presented here shows that the Tilt model provides a suitable abstract description which can be adequately predicted from higher levels of a speech synthesis system. Both these points show how Tilt fills the important gap between higher level linguistic information and an f_0 contour.

Finally, the well-defined properties and labelling ease of the Tilt model make it portable. Building intonation models from new databases is a straightforward and easy task, in comparison to rewriting a large number of rules, as is the case with some other intonation models. The relative speed and ease with which the Tilt model can be used on different speakers and speaking styles make the model both attractive and useful.

Chapter 10

ESTIMATION OF PARAMETERS FOR THE KLATT SYNTHESIZER FROM A SPEECH DATABASE

John Coleman and Andrew Slater
Phonetics Laboratory, University of Oxford

Abstract We present a practical guide to estimating reasonable parameter values for the Klatt synthesizer, focusing on practical success, rather than theoretical finesse. We employ a combination of techniques, including automatic analysis of acoustic properties from a speech database, using commonly available speech analysis tools, as well as 'tricks of the trade' learned through the trials of bitter experience.

1. Introduction

The Klatt formant synthesizer consists of a general design for generation of speech signals having acoustic characteristics specified by a set of time-varying control parameters, together with software implementing that design. A block diagram of the components of the program is shown in Figure 10.1.

Klatt (1980) includes a complete listing of the FORTRAN programs for an early implementation. Subsequently, researchers in other laboratories have translated this software into other programming languages, often with minor extensions or alterations to the original design. Klatt and Klatt (1990) introduced several enhancements, including a choice of more sophisticated source models to enable greater flexibility and accuracy in the synthesis of a wide range of individual voice qualities. This software is marketed by Sensimetrics Corporation, Cambridge, MA, under the trademark SenSyn. Another version (KLSYN88), which includes an option for using a stored voice waveform derived from natural speech, was developed by Jon Iles of the University of Birmingham and Nick Ing-Simmons of IBM, and is obtainable from various software archives (Iles and Ing-Simons 1993). Klatt's coworkers (Stevens and Bickley 1991; Bickley, Stevens, and Williams 1997) developed

R. I. Damper (ed.), Data-Driven Techniques in Speech Synthesis, 215-238.

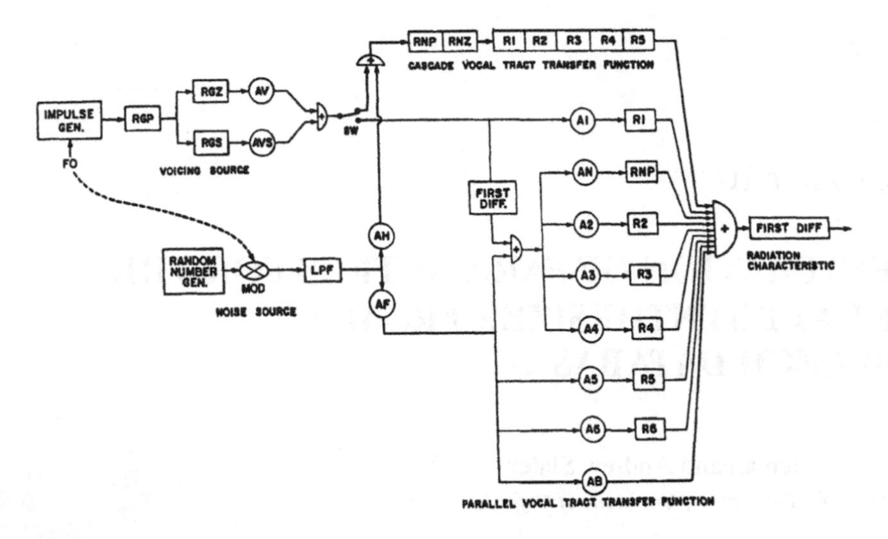

Figure 10.1. Block diagram of Klatt's (1980) cascade/parallel formant synthesizer architecture. Digital resonators are indicated by the prefix R and amplitude controls by the prefix A. Each resonator Rn has an associated resonant frequency control parameter Fn and a resonance control bandwidth Bn. (Reprinted with permission from Klatt, D.H. (1980). Software for a cascade/parallel formant synthesizer. *Journal of the Acoustical Society of America*, 67(3), 971–995, ©1980 Acoustical Society of America.)

an extension, HLsyn (also marketed by Sensimetrics), that simplifies the task of calculating values for the many parameters of Klatt's system, by employing a smaller number of higher-level parameters, which model the interrelations of various functionally-related groups of lower-level parameters. Input to the Klatt synthesizer in its standard form is via a text file of parameter values, containing one row per update interval. Several researchers (e.g. Fletcher, Local, and Coleman 1990; Simpson 1995) have developed tools to allow editing and display of parameter values using a graphical user interface.

Even the original 1980 design is capable, *if provided with the appropriate control parameter values*, of generating synthetic speech of a very high level of naturalness, as demonstrated by the high fidelity of certain carefully-produced synthetic copies of natural utterances, some of which are demonstrated on audio recordings accompanying Klatt (1987). (This replicates the earlier result of Holmes 1973, who obtained, using a somewhat different formant synthesizer, copy synthesis which was almost indistinguishable from natural speech.) Determining appropriate parameter values, however, is not easy. A careful copy usually requires long hours of painstaking parameter estimation, reestimation and fine-tuning, with repeated comparison of synthetically generated output with an original recording. Synthesis-by-rule systems that employ Klatt synthesis, such as DECTalk and MITalk, although excellent products in comparison to many other systems (see Logan, Greene, and Pisoni 1989 and van Santen 1993), produce speech which is far less natural-sounding than the theoretically possible optimum which careful copy synthesis demonstrates. The same failing is manifest in some recent models based on non-segmental phonology: YorkTalk (Coleman 1992) and IPOX (Dirksen and Coleman 1997). These systems yield some technical improvements over segmentally-based systems, by a more liberal regime for the temporal coordination of larger and smaller units of various kinds (e.g. syllables, parts of syllables, individual features and groups of features). For example, coarticulation within consonant clusters and vowels, reduction of unstressed vowels and the concomitant emergence of syllabicity of neighbouring sonorants, and rhythm are among the aspects of speech on which such models provide new insights. Nevertheless, these are still synthesis-by-rule systems (although the nature of the rules is quite different from that of many other systems), employing hand-crafted rules that may be incorrect, overly general or overly specific, or most damaging, simply lacking in the specification of certain fine details that can make the difference between almost natural synthesis and convincingly natural-sounding synthesis. Inadequate modelling of acoustic-phonetic fine detail in formant synthesis may also account for the rapid drop in intelligibility of synthetic speech in noise, compared with natural speech (Hawkins and Slater 1994). Our personal experience of using the Klatt synthesizer for over a decade is that, although on occasions it is possible to produce short stretches of synthetic speech – a few syllables, perhaps – that might deceive a naïve listener into believing it to be human speech, attaining this standard every time, for arbitrarily long stretches of speech, still evades us.

The goal of this chapter, then, is to document some of the 'tricks of the trade' of estimating, with as little work as necessary, the parameter values needed to synthesize a wide repertoire of speech sounds using the 1990 architecture. This, we hope, will form a sufficient basis on which new researchers might use the system for generation of short stretches of synthetic speech. One such group of researchers consists of those who wish to generate stimuli for use in speech perception experiments. Another group of potential beneficiaries is those who wish to study the Klatt synthesizer to employ it in a text-to-speech system, or to examine improvements in the design. To assist these researchers, we shall document a number of suggestions for the rapid prototyping of various synthetic speech sounds.

The schemas which we present in subsequent sections are *linear approximations* to observed acoustic patterns. For example, the English diphthong /ɑɪ/ (see Figure 10.2(a) for the wideband spectrogram of a typical example), can be imitated synthetically by employing the time-varying parameter pattern shown in Figure 10.2(b). (In addition, many parameter values which are fixed for the duration of this vowel must also be specified.) The schema has the benefit of

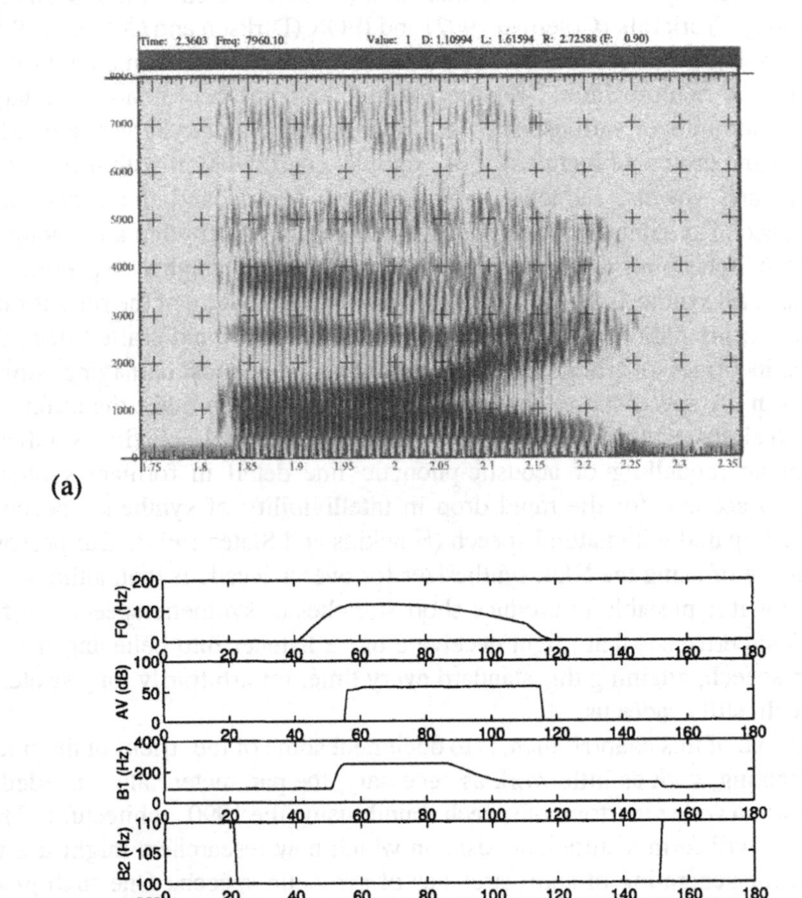

Figure 10.2. (a) Wideband spectrogram of the vowel [ɑɪ] as in the word *eye*; (b) Some of the time-varying parameter patterns required to be defined for synthesis of this vowel.

simplicity: it is not necessary to specify a value for every parameter at every time. Only a few points are specified (points at which the rate of change of the parameter is beyond a certain threshold), with linear interpolation providing all the intermediate values. Such models of parameter dynamics depart from the behaviour of natural acoustic parameters in two ways: (1) natural parameter dynamics are rarely linear, in fact; (2) natural parameter dynamics are rarely smooth, either. Hence, although problem (1) can be alleviated by employing nonlinear interpolation functions (e.g. cubic functions, arc tangents, or other sigmoids), solving problem (2) requires an element of apparently unpredictable variability to be introduced. (We do not know if such small unpredictable variations arise from stochastic processes, or simply reflect the great complexity of the neural, biomechanical, aerodynamic and acoustic systems involved in the generation of natural speech.) In practice, we have found that nonlinear interpolation results in trajectories that approximate some properties of natural speech (such as undershoot in unstressed vowels), whereas linear smoothing allows more precise control over parameter trajectories, but requires more complex rules.

The structure of the rest of this chapter is as follows. Next, in Section 2, we detail the settings of global (i.e. non-time varying) parameters which are usually kept fixed during synthesis. Section 3 is concerned with the estimation of dynamic (time-varying) parameters for synthesis of vowels, diphthongs, and glides, as well as introducing the use of schemas. The main parameters to be estimated are the amplitude of voicing (`AV`), fundamental frequency (`F0`), and formant frequencies and bandwidths, as well as various aspects of timing. Section 4 is on stop consonants, and introduces consonant-to-vowel and vowel-to-consonant transitions. The amplitudes of the principal resonances, and of the noise sources for friction and aspiration are discussed. Section 5 concerns voiceless and voiced fricatives, and the estimation of the amplitude of sinusoidal voicing (`AVS`) and of friction (`AF`). Section 6 is about affricates, nasals and liquids, and Section 7 rounds off with some observations concerning the design and construction of a monosyllable database used to build up a library of synthesis parameters, before concluding (Section 8).

2. Global Parameter Settings

The Klatt synthesizer employs two kinds of control parameters: *global* parameters, which are not time-varying, and *dynamic* parameters, which encode the time-varying acoustic properties of an utterance. Global parameters are fixed for each utterance, and control such general aspects as overall speaker voice quality, and fixed aspects of the signal such as sampling rate. To model specific speakers' voice qualities, some of the global parameters must be set carefully. However, we have no research experience of modelling different voice qualities: hence, the following suggestions represent a safe choice of settings for many circumstances.

2.1. Gain

Although the gain parameter adjusts overall loudness, it is best to set it to a constant 70 dB, and not to use it to modulate loudness. In general, modulations of loudness are better effected by alterations to the formant bandwidths, especially B1, and the amplitudes of the periodic and aperiodic source parameters, AV, AH and AF, as appropriate. These parameters are discussed further below.

2.2. Number of Formants

This is set to 5 for a male voice and to 4 for a female voice, assuming a sampling rate of about 10 kHz. Higher sampling rates require the inclusion of additional formants, although only F1–F3 (and sometimes F4) need to be varied dynamically.

2.3. Cascade-Parallel/All-Parallel Configuration

Although the all-parallel configuration may seem, on the face of it, to be the simpler control option, in our experience the mixed cascade-parallel configuration is easier to use, and is better motivated theoretically. For the synthesis of vowels, the formant amplitude parameters do not need to be controlled: formant frequencies and bandwidths suffice. When the formant amplitudes are not specified, the cascade part of the synthesizer is used to generate vowels, and the relative amplitudes of formants peaks closely approximate those found in natural speech. The cascade configuration models the vocal tract transfer function of (non-nasal) vowels more closely than the parallel configuration, and natural-sounding vowels are quite rapidly and easily obtained with this configuration. The parallel branch is, of course, essential for high-quality synthesis of fricatives and plosives, since the cascade configuration is not an appropriate model of the vocal tract transfer function when the sound source is above the larynx.

2.4. Sampling Rate

The value of this parameter depends on the digital-to-analogue conversion rates in the user's computing environment. Rates lower than 10 kHz are inadvisable, as naturalness and the intelligibility of certain sounds, such as [s], are impaired by the loss of high-frequency components. By the same token, it is inadvisable to use rates above 16 kHz without estimating and controlling values for additional formants.

2.5. Frame Rate

The usual frame rate (the rate at which control parameters are updated) is every 5 ms. This has been found to be generally adequate for formant synthesis. Using a slower frame rate (e.g. every 10 ms) can cause problems in modelling rapid speech events such as plosive bursts.

3. Synthesis of Vowels, Diphthongs and Glides

This section is concerned with the synthesis of vowels and vowel-like sounds, as well as introducing the use of schemas.

3.1. Estimation of Parameters

Vowels, diphthongs and glides ('semivowels') such as [j], [w] and [ɹ], as well as other frictionless continuants in other languages, are acoustically a uniform class, and are all synthesized in a similar way. We shall refer to them below by the more general phonetic term *vocoids*. In the Klatt synthesizer, they may be generated using the cascade configuration of resonators, which means that it is not necessary to specify the formant amplitudes: it suffices to define the formant frequencies (`F1` to `F3` or `F4`) and bandwidths (`B1` to `B3` or `B4`). In addition, only two source parameters, the amplitude of voicing (`AV`) and the fundamental frequency (`F0`), need specifying. All of these parameters are easily estimated using standard speech analysis tools. We employ the widely-used ESPS/Waves software (Entropic Research Laboratory Inc., Washington, DC) for this purpose. In this package, the `formant` program can be used to estimate formant frequencies and bandwidths of an input signal file, as well as `F0` and `prob_voice`, an estimate of the probability of voicing, which is discussed further below. It is necessary to set the frame rate `formant` uses to the same rate used for synthesis (typically 5 ms), and to choose a sufficiently high order for the linear prediction coding (LPC) analysis that `formant` uses for formant frequency and bandwidth estimation, which depends on the synthesis signal sampling rate (we typically use 11025 Hz). For an adult male voice, the Klatt synthesizer typically uses 5 or 6 resonators, requiring 12 LPC coefficients, plus two more for the source – a total of at least 14 predictor coefficients (`formant`'s default is only 12). We have found 16th or 18th order prediction to be sufficient. In practice, determining the optimum number of predictor coefficients for a given speaker requires some experimentation. An example of the complete `formant` command syntax is:

```
formant -i0.005 -o18 input_signal_filename
```

Among the several data files generated by `formant`, two of these – `input_signal_filename.f0` and `input_signal_filename.fb` – are necessary for estimating Klatt parameters. As they are binary data files, we find it convenient to convert them to ASCII using the ESPS/waves pplain function, for example:

```
pplain input_signal_filename.f0 > input_signal_filename.sourcepars
pplain input_signal_filename.fb > input_signal_filename.fbpars
```

Each line in the file input_signal_filename.sourcepars contains 5 fields, of which only the first 2 are of interest here. The first is f0, the estimate of fundamental frequency and the second, prob_voice, is the probability of voicing. This file also provides various measures of signal amplitude (rms and ac_peak). However, for Klatt synthesis of vocoids, it is unnecessary and inadvisable to try to control the output signal amplitude directly. Rather, one should set the overall gain parameter GAIN at a fixed value, e.g. 70 dB, and keep the amplitude of voicing AV at 60 dB throughout voiced intervals. Fine variations of signal amplitude are derived naturally by accurate estimation of formant frequencies and bandwidths alone. Reasonably accurate estimation of formant bandwidths in particular has a large effect on output quality, even though they do not change rapidly during the production of vocoids. Since the amplitude of formants in the cascade model is inversely proportional to their bandwidths, if estimates of B1 and B2 are too low (i.e. too narrow), the amplitudes of F1 and F2 will be unnaturally high, relative to the higher formant peaks.

Inaccuracies in bandwidth estimation of this kind can yield synthesis with too little or too much low-frequency energy in the output, making it sound 'tinny' (too little bass) or 'muffled' (too little treble). As AV is practically like an on/off switch for voicing (on = 60 dB, off = 0 dB), with very little dynamic variation, we have found it easy to estimate from prob_voice. As prob_voice is quite a reliable estimator of voicing, we set the 95% probability level as the threshold of the voiced/unvoiced boundary, so that if prob_voice $<$ 0.95, AV is set to zero. For prob_voice $>$ 0.95, we set AV $= 60 \times$ prob_voice. If small discontinuities (i.e. short silent periods) are found in the vocoids generated in this way, the 95% threshold is probably too high, and should be reduced a little. The estimate of F0 can be used without modification in the original Klatt synthesizer architecture, but in the revised version, F0 is specified in 0.1 Hz steps. This finer scaling of F0 (together with reduced period quantisation) in KLSYN88 helps reduce unnatural 'staircase' percepts in slowly-changing pitch glides. Consequently, in this architecture, F0 estimates generated using formant need to be multiplied by 10. For some speakers (particular female speakers), we have found ESPS get_f0 to give better results, with fewer pitch halvings and doublings than formant.

The estimates of formant frequencies and bandwidths generated by formant tend to be a little too 'jittery' for the Klatt synthesizer, and need to be smoothed a little. For this, we use a 10-point moving average (assuming that the synthesis frame rate is 5 ms^{-1}). During the steady state portion of consonants, bandwidths are also fairly stable and formant frequency changes slower than those generated by `formant`, so smoothing helps again here. At consonant-vowel boundaries, however, rapid changes in formant frequency and bandwidth are to be expected, so it is important not to smooth these parameters too much. Smoothing of bandwidths at these transitions adversely affects the perception of these consonants. Consonants also introduce short drops and peaks in fundamental frequency (f_0) – so-called microprosodic perturbations, for details of which see Bruce (1977) and Kohler (1990). Naturalness is improved by modelling these rapid f_0 changes. During vocoids, however, `F0` should change more slowly: any jitters are probably tracking errors, and should be removed by smoothing. We shall say nothing about the linguistic functions of intonation, and how abstract specifications of intonation may be used in the generation of f_0 contours, but refer to Pierrehumbert (1981) and the (extensive) associated literature for further details.

3.2. Synthesis of Vocoids

In speech, although the parameters of vocoids change relatively slowly, compared with consonant-vowel transitions, there are rarely any purely monophthongal sounds: even simple short vowels exhibit some changes (spectral changes, in particular) during their production. For example, the English phonemes /ɪ/, /ɛ/, /a/, /ʌ/, /ɒ/ and /ʊ/ are realised with slightly diphthongal qualities, with schwa-like offglides at their end. In many American English dialects, including the one for which Klatt (1980) provides some parameter data, this can be quite marked, so that even these short vowels must be modelled dynamically, like diphthongs. Furthermore, a vowel that is preceded by a consonant will exhibit the quickly-changing formant movements around the consonant-vowel transition that characterise the consonant's place and manner of articulation, although these formant movements mostly occur during the vocoid portion (i.e. during voicing). The formant estimates derived as described above, therefore, also provide an important aspect of the parameter values required for synthesis of consonants. Consequently, if the method described above is used to estimate synthesis parameters for a complete utterance consisting of vocoids and stop consonants, such as *a baby bear*, an incomplete but intelligible natural-sounding copy can be generated. The consonants require a number of additional parameters to be specified, especially noise source parameters for the burst and (if present) aspiration of stops, and fricatives require friction to be generated, of course.

The guidelines for the synthesis of vowels includes diphthongs too, such as English /eɪ/ (as in *a*), /ɑɪ/ (as in *eye*), /aʊ/ (*now*) and British English /ɪə/ (*ear*), /ɛə/ (*air*) etc. Acoustically, these are all vocoids. It is not necessary

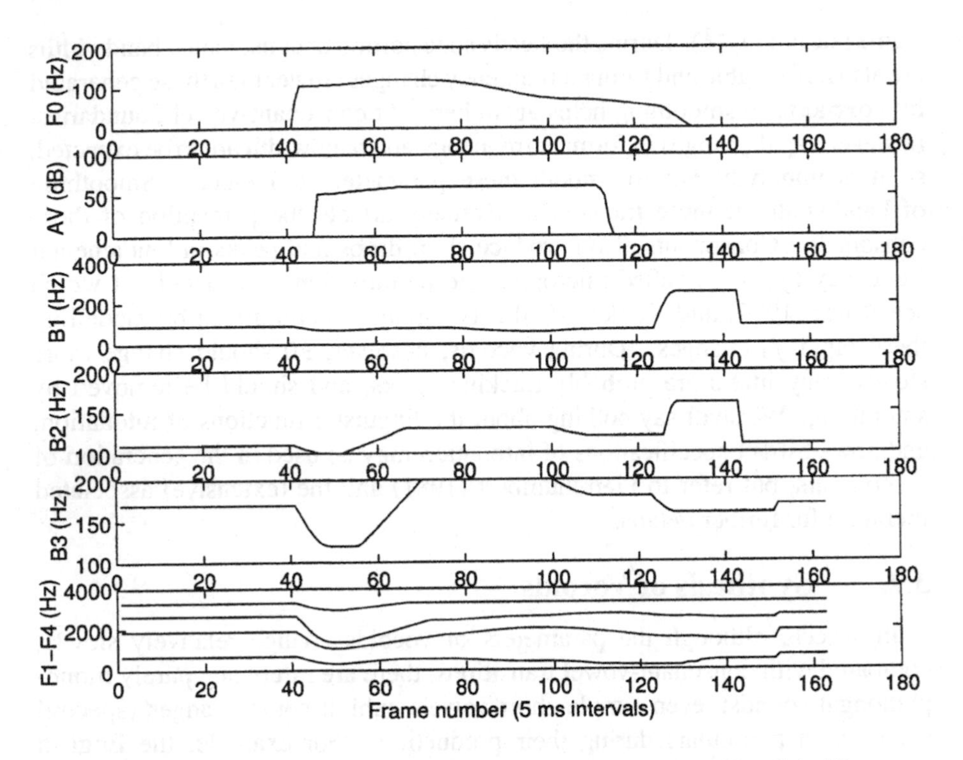

Figure 10.3. A schema for the synthesis of the vocoid portion of [ɹeɪd] (*raid*).

or relevant to pay any heed to discussions in theoretical phonology as to whether e.g. /ɑɪ/ should be analysed as one vowel, two vowels or a vowel and consonant (i.e. /ɑ/ + /j/). While such debates are of interest in respect of the abstract structure of phonological representations in the mental lexicon, or in pronouncing dictionaries etc., for synthesis purposes the nature of vocoids is clear: short vowels, long vowels, diphthongs and the glide consonants /j/, /w/ and /ɹ/ are all made of slowly changing resonances, and are all synthesized in the same way. To illustrate the timing pattern of diphthongs, a schema for the synthesis of *raid* is shown in Figure 10.3. The dipthong [eɪ] is around frames 70–100.

Although phonological theory draws a simple distinction between diphthongs and monophthongs, the parameter values given in Tables 10.1 and 10.2 illustrate how some monophthongs require some parameters to be defined with a changing sequence of target values, like a diphthong (but only in respect of those parameters). For example, [iː] has diphthongal F1 and B1, and [uː] has diphthongal F1, B1 and F2. Likewise, some parameters of some diphthongs (such as B2, F3 and B3 of [əʊ]) can be set to a constant target value. Where two values are given, the vowel has a diphthongal quality (for the parameter in question): the first value describes the early part of the diphthong; the second value, the later part. For diphthongs, the three values listed in Table 10.2 are: (a) duration of the first part, (b) duration of the transition, and (c) duration of the second part, according to the model of diphthong dynamics of Collier and t'Hart (1983) and Peeters (1987). The durations given are for vowels in isolated, closed monosyllables, and are approximate.

4. Stop Consonants (and Voiceless Vowels)

The estimation of `F0`, `AV`, formant frequencies and bandwidths discussed in the previous section also provides some of the parameters needed for stop consonants. Additional parameters need to be modelled for synthesis of stop consonants, however. First, noise source parameters `AF` (amplitude of friction, i.e. noise produced at a supralaryngeal constriction) and `AH` (amplitude of aspiration i.e. noise produced at the larynx) must be estimated, as discussed in the following section. The magnitude and time-course of these parameters must be estimated quite carefully, as small differences in these details are important perceptual cues to the voicing and place of articulation of stop consonants (Stevens and Blumstein 1978; Kewley-Port 1983). Second, the extreme constriction of the vocal tract gives rise to turbulent airflow at the place of the constriction during stop bursts and fricatives. Since, with the exception of /h/, the noise source is no longer at the larynx, the all-pole model used to model vocoids is not appropriate, as the transfer function now contains both poles and zeroes. The Klatt synthesizer approximates the effect of the zeroes via amplitude controls (`A1`–`A6`) of the six resonators in the parallel branch of the synthesizer. Fortunately, not all resonator amplitudes need to be specified for every consonant and, also, the amplitude and bandwidth specification may often be set at a single value throughout the consonant. Estimation of noise parameters helps to determine the temporal location of stops and fricatives in the signal, which enables start and endpoints to be determined within which `A2`–`A6` may then be specified.

4.1. Estimation of Noise Source Parameters

The Klatt synthesizer contains two noise sources. One, whose amplitude is determined by the value of `AH`, the amplitude of aspiration, allows noise to be passed through the cascade branch of the synthesizer, to simulate the filtering of noise produced at the glottis by the entire supralaryngeal vocal tract, irrespective of the place of articulation of the consonant. The second noise source, whose amplitude is set by the value of `AF`, the amplitude of friction, passes only through the parallel branch of the model, and thus only through those resonators in the parallel branch whose amplitude is nonzero. This aspect of the synthesizer design mimics the fact that the spectrum of friction depends

Table 10.1. Selected formant frequencies/bandwidths for the synthesis of British English vowels.

Vowel	F1 *(Hz)*	B1 *(Hz)*	F2 *(Hz)*	B2 *(Hz)*	F3 *(Hz)*	B3 *(Hz)*
[ɪ]	400	50	1900	100	2570	140
[ɛ]	620	70	1660	130	2430	300
[a]	700	70	1560	130	2430	320
[ʌ]	700	70	1220	50	2570	140
[ɒ]	620	70	850	50	2570	140
[ʊ]	400	50	890	100	2100	80
[ə]	460	90	1400	110	2570	80
[iː]	310–300	50–60	2020	80	2960	400
[eɪ]	540–380	80–60	1720–2020	130	2430–2600	200
[ɑɪ]	710–370	250–120	1180–1980	70–100	2550–2500	200
[ɔɪ]	540–380	150–60	850–1980	110	2150–2400	130–160
[aʊ]	730–380	250–60	1325–940	150–70	2590–2350	320–80
[əʊ]	540–380	150–60	1270–900	70	2300	70
[uː]	310–300	50–60	900–950	110	2200	80
[ju]	265–330	50–330	2010–1000	100–110	2430–2100	120–80
[ɪə]	400–470	50	2020–1600	100	2570–2600	140
[ɛː]	620	70	1660	130	2430	300
[ɑː]	700	80	1000	70	2570	140
[ɔː]	500	90	690	70	2570	140
[əː]	500	90	1270	60	2570	140
[ʊə]	400–470	50	1250–1180	100	2200	80

Table 10.2. Selected durations for the synthesis of British English vowels.

Vowel	*Duration (ms)*
[ɪ]	200
[ɛ]	300
[a]	300
[ʌ]	220
[ɒ]	220
[ʊ]	200
[ə]	300
[iː]	240
[eɪ]	80–140–60
[ɑɪ]	130–100–90
[ɔɪ]	150–90–80
[aʊ]	140–130–30
[əʊ]	50–150–60
[uː]	250
[ju]	80–100–120
[ɪə]	90–85–60
[ɛː]	400
[ɑː]	300
[ɔː]	300
[əː]	260
[ʊə]	350

almost entirely on the shape of the vocal tract downstream from the point of constriction, the so-called *front cavity resonances*, which vary from one consonant to another (Stevens 1972).

In speech signals, however, noise is noise: it is not easy to determine which source to use from numerical analysis of the signal alone, even though the fact that the noise has been shaped by a different number of resonators means that aspiration noise and friction noise are spectrally different. Spotting such spectral differences is not easy. However, a little knowledge about the formation of stop consonants helps to distinguish the two kinds of noise, as we shall see.

In English, voiceless stops are quite strongly aspirated, especially before vowels, but also even after vowels (contrary to what is stated in numerous phonetics textbooks): for clarity, therefore, we shall transcribe them below with aspiration, e.g. [t^h]. In phrase- and utterance-final position, the aspiration of word-final voiceless stops may be quite strong. In any of these cases, failure to model the aspiration leads to unnaturalness. English voiceless stops are really only unaspirated when they follow /s/ in the same syllable. The aspiration follows the release of the stop, and co-occurs with the beginning of the vocoid (if one follows). This means that the voicing of the vocoid may start before the aspiration has finished, in which case there will be a portion of simultaneous voicing and aspiration at the end of the aspiration interval. Often, however, voicing begins at the end of this interval. Also, where there is aspiration, it typically overlaps the formant frequency transitions between the constriction portion of the consonant and the more open vocal tract configuration of the vocoid which follows. We have observed short intervals of aspiration even in voiced stops, especially when they occur before a pause: values for the peak amplitudes of `AH` for voiced stops and affricates are therefore also included in Tables 10.3 and 10.4 below.

To become familiar with the nature of aspiration and how to model it, it is best to begin by modelling sequences of /h/ followed by a vowel (as in *he*, *her*, *hair*, *hear* etc.). (See Olive, Greenwood, and Coleman 1993, Section 9.1.1 for more details.) In such sequences, the supralaryngeal vocal tract configuration of /h/, and hence its formant frequencies and bandwidths, is the same as that of the vowel which follows. In fact, /h/ is just a cover symbol for the entire class of voiceless vowels. This class differs from the regular vowels that follow by virtue of being voiceless, (`AV` = 0 dB), with an aspiration noise source (`AH` > 0 dB), with some concomitant deviation in `F0`. In this simple case, it is easy to determine the extent and amplitude of `AH`, as we now show.

For signal portions with a magnitude greater than zero (i.e. non-silent portions), the energy comes from a combination of two sources: (a) the voice source; (b) the noise source. Where the signal is voiceless (e.g. as estimated using `prob_voice`, as above), therefore, all the energy in the signal is from the noise source. Thus, the short-time integral of the nonzero, unvoiced portions of

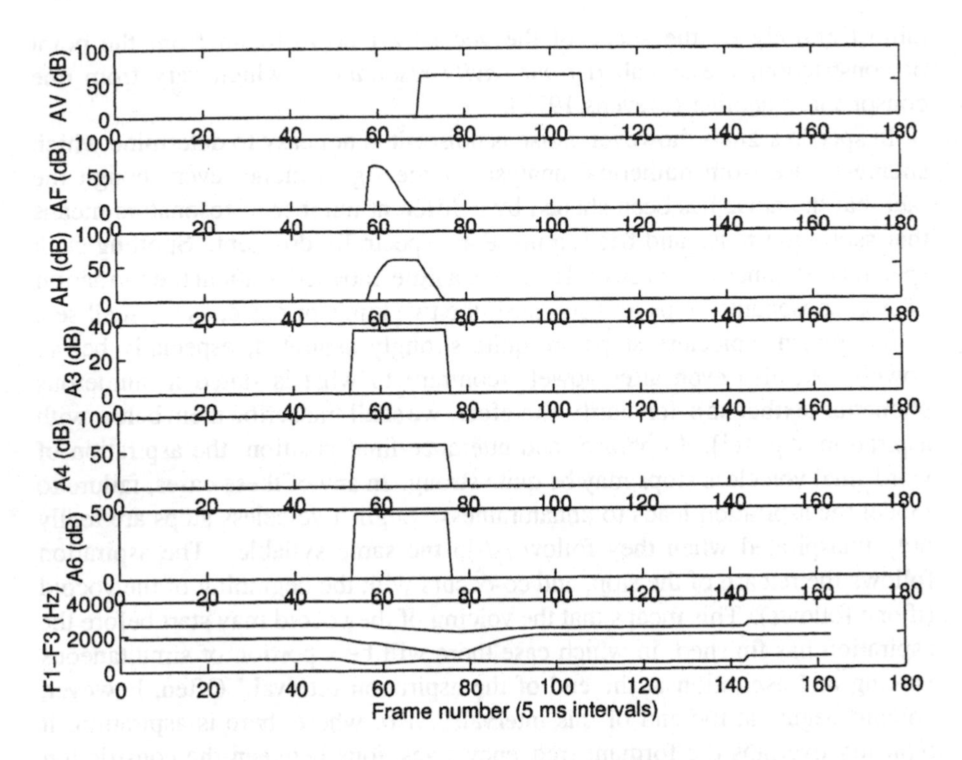

Figure 10.4. A schema for the synthesis of [tʰɔɪ] (as in *toy*).

a natural speech signal estimates the extent and amplitude envelope of friction. For /h/, that friction is specified by the AH parameter alone. To calculate the short-time integral, we square the signal amplitude at each sample to make all values positive, and then run a 50-point moving average smoother through the unvoiced portion. Alternatively, the rms amplitude of the unvoiced portion of the signal, suitably scaled, provides another estimate.

In /h/ and other fricatives, however, it is possible for the noise to overlap with voicing, in which case these methods for estimating the noise amplitude do not work, as the signal amplitude is dominated by the energy provided by the voice source. In this case, it is necessary to factor the signal into separate voice and noise components, and estimate the amplitude of each component separately. A method of achieving this is described in Section 5.

A schema for the synthesis of [tʰɔɪ] (as in *toy*) is shown in Figure 10.4.

4.2. Estimation of Resonator Amplitudes

Empirical estimation of resonator amplitudes is a little tricky, although it is sufficient to determine a single, average value of the relevant resonator amplitudes for the whole consonant. Klatt (1980) suggests a manual, trial-

Table 10.3. Amplitude parameters for consonants in prevocalic position, before front vowels.

	A2	A3	A4	A5	A6	AB	*max* AF	*max* AH
Fricatives								
[f], [v]						57	60	
[θ]]	13				29	48	65	
[ð]					27	48	50	
[s]					52		60	
[z]					52		55	
[ʃ]		57	48	48	46		55	
[ʒ]	48	48	48	41	53		53	
[h]								59
Affricates								
[tʃ]		57	48	48	53		53	58
[ʤ]		57	48	48	53		53	
Stops								
[p]						63	63	65
[b]						63	63	
[t]		30	45	57	63		53	58
[d]		47	60	62	60		53	
[k]	54	53	43	55	27		49	58
[g]	54	53	43	43	32		53	

and-error method, taking linear prediction spectra of the original signal every 10 ms, using a 25.6 ms Kaiser window centred on the time of each frame. (For fricatives, a single spectrum using a 40 ms window is suggested.) However, it is not appropriate to determine an amplitude setting for every resonance peak in such a spectrum, as only front-cavity resonances are strongly excited and require amplitude specifications. Klatt (1980, p. 987) provides a table of amplitude settings (his Table III) appropriate to various American English stops, affricates and fricatives. A version of the table for amplitude parameters pertaining to the British English pronunciation of consonants in prevocalic and postvocalic consonants is provided in Tables 10.3 and 10.4. We have found it easier to take values from the tables and specify them as constants holding for the entire duration of a consonant, rather than attempting to estimate them from natural signals. For dynamically-changing amplitudes, such as AF and AH, the table values are peak amplitudes, which can be used to scale dynamic estimates appropriately. The same tables also suggest values for AB, the amplitude of a path that bypasses the parallel configuration of resonators, circumventing the spectral shaping effect of the A2–A6 settings and resulting in a flattening of the spectrum, a characteristic of labial and other spectrally 'flat' consonants (i.e. English /θ/ and /ð/).

Table 10.4. Amplitude parameters for consonants in postvocalic position, after front vowels.

	A2	A3	A4	A5	A6	AB	*max* AF	*max* AH
Fricatives								
[f]						57	60	
[v]						57	52	
[θ]	25				27	48	60	
[ð]	25				27	49	57	
[s]					48		60	
[z]					52		50	
[ʃ]	45	48	48	22	46		47	
[ʒ]				41	44		47	
Affricates								
[tʃ]	35	40	29	42	39		55	61
[ʤ]	35	40	29	42	39		50	54–59
Stops								
[p]						60	53	57
[b]						52	49	54
[t]		30	45	56	62		47	54
[d]		35	50	53	53		50	50
[k]	54	52	38	43	32		48	54
[g]	54	33	46	48	40		48	48

4.3. Observations on the Time-Course of Stop Consonants

In addition to estimating values of the many parameters required for the synthesis of stop consonants, close attention must be paid to the relative temporal arrangement of events (e.g. peaks) in the various parameters. In human speech production, for example, the production of a voiceless stop such as the [tʰ] of [tʰɔɪ] (*toy*) involves the following events (see Figure 10.4).

First, there is an interval during which the vocal tract is completely closed, so that air cannot exit the vocal tract at the lips. In the case of oral stops, the velum is raised preventing the egress of air by way of the nose either. Since in this case the stop is voiceless, sound is not even transmitted through the skin, as is the case with strongly voiced stop closures. In consequence, there is no airflow, and hence no acoustic signal during the closure interval. In the Klatt synthesizer, this result is modelled by setting all the source amplitude parameters (AV, AH and AF) to zero. Under this condition, the settings of other parameters do not matter. It is possible, and for reasons of interpolation may actually be convenient, to set non-source parameters to the values that they must have in the subsequent interval, in readiness.

Second, when the oral closure is released, there is a short interval of friction (the *burst*), arising from the fact that the active articulator cannot move from a position of complete closure to the position required for the following vowel without passing, however rapidly, through an intermediate position in which there is a very small opening in the vocal tract. The flow of air through this narrow opening, resulting from the build-up of air pressure behind the constriction, causes the burst. In the synthesizer, this is modelled by a single, short, peak in the AF parameter (see Fig. 10.4). The overall duration of this peak varies according to the place of articulation of the stop. For example, the burst of [p^h] is perhaps only *circa* 5 ms long, whereas [t^h] and [k^h] have longer bursts, of the order of 40 ms. Burst duration is also conditioned by segmental context. For example, we have found longer velar bursts before high, front vowels than before other vowels. However, as the time-course of parameters is quantised to the frame rate (e.g. 5 ms), the timing of synthesis parameters for the burst may differ somewhat from measured burst durations. For example, a measured burst of 3 ms duration should be rounded up to one frame at least, rather than rounded down and thereby obliterated. Measured burst durations that are intermediate between multiples of 5 ms, such as 7 ms or 12 ms, should be modelled with shorter or longer durations in synthesis, to fit the frame rate. In such cases, experimentation will determine whether it is better to round up or down. Bursts of friction occur in voiced and voiceless stops alike.

Third, in voiceless stops, there is an interval of aspiration, modelled with the AH parameter. Where the preceding burst decays relatively slowly, as in [t^h] or [k^h], its tail may overlap the rise part of the aspiration interval. This fact makes it difficult to analyse AH and AF from a single estimate of overall signal noise in that portion. To get around this problem, it is helpful to estimate the amplitude of aspiration noise in /h/ alone (i.e. not as part of an aspirated stop). Because of the difficulty of temporal resolution in modelling bursts, a certain amount of trial-and-error adjustment in estimating the magnitude and time-course of AF is unavoidable for natural-sounding synthesis.

Fourth, in voiced stops, voicing begins shortly after the burst. In voiceless aspirated stops, the onset of voicing is delayed until shortly before the end of aspiration. During the interval following the burst, whether voiced or voiceless, the formant frequency and bandwidth parameters must be dynamically varied to model the changing shape of the vocal tract as the articulators move from the position appropriate to the stop closure to the target position of the following vowel.

Approximate durations of the various phases of stops in prevocalic and postvocalic position in isolated monosyllables are given in Table 10.5.

5. Estimation of Fricative Parameters

This section is concerned with the synthesis of voiced and unvoiced fricatives.

Table 10.5. Approximate durations (in ms) of phases of stops in isolated monosyllables.

Prevocalic position				
	Burst	*Aspiration*	*VOT*	*Stop-Vowel Transition*
[pʰ]	10	90	65	60
[tʰ]	55–60	90	65	55
[kʰ]	65	100	65	70
[b]	10		20	60
[d]	15		20	55
[g]	25		20	70
Postvocalic position				
	Vowel-Stop Transition	*Stop Closure*	*Burst*	*Aspiration*
[pʰ]	30	120	40	55
[tʰ]	30	85	40	100
[kʰ]	30	120	40	100

5.1. Estimation of Noise Source Amplitudes

Techniques for estimating most of the dynamic parameters have now been discussed, with two exceptions: (a) the amplitude of friction noise *even during voicing*; (b) the quasi-sinusoidal voicing characteristic of voiced fricatives.

The fact that the noise and voice sources may both be turned on at the same time means that it is not possible simply to segment the signal into voiced and unvoiced segments, as is done in some synthesizer designs. Two commonly-used noise estimation techniques in particular, zero-crossing rate and high/low energy ratio, are inappropriate for accurate estimation of AF (or AH). The zero-crossing rate is much higher in voiceless fricatives than in neighbouring vowels, and for this comparison is a useful measure for determining the presence of voiceless friction. But in voiced fricatives, the zero-crossing rate is determined by the frequency of voicing: the much higher frequency noise components are exhibited as small, rapid variations of an otherwise fairly sinusoidal signal, well away from the zero level. Although both noise and voice sources are involved, zero-crossing rate only characterises the voicing of such portions, missing the fact that noise is also present. This quasi-sinusoidal voicing differs from normal voicing by the absence of higher harmonics in the source spectrum. The Klatt synthesizer allows the user to model this type of voicing (found also in phonetically voiced stop consonants) via the AVS voice source parameter, usually employed together with AV.

Another technique for estimating the presence of fricatives compares the energy at lower and higher frequency bands of the signal spectrum. If there is less energy in some lower frequency band than in some higher frequency band, the relevant interval could be classified as having a high-frequency noise component. This metric was employed in several, early knowledge-based speech recognition systems (e.g. de Mori, Lam, and Gilloux 1987; Mercier, Bigorgne, Miclet, le Guennec, and Querre 1989). In practice, however, this technique is not robust. It is difficult to determine appropriate frequency bands for the analysis: in experiments, we found that the higher frequency components of some vowels caused them to be sometimes misclassified as (non-strident) fricatives. The metric does not get at the most important characteristic of fricatives, which is not that they contain energy at high frequencies, but that they have a strong *aperiodic* component. Aperiodicity, not frequency, is the defining characteristic of friction, voiced or voiceless.

To determine the strength of the aperiodic component, therefore, we attempt to factor the signal into separate (quasi-)periodic and aperiodic components. The obvious way to do this is to use linear prediction, which models each sample s_n of the signal as the sum of the linear combination of the previous p samples plus a residual term r_n:

$$s_n = -a_1 s_{n-1} - a_2 s_{n-2} - \cdots - a_p s_{n-p} + r_n$$

where the first p terms on the RHS represent the quasi-periodic part, and the residual r_n is the noise part.

The residual signal derives from three sources:

1 rapid changes in the signal deriving from rapid changes in the vocal tract, either from (a) the closure phase, or release ('stop transients') of plosive consonants, or (b) rapid aerodynamic changes as the vocal cords part or snap together;

2 friction noise, which is inherently random in character and, hence, cannot be predicted;

3 numerical error, e.g. from rounding.

The errors from 1(a) occur infrequently – at most two for any consonant – making it the least serious source of error. Errors of type 1(b) are correlated with the fundamental frequency: for this reason, the error signal has a pattern of pitch pulses that corresponds to pitch pulses of voiced portions of the original signal. We can reduce this component of the residual signal by taking account of the fundamental frequency, as discussed below. When this is done, errors of type 2 and 3 remain, with the 'intentional' unpredictability of noise dominating the accidental errors of numerical rounding.

To remove type 1(b) errors (pitch-pulse errors) we work out, for each pitch period, the time delay over which the signal is most like itself in an earlier period. This is done using the autocorrelation function, calculated over a moving window of samples that is approximately equal to the period of the lowest fundamental frequency. (There is no point looking for pitch-pulse correlations over a longer interval than this.) In our preferred signal processing environment, MATLAB (The MathWorks Inc., Natick, MA), the autocorrelation function is slow to calculate, but yields a very accurate measure of the duration of each pitch period, the autocorrelation lag L. The autocorrelation lag for each sample is the time between that sample and the most similar earlier sample within the preceding pitch period or two. The graph of the autocorrelation lag is very flat and smooth for voiced portions (where the similarity between one pitch period and the next is clear and well-defined), but is very variable during voiceless portions. It is easy to divide the signal into voiced and voiceless portions: since L changes little during voiced speech, its (smoothed, scaled) first difference, L', provides an estimate of the likelihood of voicing. Its inverse, $1/L$, suitably scaled, provides a sample-by-sample estimate of F0.

By subtracting the most similar earlier sample from the current sample, for each sample of the residual signal in turn, errors arising from pitch pulses can be reduced. For each sample n, L_n is the distance in sample periods between n and the most similar recent sample, and $r_n - r_{n-L_n}$ is the difference between each error sample and the most similar sample earlier in the residual signal, i.e. the sample L_n periods earlier. If r_n and r_{n-L_n} were the same size, their difference would be zero, i.e. subtraction would reduce the linear prediction residual to zero. In practice, this is hardly ever the case, so instead of subtracting the whole of r_{n-L_n}, it can only be partly taken account of. In our implementation, we have found $e_n = r_n - 0.65r_{n-L_n}$ to give the best overall reduction of pitch pulse error. The constant 0.65 is, of course, a 'magic number', with no theoretical basis: it has just been found to work well empirically. The error signal, e, that is computed in this way results mainly from acoustic noise, together with a small contribution from pitch pulse and rounding errors. The signal bears a resemblance to the original speech, but without voicing: consequently the portions corresponding to vowels sound like whisper and only the friction and aspiration components of the consonants remain. The error reflects the noise component of the original signal, and is thus an excellent basis on which to estimate the amplitude of friction or aspiration.

5.2. Observations on the Time-Course of Fricatives

Fricatives may defined acoustically as segments containing relatively long intervals of aperiodic noise. The place and manner of articulation of the fricative are characterised by its spectral shape, as well as associated formant transitions, as for stops. Phonological voicing of fricatives is cued by some combination of (a) quasi-periodicity, (b) the duration of the fricative, and (c) the duration of any preceding vocoid. To avoid confusions with stops or affricates, it is important to synthesize the amplitude envelope of the fricative by accurate estimation of AF. For example, we have found that a rise-time of around 40 ms

Table 10.6. Approximate durations (in ms) of fricatives in isolated monosyllables.

Fricative	*Prevocalic duration*	*Postvocalic duration*
[f]	185	200
[θ]	205	200
[s]	195	225
[ʃ]	185	275
[h]	210	
[v]	135	200
[ð]	135	180
[z]	145	190

is about right for the alveolar fricatives /s/ and /z/ in English. Dynamic changes in resonator frequencies need to be modelled to reflect the coarticulatory efforts of neighbouring sounds (such as darkening of friction before rounded vowels). As with stop consonants, continuity of resonance patterns between the fricative and its adjacent segments contributes to an overall percept of coherence and fluency. Table 10.6 gives approximate durations of fricatives in pre- and post-vocalic position in isolated monosyllables.

6. Other Sounds

Here we consider the synthesis of other sounds: affricates, nasals and liquids.

6.1. Affricates

Affricates consist of a stop closure followed by a friction interval. Formant frequency, bandwidth and amplitude parameters for the stop closure are estimated as for stop consonants (see Section 4 above). In particular, the movements of the formants in the transitional interval at the end of the preceding vowel (if there is one), provide information about the place of articulation of the affricate (always post-alveolar in English). For the frication portion, formant frequencies and bandwidths can be estimated using the ESPS `formant` program, and amplitude parameters taken from Klatt's (1980) Table III. Approximate durations of closure and frication portions of affricates are given in Table 10.7. Note that in prevocalic position the closure portion is not distinguishable from silence, and therefore has no measurable duration: that is, shorter or longer silences are equally acceptable. (A similar consideration applies to prevocalic stop closures.)

6.2. Nasals and Liquids

Close spectral modelling of nasal consonants and nasalised vowels is difficult, because of the extra poles and zeroes present in their natural spectra, resulting

Table 10.7. Approximate durations (in ms) of closure and frication portions of affricates.

Affricate	*Prevocalic frication*	*Postvocalic closure*	*Postvocalic frication*
[tʃ]	100	75	100
[dʒ]	75	50	160

Table 10.8. Approximate durations (in ms) of sonorants in isolated monosyllables.

Sonorant	*Prevocalic duration*	*Postvocalic duration*
[m]	120	210
[n]	110	210
[ŋ]		200
[l]	125	250
[ɹ]	115	
[w]	130	
[j]	150	

from the coupling of the nasal side branch. For semiautomatic estimation of parameters, we have found that the formant frequency and bandwidth estimates provided by using `formant` to analyse nasals and liquids yield all-pole approximations to the natural spectra, so that it is not necessary to employ the nasal branch amplitude parameter `AN`. In this case, parameters for nasals can be estimated just like other sonorants, i.e. the vocoids, as described in Section 3 above. However, with this approach it is hard to get natural-sounding nasals that give high intelligibility results. Use of the nasal pole-zero pair (`FNP` and `FNZ`) and amplitude of the nasal pole (`AN`) yield better results (Klatt 1980), but automatic estimation of the additional pole and zero frequencies and bandwidths is more difficult. Approximate durations of sonorants are given in Table 10.8.

7. Application: A Database of English Monosyllables

We have recently developed a phonologically rich database of English monosyllables, together with their associated parameter tracks using the techniques described in this paper. We have used the database in experiments in non-segmental speech analysis and synthesis (Slater and Coleman 1996), but the comprehensive set of parameters it contains should be of general interest to other researchers using the Klatt synthesizer. In Section 7.1, we describe the acoustic recordings and their associated token parameter files, and in Section 7.2 we outline a technique for generating parameter prototypes.

7.1. Recordings

All 8444 monosyllabic words from the computer-readable Oxford Advanced Learners' Dictionary (Hornby 1974) were parsed into the syllable constituents onset, nucleus and coda. From these, a subset of 1066 words were chosen to include almost every onset, nucleus and coda, and almost all onset-nucleus and nucleus-consonant combinations. Some categories were excluded because of their marginal status in British English (e.g. onset /ʃm/). Each word was embedded in a carrier phrase and five tokens of each were read by a male speaker of British English. (For full details of the methodology, see Slater and Coleman 1996.) For each token, several acoustic parameters were extracted semiautomatically, using the techniques outlined in this paper. Formant tracks were visually inspected and tracking errors manually corrected using wideband spectrograms and LPC spectra. The acoustic parameters were recorded at a 5 ms frame rate.

7.2. Parameter Prototypes

We have found it useful to generate a single set of parameters, or *parameter prototype*, for each word stored in the database. This is done to minimise the noise present in individual tokens. Parameter prototypes are generated from centroid parameter trajectories, as follows. Dynamic time warping (DTW) techniques (Sakoe and Chiba 1978) are used to construct centroid trajectories from the five tokens of each word in the database. We use a slightly modified version of the usual DTW algorithm, using the distance metric $|\log(y/x)|$ where samples x and y are single parameter values from a given pair of words.

Centroid trajectories are computed for each word, for each parameter separately, as follows: (a) pairwise time warps of the five recorded trajectories are computed, giving 25 new trajectories, and (b) a set of five new trajectories – candidate centroids – is computed from the medians of the trajectories obtained in (a). The candidate centroids nearly converge after just two iterations. When candidate centroids are computed for all acoustic parameters, a single centroid is chosen from each set of five candidates, using the lowest median sum distance of the candidates from the original trajectories. This choice is cast as a vote for a set of related parameter centroids of the same length. Thus, for each word in the dataset, we have a parameter prototype, consisting centroid trajectories, one per acoustic feature. The quality of the parameter prototypes was confirmed impressionistically by listening to resynthesis of each word using the prototypes. The normalisation of parameter trajectories and the selection of a parameter prototype (centroid) are illustrated in Figure 10.5.

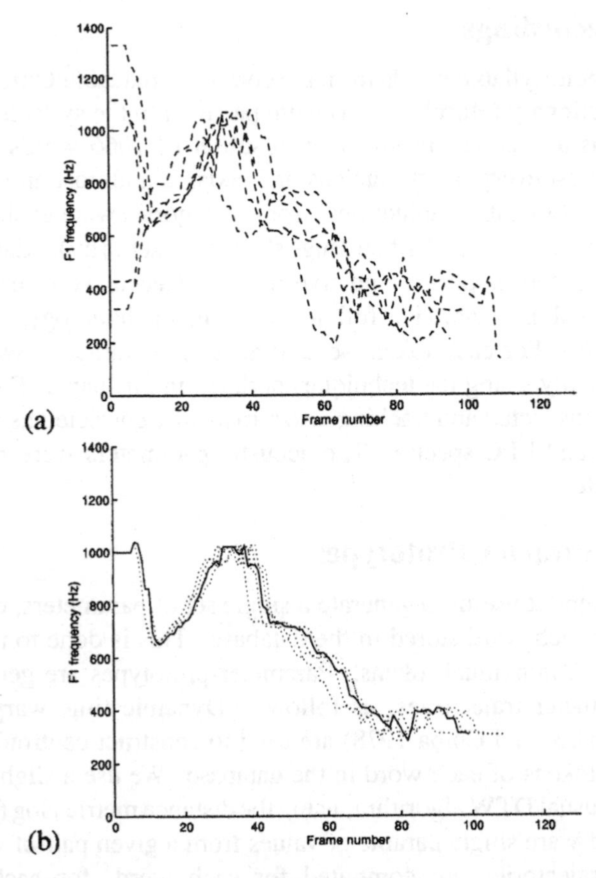

Figure 10.5. (a) Normalisation of F1 parameter values based on recordings of the word *height*. (a) original unnormalised trajectories. (b) candidate centroids. The solid line is the centroid selected to be the prototype.

8. Conclusion

In our experience, learning how to determine parameters for any formant synthesizer was a rather slow, laborious process of trial-and-error. Along the way, we have accumulated various ideas and methods, which we have attempted to set out clearly above in order to pass on some hard-won but (we hope) useful tips.

Chapter 11

TRAINING ACCENT AND PHRASING ASSIGNMENT ON LARGE CORPORA

Julia Hirschberg
AT&T Labs–Research

Abstract This chapter describes techniques for acquiring intonational phrasing rules and prominence assignment rules for text-to-speech synthesis automatically from labeled corpora or from annotated text together with some evaluation of these procedures for Standard American English. The procedures employ decision trees generated automatically using classification and regression tree (CART) machine learning techniques, from audio corpora labeled for pitch accent and phrase boundary location or from text corpora labeled by native speakers with likely locations of intonational features. Both types of corpus are used as training material, together with information available about the text via simple text analysis techniques, to produce decision trees, which in turn are used to predict accent and phrasing decisions for text-to-speech. Rules generated by these methods achieve more than 95% accuracy for phrasing decisions and 85% for prominence assignment.

1. Introduction

The association between prosodic variations and semantic, syntactic and discourse features of utterances has been a long-term issue in theoretical studies of language as well as applications to speech synthesis and speech interpretation. Understanding the associations between the prosodic choices speakers make and the structure and meaning of the utterances they generate, together with the context in which they are generated, can help to produce more natural sounding synthetic speech and to interpret the full meaning of natural utterances. Currently, however, these associations are imperfectly understood. Even when well-understood, it is often difficult to obtain the information needed to produce them in real text-to-speech systems and to identify their correlates in automatic speech recognition systems. In the past few years, however, there have been numerous attempts to learn such associations automatically from prosodically-

R. I. Damper (ed.), Data-Driven Techniques in Speech Synthesis, 239-273.

labeled corpora. This chapter reports on such work, undertaken for the purpose of generating intonational variation for text-to-speech synthesis.

Human speakers use pitch contours to convey part of the overall meaning of their speech. In varying contour, they are also varying pitch accent choice and placement, as well as deciding how to 'chunk up' words into two levels of prosodic phrasing: major and minor.

Most current text-to-speech systems use simple word-class information to assign intonational prominence, determining which words in an utterance are going to be prosodically emphasized based on a function/content word distinction. So, function words such as articles and prepositions are deaccented, while content words such as nouns and verbs are accented. Further, intonational boundaries within an utterance are located wherever non-final punctuation (commas, semicolons) occurs in text. So, for most synthesizers, words that should be emphasized are often not prominent at all (e.g most synthesizers say *My* PRINTER RAN *out of* PAPER instead of *My* PRINTER *ran* OUT *of* PAPER), while other words may be much too prominent (e.g. THANK YOU instead of THANK *you*). Long sentences that lack internal punctuation are typically uttered without 'taking a breath', so that it is almost impossible to remember the beginning of the sentence by the end. In general, the basic intonational contour of a sentence is varied only by reference to its final punctuation; sentences ending with a period, for example, are always produced with the same contour, contributing to a numbing sense of monotony. And sentences ending with a question mark are generally produced with rising intonation, whether appropriate (for a yes-no question) or not (for a *wh*-question).

While message-to-speech systems (which take as input an abstract representation of the message to be conveyed) and text-to-speech systems for restricted domains may take advantage of richer semantic, syntactic, and discourse-level information in assigning prosodic variation, such information has generally not been available for unrestricted text-to-speech systems. Although, indeed, truly natural prosodic assignment is not currently achievable for such systems, present-day machine learning techniques applied to (relatively) large prosodically-labeled corpora, or text annotated with likely prosodic information, do make it is possible to improve prosodic assignment considerably, using associations which can be learned between such prosodically annotated and other information gleaned from fairly simple text analysis. Such associations are exploited for the assignment of prosodic features in the Bell Labs Text-to-Speech System (TTS). This system is commercially available as AT&T's Watson FlexTalk and Entropic Research Laboratory's TrueTalk. Both phrasing and accent algorithms are derived from the analysis of sizable, prosodically-labeled corpora. The development of these procedures, together with some measurements of their performance, is the subject of this chapter.

2. Intonational Model

Intonational contours in TTS represent a partial implementation (Anderson, Pierrehumbert, and Liberman 1984) of Pierrehumbert's description of Standard American English intonation (Pierrehumbert 1980; Beckman and Pierrehumbert 1986; Pierrehumbert and Beckman 1988). In Pierrehumbert's model, intonational contours are represented as sequences of high (H) and low (L) targets in the fundamental frequency (f_0) contour. Intonational contours are defined over prosodic units known as *intonational phrases*. Intonational phrases are composed of one or more smaller phrasal units called *intermediate phrases*, plus a high or low boundary tone (H% or L%), which occurs at the right edge of the intonational phrase. Utterances are modeled as sequences of one or more intonational phrases. So, an utterance such as *I like apples a lot, but I really prefer oranges* might be realized as two intonational phrases: *I like apples a lot* and *but I really prefer oranges*. And each of these intonational phrases might in turn be uttered as several intermediate phrases; for example *but I really | prefer oranges* might be produced with a slight prosodic boundary between *really* and *prefer*. Intermediate phrases are themselves composed of one or more *pitch accents*, or intonational prominences, from an inventory of six possible accent types (H*, L*, H*+L, H+L*, L*+H or L+H*), plus a high (H) or low (L) *phrase accent* which controls the pitch from the last pitch accent to the end of the phrase. In theory, every intermediate phrase must contain at least one pitch accent. However, there have been observations that some tag phrases – attributions like *he said* – can be uttered as separate intermediate phrases with no perceptible pitch accent.

For both prominence and phrase boundary assignment in TTS, speech corpora were first labeled in Pierrehumbert's system for prosodic information, including location and type of pitch accents and phrase boundaries. Transcriptions of these corpora were then subjected to various forms of text analysis to identify potential predictors of the labeled intonational phenomena. Locations of accents and boundaries were then (separately) trained on a variety of corpora, with the text features used as independent variables, using classification and regression tree techniques.

3. Classification and Regression Trees

Classification and regression tree (CART) techniques (Breiman, Friedman, Olshen, and Stone 1984) permit the automatic generation of decision trees from sets of continuous or discrete variables. At each node in the generated tree, CART selects the statistically most significant variable to minimize prediction error at that point. In the implementation of CART used in this study (Riley 1989), all of these decisions are binary, based upon consideration of each possible binary split of values of categorical variables and consideration of different cut-points for values of continuous variables.

Generation of CART decision trees depends on a set of splitting, stopping and prediction rules. These affect the internal nodes, subtree height, and terminal nodes, respectively. At each internal node, the program must determine which factor should govern the forking of two paths from that node. Furthermore, the program must decide which values of the factor to associate with each path. Ideally, the splitting rules ought to choose the factor and value split which minimize the prediction error rate. The splitting rules described by Riley (1989) use a heuristic which approximates optimality by choosing at each node the split which minimizes the prediction error rate on the training set of data.

In this study, following Riley (1989), all of these decisions are binary, based upon consideration of each possible binary split of values of categorical variables and consideration of different cut-points for values of continuous variables. CART's cross-validated estimates of the generalizability of the trees it produces have proven quite accurate for the current task when compared with tests on separate datasets; in every case, CART predictions for a given prediction tree and that tree's performance on a hand-separated test set fall within a 95% confidence interval. Cross-validation estimates are derived in (roughly) the following way: CART separates input training data into training and test sets (90% and 10% of the input data in the implementation used here), grows a subtree on the training data and tests on the test data, repeats this process a number of times (five, in the implementation used here), and computes an average result for the subtrees.

Stopping rules are necessary for terminating the splitting process at each internal node. Thus, these rules govern the height of each subtree. To determine the best tree, this implementation uses two sets of stopping rules. The first set is extremely conservative, resulting in an overly large tree. The problem with this type of tree is that it usually lacks the generality necessary to account for data outside of the training set. To compensate for this difficulty, a second set of rules is used. First, a sequence of subtrees is formed. Then, each tree is grown on a sizable fraction of the training data and tested on the remaining portion. This is repeated until the tree has been grown and tested on all of the data. By performing this step, the stopping rules have access to cross-validated error rates for each subtree. The subtree with the lowest rates then defines the stopping points for each path in the full tree. Trees presented below were all obtained using this cross-validation procedure.

The prediction rules work in a straightforward manner to add the necessary labels to the terminal nodes. In the case of a continuous variable, the rules calculate the mean of the data points that have been classified together at that node. In the case of a categorical variable, they simply choose the class that occurs most frequently among the data points. The success of these rules can be measured through estimates of deviation. In this implementation, the deviation for continuous variables is the sum of the squared error for the observations. The deviation for categorical variables is simply the number of misclassified observations.

4. Predicting Pitch Accent Placement

In natural speech, some words appear more intonationally prominent than others. Such words are said to be stressed, or to bear pitch accents. While, in English, each word has a characteristic (lexical) stress pattern, not every word is accented. For example, *telephone* has three syllables and a 1-0-2 stress pattern; that is, its first syllable exhibits primary stress, its second no stress, and its third, secondary stress. When uttered, *telephone* may be accented or not. If accented, the pitch accent is associated with the (primary) stressed syllable, *tel–*. Thus, *lexical stress* may be distinguished from pitch accent. For example, note that the nominal and verbal homonyms of *object* are distinguished in speech by differences in lexical stress, but either may be uttered with or without a pitch accent with this difference preserved.

Although pitch accent is a perceptual phenomenon, words that hearers identify as accented tend to differ from their deaccented versions with respect to some combination of pitch, duration, amplitude, and spectral characteristics. Accented words are usually identifiable in the f_0 contour as local maxima or minima, aligned with the word's stressed syllable; deaccented items do not exhibit such pitch excursions. Words perceived as accented also tend to be somewhat longer and louder than their deaccented counterparts. The vowel in the stressed syllable of a deaccented word is often reduced from the full vowel of the accented version. Words that are deaccented may or may not be *cliticized*; that is, they may lack adjacent word boundaries as well as exhibit vowel reduction (e.g. *to* in *gotta* versus *got to*).

How humans decide which words to accent and which to deaccent – what constrains accent placement and what function accent serves in conveying meaning – is an open research question in linguistic and speech science. While syntactic structure was once believed to determine accent placement (Quirk, Svartvik, Duckworth, Rusiecki, and Colin 1964; Crystal 1975; Cruttenden 1986), it is now generally believed that syntactic, semantic, and discourse/pragmatic factors are all involved in accent decisions (Bolinger 1989). Currently, theoretical and empirical investigations of accent include word class, syntactic constituency, and surface position, as well as less easily defined phenomena falling into the broad category of information status, including contrastiveness (Bolinger 1961), focus (Jackendoff 1972; Rooth 1985), and the given/new distinction (Chafe 1976; Halliday and Hassan 1976; Clark and Haviland 1977; Prince 1981).

Most current text-to-speech systems use simple word-class information for input text to determine which items to accent and which to deaccent. In such systems, function (or closed class) words, such as prepositions and articles, are deaccented while content (or open class) words, such as nouns and verbs, are accented. It is widely believed that more detailed syntactic, semantic, and discourse-level information is needed to model human performance accurately, particularly for the synthesis of longer texts, but the problem of obtaining such information for unrestricted text appears unlikely to be solved in the foreseeable future. Yet considerably more detailed syntactic information than the simple function/content word distinction is already available for unrestricted text through improvements in part-of-speech tagging. And simple sorts of 'higher level' information can also be inferred about a text, based upon results of a number of psycholinguistic studies of information status, taken in conjunction with proposals advanced by computational linguists for modeling discourse structure. In short, fairly minimal sorts of text analysis can provide much richer sources of information to guide prosodic prediction than are currently being exploited.

The following section describes the development of procedures for the assignment of pitch accent automatically in synthetic speech from a minimal analysis of unrestricted text. These procedures are based on results of a series of experiments in the modeling of pitch accent assignment from analysis of prosodic features of recorded speech in several speech corpora. In these experiments, both hand-derived models and automatically-derived models were generated, the latter using the CART techniques described in Section 3. These experiments focused upon discovering answers to the questions: What sorts of information which can be derived currently from text analysis are useful in the prediction of pitch accent assignment? What sort of discourse representation permits the most accurate modeling of human accent strategies? How does automatic prediction from obtainable information compare with prediction from the fuller (and presumably more accurate) information obtained by hand?

The hand-crafted rules used to address this question (and described in detail below) were developed on a corpus of radio speech and modeled human accent decisions on this data with 82.4% accuracy. This performance compares quite favorably with previous results reported by Altenberg (1987) of 57% on partly scripted monologue. However, when tested on a corpus of citation-format utterances, the hand-written rules performed at 98.3% accuracy, a quite impressive success rate, albeit on a simpler task. Procedures developed automatically via CART techniques produced decision trees which predicted accent in the radio speech with a (cross-validated) success rate of only 76.5%, but whose performance improved to 85.1% when trained on larger corpora.

4.1. Speech Corpora for Testing and Training

Several speech corpora of American English were examined in the course of developing the accent prediction procedures discussed below, including citation-form sentences spoken by each of three speakers, news stories read by a single speaker, multi-speaker broadcast radio speech, and multi-speaker elicited spontaneous speech. Such variety permitted comparison of prediction performance for single versus multi-speaker speech, sentence versus discourse length speech, and planned versus spontaneous speech.

The citation-form sentences examined for this study represent a portion of the utterances elicited from each of three speakers (one male and two female) for the acoustic inventory used in concatenative synthesis by the Standard American English version of TTS. The 1380 utterances selected for analysis below consist of fairly short sentences and phrases, which are nevertheless all over 4 words long, with an average length of 5.5 words per sentence. Some examples are: *Plow a young man* and *The large cow accidentally brushed against the fence.* While such sentences may be difficult to interpret without constructing a rather specialized context, none of them appeared to present the speakers with prosodic challenges. The productions are regular and monotonous, without hesitations or disfluencies. The total duration for this corpus is approximately 45 minutes for each of the three speakers.

The single-speaker corpus of read news stories is the Audix Speech Corpus, consisting of ten news stories recorded by a female professional newscaster under laboratory conditions. The stories were selected from the Associated Press (AP) newswire and produced by the speaker (after familiarizing herself with the text) largely as written, with minor omissions and substitutions. While this database was collected for potential use in concatenative speech synthesis, the productions are quite natural for professional radio speech. This corpus currently has a total duration of approximately half an hour, with each story averaging about five minutes in length. Sentences (from the materials read) averaged 23.4 words in length. There are virtually no hesitations or disfluencies in the speech analyzed, since disfluent material was re-recorded.

The multi-speaker broadcast radio speech used in this study is a portion of the FM Radio Newscasting Database, a series of studio recordings of newscasts and other material provided by National Public Radio Station WBUR in association with Boston University, collected by SRI International (Patti Price), Boston University (Mari Ostendorf), and MIT (Stefanie Shattuck-Hufnagel). Prosodic labeling used for the current analysis was done at Bell Laboratories. Two news stories were used, representing approximately nine minutes of speech, produced by a total of nine speakers. While, again, these stories appear typical of fluent 'radio speech', the inclusion of 'sound bites' from interviewed speakers makes them somewhat less useful as exemplars of coherent read text. There are some minor disfluencies in the recordings.

The multi-speaker spontaneous speech database investigated here included 298 sentences (representing 24 minutes of speech by 26 speakers, with an average length in words of 11.5) from the DARPA Air Travel Information Service (ATIS) database. These utterances were taken from the early Texas Instruments portion of the ATIS corpus, and consisted of 774 sentences elicited in a wizard-of-Oz simulation by Texas Instruments to provide a common database for the DARPA Spoken Language Systems task (Hemphill, Godfrey, and Doddington 1990). To simulate interactions between a caller and a travel agent, subjects were given a travel scenario and asked to make travel plans accordingly, providing spoken input and receiving visual output at a terminal. They were told that their utterances were being recognized by machine; vocabulary constraints were enforced by error messages.

All but the citation form sentences were assigned prosodic labels by hand, using speech analysis software (WAVES) developed by Talkin (1989). Prosodic labels were those described in Section 2.

4.2. Variables for Predicting Accent Location

As noted above, syntactic, semantic, and pragmatic/discourse factors have been proposed as predictors of pitch accent. Of course, not all the types of information proposed in the literature can be obtained automatically from text with current natural language processing capabilities. Complete and accurate information about syntactic structure is an obvious example of information believed essential to 'natural' pitch accent assignment which is not available computationally for unrestricted text. However, the goal of the current study is to determine what sort of success can be achieved in predicting pitch accent in natural speech simply from information which *can* be obtained automatically from text with current computational techniques – and which types of information are most useful in this endeavor.

As noted above, part-of-speech information is perhaps the most commonly employed predictor of pitch accent. Most text-to-speech systems assign accent purely based on simple function word/content word distinction, with the consequence of sometimes accenting words human speakers would deaccent, especially in the synthesis of longer texts. Also, items that are ambiguous with respect to the function-content distinction, as in preposition (commonly deaccented) versus verb particle (commonly accented) ambiguities, are rarely disambiguated, since text-to-speech parsers and taggers tend to be extremely simple. Finer-grained distinctions among word classes have, however, been explored empirically, with rather limited results to date. Altenberg (1987), for example, examined the extent to which pitch accent (as a binary feature) could be predicted in a 48-minute partly-read, partly-extemporaneous public talk by a non-professional speaker. From hand-tagged data, Altenberg found that accent could be predicted on the basis of word class in this (training) data only 57% of the time (with 92% correct and 62% coverage of the corpus). While part-of-speech alone clearly cannot account for all accent behavior, however, it is nonetheless an important factor. For the current study, part-of-speech information was obtained using the tagger described by Church (1988), which has a success rate of around 98%.

A special difficulty for assigning pitch accent from part-of-speech information is the problem of stressing complex nominals – items such as *city hall* and *parking lot*. Liberman and Sproat (1992) estimate there are approximately 70,000 complex nominals per million words from their examination of the Brown Corpus (`http://www.hit.uib.no/icame/brown/bcm.html`). While both these phrases are commonly analyzed in citation form as noun-noun sequences, the first commonly has its most prominent accent on its right-hand element, *hall*, while the second has its on the left element, *parking*. (Liberman and Sproat (1992) find that 75–95% of complex nominals are stressed on the left, depending upon genre.) But when the two are combined to form the complex nominal *city hall parking lot*, the stress on *hall* is normally shifted to the left, so that *city* is now more prominent than *hall*. So, while knowing that items are tagged as sequences of nouns is essential to identifying them as complex nominals, this information does not tell us how they will be accented. Stress patterns in complex nominals appear to be predicted in part by the semantic roles filled by the nominal's components, but are also to some extent idiosyncratic. In the work described below, citation form stress patterns of complex nominals are obtained from a complex nominal parser, NP, described by Sproat (1990), which employs semantic rules together with table lookup.

Surface order is also mentioned in the literature as a factor in predicting pitch accent. While this feature is most often cited in schemes to predict the location of nuclear stress, it has also been observed informally that items which are preposed, for example, will be accented more frequently than the same item in non-initial position. Compare TODAY *we will* BEGIN *to* LOOK *at* FROG *anatomy*, for example, with *We will* BEGIN *to* LOOK *at* FROG *anatomy today*.

Even prepositions in preposed prepositional phrases will sometimes receive an L* accent, although they are rarely accented when not preposed. In certain genres, too, the importance of surface order in combination with word class in determining pitch accent has been noted. For example, in broadcasters' speech, Bolinger (1982, 1989) has noted the tendency to place nuclear stress as close to the end of a phrase as possible, even when this decision leads to accenting items that, in ordinary speech, would probably be deaccented.

The remaining predictors of pitch accent commonly noted in the literature may be broadly characterized under the rubric *information status*. In this category, we include an item's status in the discourse as *given* or *new*, that is, whether the item represents 'old' information, which a speaker is entitled to believe is shared with a hearer, or not (Prince 1981), as well as whether the item is being linguistically *focused* or is being used *contrastively*.

Experimental evidence indicates that speakers tend to introduce new items in utterances by accenting them, and tend to deaccent items representing given information (Brown 1983). Using a taxonomy based on the Prince (1981) classification of given/new information, Brown (1983) found that 87% of *brand new* entities (assumed by the speaker not to be known by the hearer) and 79% of *new inferrable* entities (assumed by the speaker to be inferrable by reasoning from previously evoked entities) were in fact accented by subjects, while 96–100% of *evoked* entities (those already mentioned or situationally salient) were deaccented. Perception studies also indicate that hearers appear sensitive to this association of givenness with deaccenting and newness with accent (Terken and Nooteboom 1987). For example, in *There are* LAWYERS, *and there are* GOOD *lawyers*, the token *lawyers* is likely to be accented in its initial occurrence, but it is likely to be deaccented in its second occurrence.

But how items come to be perceived as given – and how they lose that 'givenness' – are open research questions. For example, a concept need not have been explicitly evoked in prior discourse to be taken as given; in EACH NATION DEFINES *its* OWN *national* INTEREST, mention of *nation* can make the subsequent mention of *national* deaccentable, as given information. However, mention of *sleepy* does not seem to make *slept* deaccentable in *The* SLEEPY SOLDIER SLEPT *for an* HOUR. Mention of one item in a contrast set can license the givenness of other members of a set, although sometimes with different consequences for accentuation; for example, *unhelpful* in JOHN TRIED *to be* HELPFUL, *but* ONLY SUCCEEDED *in being* UN*helpful* is made given by mention of *helpful*, and, in consequence, only the negative prefix is stressed. Mention of subordinate categories can make superordinates given and, hence, deaccentable, as illustrated in I LIKE GOLDEN RETRIEVERS, *but* MOST *dogs* LEAVE *me* COLD, where mention of *golden retrievers* makes *dogs* deaccentable.

While rules encapsulating the full complexities of givenness may be far in the future, it is nonetheless possible to model the simpler aspects of this phenomenon. Silverman (1987) proposed using a first-in first-out (FIFO) queue of word roots of mentioned open-class items in a text-to-speech system to determine whether subsequent items should be deaccented, clearing the queue at paragraph boundaries. (This proposal was not actually tested by Silverman, although certain aspects of the proposal appear to be supported by experimental findings of the current study discussed below.) In contrast to this linear approach, the study described below makes use of Grosz and Sidner's (1986) hierarchical notion of a discourse's *attentional structure* to define the domain of givenness. In this model of discourse, discourse structure itself comprises three structures, a *linguistic structure*, which is the text/speech itself; an *attentional structure*, which includes information about the relative salience of objects, properties, relations, and intentions at any point in the discourse; and an *intentional structure*, which relates the intentions underlying the production of speech segments to one another. The attentional structure is represented as a stack of *focus spaces*, which is updated by pushing items onto or popping them from the stack. The notion of *focus* in this tradition as 'what is currently being talked about' should not be confused with the linguistic notion of *focus* as 'what the speaker wishes to make specially prominent'.

For this study, no attempt has been made to model Grosz and Sidner's attentional structure entirely. In particular, no aspect of intentional structure is included, and objects, properties and relations are represented here by their roots rather than by some more abstract conceptual representation. At any point in the discourse, the current focus space contains some representation of items that are currently salient in the discourse, or in *local focus* (Grosz, Joshi, and Weinstein 1983; Sidner 1983; Brennan, Friedman, and Pollard 1987). The relationship between items in a particular focus space – as well as the relationship between items in different focus spaces – is not well understood. However, some hierarchical relationship between focus spaces is usually assumed in the literature. Specifically, items in a focus space associated with a portion of the discourse that is understood as superordinate to another portion (e.g. a super-topic perhaps) are generally assumed to be available during the processing of the subtopic. Side by side with this notion of local focus is a notion of *global focus*, in which concepts central to the main purpose of the discourse remain salient throughout the discourse.

Local focus is modeled here as a stack of focus spaces, each of which contains the roots of some set of lexical items contained in the (orthographic) phrase currently being processed. These roots substitute for the concepts which would ideally be represented in the focus stack. Hierarchical relationships between the discourse segments associated with these focus spaces – and, thus, hierarchical relationships between the focus spaces themselves – are inferred from orthographic cues, such as paragraphing and punctuation, and lexical cues, such as *cue phrases*. Cue phrases are words such as *now*, *well*, and *by the way*, which provide explicit information about the structure of a discourse. Each cue phrase has a *push* or *pop* operation associated with it for this purpose (Litman and Hirschberg 1990). In the current study, roots of members of certain word classes mentioned in the discourse are added to the focus space; the classes that are to be considered were varied experimentally, to see which strategy best supported observed accent behavior. While local focus is continually updated during the discourse, items in global focus are fixed early in the analysis of the text. The procedures by which local focus is updated and the point at which global focus is fixed were also varied experimentally. Positional relationships between items in local focus was also examined and the content of both global and local focus spaces were varied systematically by word class.

Linguistically focused or *contrastive* commonly receive special intonational prominence (Bolinger 1961; Lakoff 1971; Jackendoff 1972; Schmerling 1976; Rooth 1985; Horne 1987). (To avoid confusion with the concepts of local and global focus discussed above, these will be termed 'contrastive', although this term too has its problems. In particular, it is often impossible to find a contrast set for many items receiving so-called *contrastive stress*.) In some cases, such prominence is the sole means by which contrastiveness can be inferred. Suggestions for how to predict this status have been many and varied, including the identification of preposed constituents (Prince 1981; Ward 1985) and of semantic or syntactic parallelism. In addition, a notion of *proper-naming*, involving the use of proper names for individuals who might otherwise be differently referenced (e.g. by common nouns or pronouns) and derived from experimental results discussed in Sanford (1985), provides another method for predicting contrastive stress. The discourse model described above has been employed to model this phenomenon. For example, items in global focus but not currently in local focus were frequently observed to be uttered with special emphasis, as if these items were being reintroduced into the discourse. This observation was tested empirically by keeping track of items in both local and global focus and comparing their status along both dimensions.

4.3. The Rule-Based System

The first stage of development of the rule-based accent assignment system was done on the FM Radio database described in Section 4.1. These rules were implemented in Quintus Prolog in a program which took unrestricted text as input and assigned accent based upon values of the variables discussed in Section 4.2. This system was employed to analyze transcriptions of recorded, prosodically-labeled speech, and predictions were compared with hand-labeled observations to retrain the system. Experimentation with this paradigm included varying both the text analysis and the structure of the decision procedure to minimize error rates when predictions were compared with observed accent values in the prosodically-labeled speech. So, for example, composition of broad word classes was varied systematically, and so was the use of the part-of-speech variable in the decision procedure. Best results for each of the FM news stories examined (79% correct pitch accent prediction for one and 85% for the other) occurred for the same variations of individual variable treatment and decision tree composition. Combined results with a success rate of 82.4% are presented in Table 11.1, where the procedure's decision is based on the feature specified.

Table 11.1. Accent prediction for the FM Radio database ($n = 1214$).

	Total correct/1000 (82.4%)	
Correctly predicted by:		*Total error (%)*
new	418	42
given	32	3
closed-accented	60	6
closed-cliticized	283	28
closed-deaccented	119	12
compound-accented	0	0
compound-deaccented	19	2
cue-phrase	16	2
contrast	59	6
	Total incorrect/214 (17.6%)	
Incorrectly predicted by:		*Total error (%)*
new	63	30
given	35	16
closed-accented	11	5
closed-cliticized	30	14
closed-deaccented	36	17
compound-accented	0	0
compound-deaccented	13	6
cue-phrase	6	3
contrast	18	8

The association between word class and accent decision was initially modeled from the data of Altenberg (1987), translating from Altenberg's tag set to that of Church (1988) and filling in gaps in Altenberg's coverage where necessary based upon the FM corpus. Such observation had previously suggested that a simple division by part-of-speech into open and closed classes was insufficient for making accurate pitch accent predictions. So, from observed pitch accent decisions, finer-grained distinctions were developed. Optimal predictive power was obtained for this corpus when four broad classes were assumed: *open class*, *closed-cliticized*, *closed-deaccented* (but not cliticized) and *closed-accented*. Membership in the four classes defined was not always possible from the tag set used by Church (1988), so exceptions were further specified by lexical item. In the final version of the hand-crafted rule set, closed-accented items include the negative article, negative modals, negative *do*, most nominal pronouns, most nominative and all reflexive pronouns, pre- and post-qualifiers (e.g. *quite*), pre-quantifiers (e.g. *all*), post-determiners (e.g. *next*), nominal adverbials (e.g. *here*), interjections, particles, most *wh*-words, plus some prepositions (e.g. *despite*, *unlike*). Closed-deaccented items include possessive pronouns (including *wh*-pronouns), coordinating and subordinating conjunctions, existential *there*, *have*, accusative pronouns and *wh*-adverbials, some prepositions, positive *do*, as well as some particular adverbials like *ago*, nominative and accusative *it* and nominative *they*, and some nominal pronouns (e.g. *something*). And closed-cliticized items include the definite and indefinite articles, many prepositions, some modals, copular verbs and *have* auxiliaries. Other items (adjectives, adverbials, nouns, verbs) are deemed 'open'.

Forty-six percent of correct accent predictions for the FM Radio corpus were made simply by distinguishing which closed-class items were likely to be accented and which were not: 86% of closed class items were correctly predicted. Most errors for closed class items (83%) involved instances of function words which were deaccented in the majority of cases but which were sometimes accented by the speakers, such as coordinate conjunctions, copular verbs, and some prepositions and determiners. The simple division into broad word classes would in fact correctly predict 77% of observed pitch accents in the corpus. Note, however, that the simple traditional association between function word status and deaccenting and content word and accenting would correctly predict only 68% of items in the corpus. So, the finer-grained model employed here appears to produce an improvement.

Citation-form stress patterns predicted by NP account for only 2% of correct accent predictions and 6% of incorrect predictions; on balance, 59% of predictions based on citation-form compound stress assignment proved to be correct. Errors in every case arose from incorrectly predicting deaccenting of an element. In half these cases, however, the algorithm predicted accent strategies which were in fact clearly acceptable to native speakers, even though not chosen by the speakers in the corpus.

The collection and manipulation of the attentional state representation was also varied experimentally in the following ways: both global and local focus representations were manipulated independently such that the global focus space was set, and the local focus spaces updated, by the orthographic phrase, the sentence, or the paragraph. So, for example, the global space was considered to be defined after the (items mentioned in the) first phrase, first sentence or first paragraph of a text. The local stack was updated independently at the end of each phrase, sentence or paragraph – although cue phrases could also motivate a push or a pop of the stack under any of the experimental conditions. For the FM database, the best results were obtained when the global space was defined as the first full sentence of the text and the local attentional stack updated by paragraph; the latter finding to some extent appears to bear out Silverman's (1987) proposal for updating a set of roots at paragraph boundaries, although the updating for the current study was also sensitive to the presence of cue phrases.

The potential content of both global and local focus spaces was also varied, so that, for example, all open-class words, nouns only, or nouns plus some combination of verbs and modifiers were allowed to affect and be affected by changes in the attentional state representation. For the FM data, best results were obtained when focal spaces included roots of all content words (nouns, verbs and modifiers) rather than some subset. However, overall, distinguishing between given and new for this corpus proved to be of limited use: while 42% of total correct predictions were due to associating new information with accent, only 3% of correct predictions were those associating given items with lack of accent. Thirty percent of the total error was due to predicting incorrectly that new items were accented, and 16% to predicting incorrectly that given items were deaccented. Overall, 87% of items labeled new were correctly predicted to be accented, but only 48% of given items were correctly predicted to be deaccented. Incorrect accent predictions based on inference of given/new status account for 31.6% of the overall error on this corpus.

It is possible that the inaccuracy of given/new predictions might be due to the fact that interview portions of these news stories have been spliced into the newsreaders' commentary. So, it is not clear how information status should be assigned, when some speakers are not in fact participants in the conversation in which their speech appears. That is, for the newsreader, what is said in an interview presumably becomes given information; but clearly what represents given information for the interviewee is not always discernible from a fragment of the interview. It is also possible, of course, that givenness was not correctly modeled, or that other factors should have been considered in evaluating the influence of this type of information status on pitch accent. These questions will be pursued below.

Finally, an attempt was made to infer local and global focus and contrastiveness from surface position, part-of-speech, and the state of the local and global focus spaces. In earlier experiments, it was noted that, in the FM corpus, speakers often placed particular intonational prominence on proper names which had previously been introduced into the text, and which thus seemed suitable for deaccentuation as given. The referential strategy of proper-naming (Sanford 1985), in which the use of proper names was noted as a mechanism to focus attention, was hypothesized as a means of accounting for this tendency. That is, it was conjectured that such referential behavior might indicate the speaker's attempt to refocus attention upon recently-mentioned persons when other persons had been mentioned even more recently. It was further hypothesized that this re-focusing accounted for the special intonational prominence which appeared to be assigned to these items. While there were not enough data in the FM corpus to test this hypothesis in much detail, it was found to predict correctly in all cases where a proper name was employed to identify a person whose name was already represented in local focus. Similarly, it was noted through experimentation that speakers tended to accent items that were predicted to be in global but not in local focus.

Again, the general notion of reintroducing an entity mentioned earlier in the text, when the discussion had shifted to other topics, seemed related to the assignment of intonational prominence, although there were relatively few instances in which to test the hypothesis. Other approaches to modeling contrastive stress included the prediction of accent on preposed adverbials and prepositional phrases and the proposal that in complex nominals, when several elements are given and at least one is new, the new item(s) receives special prominence. In general, accent for items identified as contrastive was correctly predicted 77% of the time. However, such items account for only 6% of all items in the data.

The hand-crafted rules were implemented in TTS (Sproat, Hirschberg, and Yarowsky 1992). They operate on text which is first tagged for part-of-speech, and in which standard orthographic indicators of paragraph and sentence boundaries are preserved. Tables are maintained of word classes and individual items divided into four broad classes – closed-cliticized, closed-deaccented, closed-accented and open (as above) – where distinctions among the first three groups are based upon frequency distributions in the training data. As processing of the labeled text proceeds from left to right, the following information is added to a record maintained for each item:

1 Preposed adverbials are identified from surface position and part-of-speech, as are fronted PPs, and labeled as *preposed*.

2 Cue phrases are identified from surface position and part-of-speech as well, and their accent status is predicted following the findings in Hirschberg and Litman (1987) and Litman and Hirschberg (1990).

3 Verb-particle constructions are identified from table look-up.

4 Local focus is implemented in the form of a stack of roots of all nouns, verbs, and modifiers in the previous text. New items are pushed on the stack as each phrase is read. Individual cue phrases trigger either push or pop operations, roughly as identified in Grosz and Sidner (1986). Paragraph boundaries cause the entire stack to be popped. As new items are read, those whose roots appear in local focus are marked as given, while others are labeled new. Additionally, items with prefixes such as *un–*, *il–*, and so on whose roots appear in local focus are marked as 'prefixed'.

5 Global focus is defined as simply the set of all content words in the first sentence of the first paragraph of the text. Nouns, verbs and modifiers whose roots appear in global focus are so marked.

6 Potential complex nominals are identified from part-of-speech and shipped to the NP complex nominal parser (Sproat 1990); their stress pattern in citation form is stored with the nominal for subsequent processing.

Other information collected for these nominals for subsequent evaluation includes whether or not they are inferred to be proper (e.g. street names, personal names that include titles, acronyms, and so on).

7 Finally, possible contrastiveness within a complex nominal is inferred by comparing the presence of roots of elements of the nominal in local focus; roughly, if some items are given and others new, the new items are marked as potentially contrastive. When each record is complete, a simple decision procedure determines whether the item will be predicted as cliticized, deaccented (but not cliticized), or accented.

In summary, experimentation with a hand-crafted rule set for pitch accent prediction on a corpus of recorded radio speech revealed the following:

- A significant improvement in accent prediction can be made by a more fine-grained use of word class information.
- Where a simple, function/content distinction correctly predicts 68% of pitch accents observed in this corpus, more sophisticated distinctions permit correct prediction of 77% of observed accents.
- Disappointingly, modeling of the given/new distinction and of contrast adds only 5.4% to the overall score of 82.4% correct. However, while the given/new predictions are often incorrect, predictions of accent from inferred contrastive status are generally correct.

The second corpus examined in developing rules for pitch accent prediction was the set of citation-form sentences. The original purpose in examining this corpus was not to test the success of the accent assignment algorithm specifically, but rather to see whether it could supply prosodic information which was not available from other hand labelings for an independent analysis of segmental durations (van Santen 1992). There is some empirical evidence that a word's accent status plays a significant role in determining the duration at least of its stressable syllable (Cooper, Eady, and Mueller 1985; Eefting 1991). So, using the hand-crafted rules described above, three levels of accent were predicted for these utterances: accented, deaccented but not cliticized, and cliticized. To obtain a measure of performance on this unlabeled corpus, these predictions were checked by hand for 302 of the utterances for one speaker – a total of 1756 words. Results from this evaluation indicate that accent predictions were correct in 98.3% of cases, a quite phenomenal success rate – especially in comparison with the earlier results on the data for which these rules were originally developed. The validity of this evaluation received further indirect confirmation by results of the durational analysis subsequently performed on the entire set of 1380 sentences for each of the three speakers (van Santen 1992).

Stressable syllables in words the accent prediction rules predicted as accented were found to be 18–26% longer than stressable syllables in words predicted to be deaccented but not cliticized, and 32–39% longer than similar syllables in words predicted to be cliticized.

It seems likely that the high success rate of accent prediction in citation-form sentences reflects the nature of the data. In these sentences, intonation is monotonous and regular, and each sentence has clearly been uttered with little influence from the semantic content of prior utterances. The 302 sentences checked indicate little effect of a given/new distinction, and almost no evidence of focal or contrastive accent. Most of the success of the prediction rules, then, appears to have come from the finer-grained distinctions made between word class and pitch accent, as discussed above. Thus, the usefulness of this simple refinement in word class use appears very promising for assigning pitch accent at the level of isolated sentences in text-to-speech.

4.4. Automatic Classification of Pitch Accent

The derivation of prediction rules by hand as was done for the FM Radio corpus is a labor-intensive and time-consuming process, with results whose generalizability to new corpora is difficult to predict. The rule set described in Section 4.3 took several months to construct and refine, for example. And the difference in performance of these rules on the citation sentence data and the FM data provides some evidence that different text genres require different rule sets, despite the interesting result that rules derived from one dataset in fact performed considerably better on another. To address these concerns, further experimentation was performed using the CART techniques described in Section 3. It was highly desirable to see whether procedures derived automatically could perform as well or better than hand-crafted rules.

To compare performance of the rule-based system with those which could be produced automatically, the same variables referenced by those rules (described in Section 4.2) were used to create input for CART classification. Data from the FM stories described in Section 4.1 were pooled for this analysis, for a total of 1214 accent decisions for testing and training – a rather small corpus for automatic prediction purposes as it turned out. In fact, the prediction tree produced for this dataset was quite degenerate, consisting of only a single split, on the function/content word distinction. The cross-validated success rate for this tree was 76.5%, the same percentage correctly predicted by this variable for the hand-crafted rules. Testing on a larger corpus produced more interesting results, suggesting that the classification procedure simply had too little data to make generalizable predictions for this corpus.

Analysis of the larger Audix corpus involved the same variables used above, for 2854 data points – more than twice the number in the FM Radio corpus.

Since the performance of the given/new distinction in the analysis of the FM Radio database had been somewhat disappointing, it was desirable to determine if the modeling of this feature was flawed in general, or if a larger corpus would provide a better testbed. This larger database also provided an opportunity to examine several variations on the discourse modeling strategy described in Section 4.2. In addition to varying the units used in updating the local focus stack and in modeling global focus, position of a lexical root within the local focus space was noted when the current word was found to be given. The notion here was that items more remote in local focus (i.e. made given earlier rather than more recently) might be less likely to impart givenness to the current item.

In analyzing the ATIS corpus, several new variables were considered as well, capitalizing upon prosodic information which had been identified for a separate analysis of intonational phrasing (Wang and Hirschberg 1991a, 1992). Since it seemed possible that the position of an item within its intonational phrase, as well as the overall length of this phrase, might be related to its likely accent status, these additional features were examined. At a minimum, most phonological theories predict that items alone in their intonational phrase will be accented. In addition, since a large number of speakers is represented in the ATIS corpus, and since accent strategy may indeed be speaker dependent, a speaker identity variable was added. Finally, to capture the intuition that a larger word class context may play a role in accent decisions, the part-of-speech of both left and right contexts of the current items, as well as the part-of-speech of the item itself, were examined.

Results of using CART techniques to model pitch accent predictions automatically for the Audix and ATIS corpora are illustrated by the cross-validated decision trees shown in Figures 11.1 and 11.2, respectively. The following variable names were used in these figures:

`sw`	distance from beginning of utterance
`ew`	distance from end of utterance
`totw`	total words in utterance
`bp`	distance in words from prior boundary
`bs`	distance in words from next boundary
`bpt`	type of prior boundary (n: none, f: minor, i: major, s: utterance)
`bst`	type of next boundary (n: none, f: minor, i: major, s: utterance)
`pos`	part of speech for prior (p), subsequent (s) and current (c) word
`v`	DO HV HVD HVZ MD SAIDVBD VB VBD VBG VBN VBZ NA
`b`	BE BED BEDZ BEG BEM BEN BER BEZ NA
`m`	ABN DT DTI DTS DTX JJ QL RB NA
`f`	* AT CC CD CS EX IN PART TO UH NA
`n`	NN NNS NP PN NA
`p`	PPL PPO PPS PPSS NA
`w`	WDT WPS WRB NA
`cl`	class: clac clcl clde op

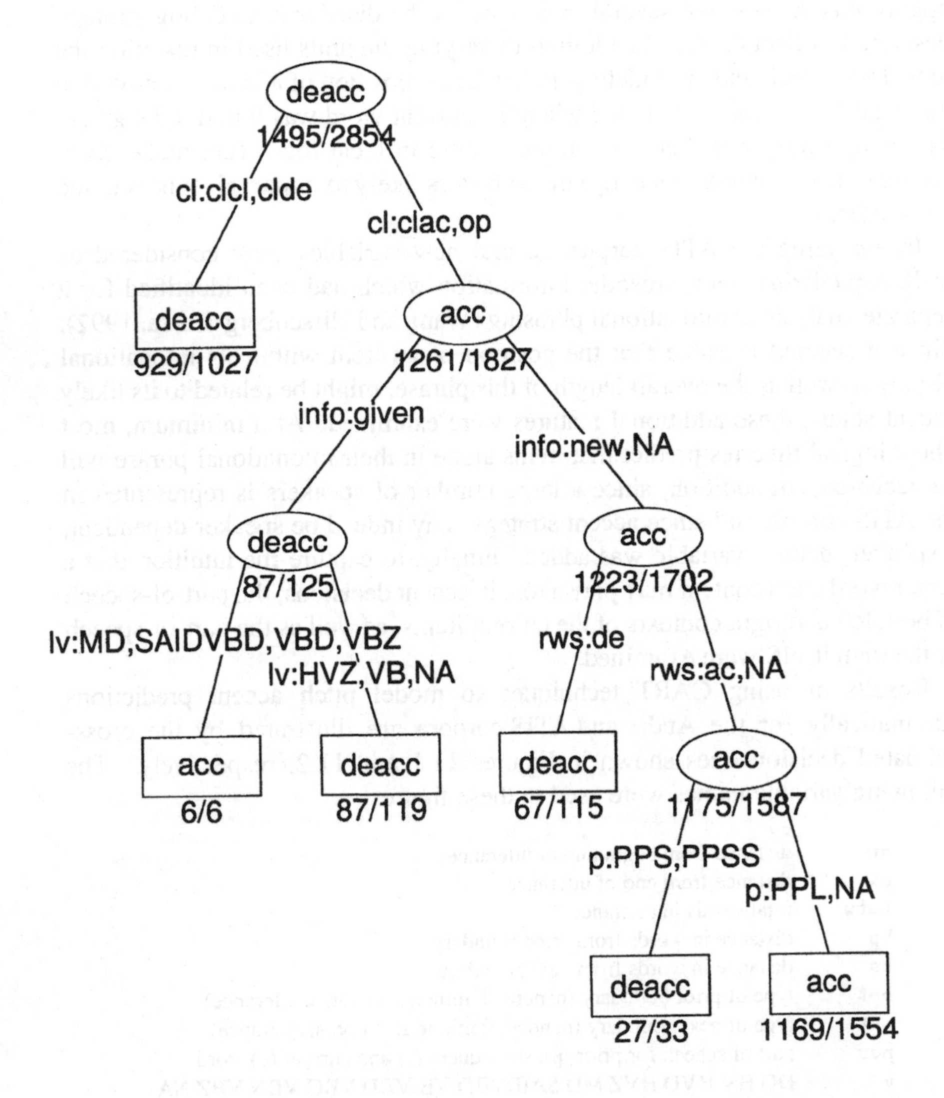

Figure 11.1. **Decision tree for pitch-accent prediction trained on the Audix corpus (80% correct).**

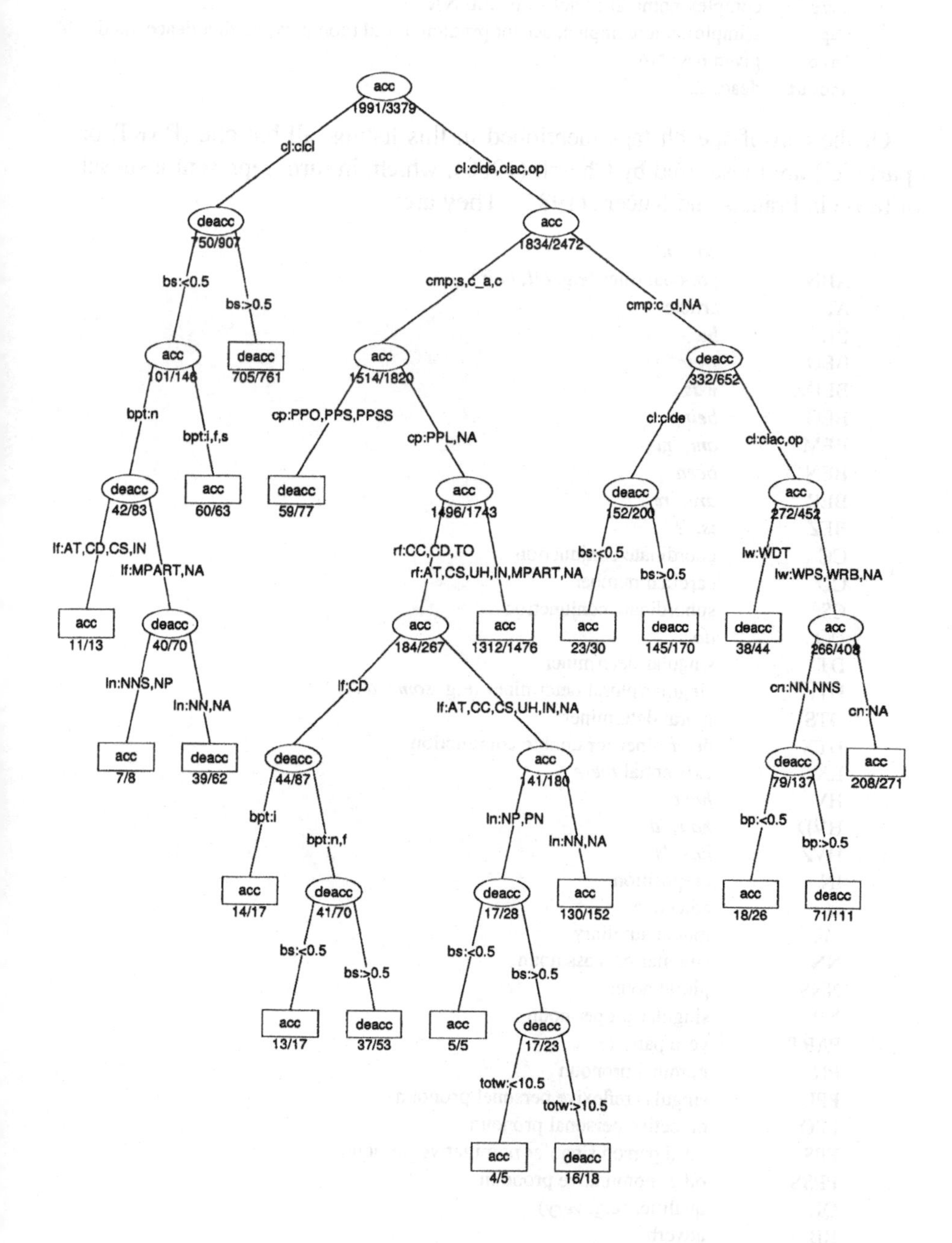

Figure 11.2. ATIS pitch-accent prediction (85.1% correct).

`cmpacc`	whether np predicts accented or deaccented (ac de NA)
`comp`	complex nominal or not smp cmp NA
`cmp`	s(imple), c_a (complex, accent predicted) c_d (complex, predict deaccented) NA
`info`	given new NA
`accent`	deacc acc

Of the part-of-speech tags mentioned in this listing, all but one (PART, or 'particle') are those used by Church (1988), which, in turn, represent a subset of those in Francis and Kučera (1982). They are:

*	*not*, *-n't*
ABN	pre-quantifier (e.g. *all*, *half*)
AT	article
BE	*be*
BED	*were*
BEDZ	*was*
BEG	*being*
BEM	*am*, *'m*
BEN	*been*
BER	*are*, *'re*
BEZ	*is*, *'s*
CC	coordinate conjunction
CD	cardinal number
CS	subordinate conjunction
DO	*do*
DT	singular determiner
DTI	singular/plural determiner (e.g. *some*, *any*)
DTS	plural determiner
DTX	determiner for double conjunction
EX	existential *there*
HV	*have*
HVD	*had*, *'d*
HVZ	*has*, *'s*
IN	preposition
JJ	adjective
MD	modal auxiliary
NN	singular or mass noun
NNS	plural noun
NP	singular proper noun
PART	verb particle
PN	nominal pronoun
PPL	singular reflexive personal pronoun
PPO	objective personal pronoun
PPS	third person singular nominative pronoun
PPSS	other nominative pronoun
QL	qualifier (e.g. *very*)
RB	adverb
SAIDVBD	*said*
TO	infinitive marker
UH	interjection

VB	verb, base form
VBD	verb, past tense
VBG	verb, present participle
VBN	verb, past participle
VBZ	verb, third person singular, present tense
WDT	*wh*-determiner
WPS	possessive *wh*-pronoun
WRB	*wh*-adverbial

As noted in Section 3, nodes in the trees are labeled with the dependent variable value for the majority of data points classified at that node. So, in Figure 11.1 the root node is labeled *deacc*, or 'deaccented', since the majority of nodes classified by the tree are classified by it as deaccented. The proportion of nodes correctly so classified is indicated by the fraction under the box or circle; e.g. for the root in Figure 11.1, 1495 out of 2854 items are correctly classified by the tree as deaccented. Arcs in the tree are labeled by the feature controlling the split from the parent node, as well as the values of the feature which the split is made on. The first split from the root node in Figure 11.1 is made on the feature *cl*, or 'class'. Values *clcl* and *clde* (closed-cliticized and closed-deaccented) of the class feature are grouped together here, both being used to predict deaccenting on the left branch, while features *clac* (closed-accented) and *op* are collapsed to form the right branch of the tree. The total number of correct predictions for a full tree may be ascertained by summing over the numerators of fractions labeling leaf nodes. As discussed in Section 3, however, it should be noted again that these numbers are somewhat higher than the cross-validated scores for these trees.

The cross-validated success rates for automatic classification are similar to those obtained for the hand-written rules described in Section 4.2 without cross-validation – about 80% of pitch accents in the Audix data are correctly predicted and 85% of the ATIS accents. The best full decision trees for both corpora cover 92–93% of the data.

The decision tree for the Audix corpus (Fig. 11.1) makes use of the same variables used in the hand-written rules created for the FM corpus. This tree successfully classifies 80% of the data. In this tree, as in the remainder of the predictions for the Audix corpus, the procedure over-predicts accent: two thirds of the errors occur when an item which the speaker has deaccented is predicted to be accented. The (cross-validated) tree itself is quite a simple one, however, and makes a good deal of intuitive sense.

The first split in the tree occurs on the partition of items by part-of-speech into open or closed, with the latter class divided into closed-cliticized, closed-deaccented, and closed-accented; 76.7% (2190/2854) of the data can be correctly classified on this distinction alone – virtually identical to the proportion of items in the FM database that could be so identified. Here, 90.5% (929/1027) of items predicted to be deaccented or cliticized by virtue of part-of-speech assignment are indeed deaccented or cliticized, while 69% (1261/1827) of items predicted by class to be accented are accented. However, the second split in the tree is on the given/new variable, which, it will be recalled, did not prove a very successful predictor of accent for the FM corpus. Here, open class and closed-accented items are further subdivided into given and new (or not applicable). Of given items at this node, 69.6% (87/125) are deaccented, while 71.9% (1223/1702) of new items are accented. The third split in the tree is on this set of new items, which is divided according to accent predictions arising from citation stress for complex nominals; 73% (1242/1702) of these items are correctly predicted by this feature. The final split in the tree separates nominative pronouns (both singular and plural) from all other parts-of-speech: these pronouns tend to be deaccented (27/33, or 81.8%).

So, the decision procedure illustrated in Figure 11.1 can be summarized as follows:

1 If an item's part-of-speech indicates it is closed-cliticized or closed-deaccented, or if it represents given information, or if it is (predicted to be) deaccented in complex nominal citation form, or if it is a nominative pronoun, predict that it is deaccented;

2 otherwise, predict accented.

This same simple result – and almost identical predictive success – appears even when the method of updating local and global focus is varied, when a 'focal space distance' measure is included, or when overall length of the current phrase and the current sentence are considered. While this result thus seems reliable, it does not suggest avenues for improving predictive power.

Examination of the full classification tree, however, suggests that obtaining more data might reduce the errors in classification, where the addition or manipulation of variables will not. In many cases, there are too few representatives of a potentially disambiguating word class to justify disambiguation on grounds of class membership, even though the existing members seem to be behaving predictably. In fact, this classification procedure does not make use of word identity, so there are no lexical exceptions to accenting likelihoods based upon part-of-speech, as there are in the hand-crafted rules.

Results from automatic classification of pitch accent for the ATIS corpus (Fig. 11.2) initially proved similar to results obtained for the Audix corpus – but improved considerably when additional features were examined. Results using the original variable set as described in Section 4.2 correctly predicted 81.9% of pitch accents. However, the decision trees produced by the learning algorithm contain a wider variety of features than any of the trees produced by CART for the Audix corpus. Another general observation we can make from comparing CART trees grown on the ATIS data to those grown on the

Audix corpus data is the more important role allocated to various measures of distance and to the part-of-speech of the right and left contexts in the ATIS trees.

The best predictions of accenting for the ATIS corpus are made when we make additional prosodic variables available to CART. The decision tree depicted in Figure 11.2 successfully predicts 85.1% of the ATIS corpus when two new variables, distance with respect to preceding or succeeding intonational boundary and the type (major or minor) of that boundary are available. In this tree, relative distance to boundary appears to subsume relative distance to beginning or end of utterance – understandably, since utterance edges are themselves boundaries – but also provides additional discriminatory power. Words predicted by overall class to be closed-deaccented are found to have their actual accentual status dependent upon position within intonational or intermediate phrase and part of speech of left context. In this way, 84% (168/200) of these closed-deaccented items are correctly predicted. Since the actual intonational phrasing of an utterance may not itself be reliably recoverable from text, it might seem optimistic to include observed values as predictors here. However, we will see in Section 5 that phrase boundary location can be predicted with even more accuracy than accent decision, so such predictions should serve as reasonable substitutes for observed values.

4.5. Discussion of Pitch Accent Prediction

The experiments described above were designed to develop procedures for modeling human pitch accent behavior in several different types of speech corpora. The most successful attempt involved predictions for citation-form sentences, where a prediction rate of 98.3% was achieved. Less successful, although still improving over the simple function/content word prediction method employed by most text-to-speech systems, were attempts to predict accent location in larger contexts. Rule systems developed by hand and others generated through CART automatic classification techniques attained success rates of 80–85% on these corpora.

Throughout the experiments, it was clear that part-of-speech plays a major role in the prediction of pitch accent. However, this feature alone reliably predicts only about three-quarters of observed pitch accents. To improve predictive power, it is thus necessary to explore additional predictive features, none of which individually may be responsible for a large number of successful predictions, but which may together improve predictive accuracy. In exploring new sources of automatic accent prediction, a hierarchical representation of the attentional structure of the discourse was employed, which supported inference of local and global focus as well as contrast. Additional sources of information tested included: word class context, citation-form compound nominal stress (derived from Sproat's NP parser), a variety of relative distance measures within intonational phrase and utterance, as well as utterance length and speaker identity. While not enough data have been examined to assess fully the usefulness of any of these additional information sources, it is notable that different collections of features can produce roughly similar success rates for a given corpus, suggesting a certain amount of redundancy in predictors of pitch accent. Larger databases will have to be analyzed in order to tease apart relationships among the variables examined, and to improve our ability to model pitch accent strategies successfully.

5. Predicting Phrase Boundary Location

Intuitively, intonational phrases divide utterances into meaningful 'chunks' of information (Bolinger 1989). Variation in phrasing can change the meaning hearers assign to utterances of a given sentence. For example, the interpretation of a sentence like *Bill doesn't drink because he's unhappy* will vary, depending upon whether it is uttered as one phrase or two. Uttered as a single phrase, this sentence is commonly interpreted as conveying that Bill does drink – but the cause of his drinking is not his unhappiness. Uttered as two phrases, with an intonational boundary between *drink* and *because*, it is more likely to convey that Bill does not drink – and that the reason for his abstinence is his unhappiness. Accent may also vary between the two readings, but, for English, this variation is not as consistent as phrasing (Avesani, Hirschberg, and Prieto 1995). In this section, we describe the use of a number of textual features to predict the location of intonational phrase boundaries, using several speech corpora for training and testing, annotated with prosodic labels by hand, as discussed above in Section 4.1. However, procedures were also developed for using hand-annotated text corpora to achieve better results with much less time and effort. Best results on the speech corpora were just over 90% CART cross-validated success rates; on the text corpora, these success rates rise to just over 95% correct. The decision trees produced using the CART techniques described in Section 3 are currently used to assign prosodic boundaries in TTS.

While we assume phrase boundaries to be perceptual categories, these have been found to be associated with certain physical characteristics of the speech signal. In addition to the tonal features described above, phrases may be identified by one of more of the following features: pauses (which may be filled or not), changes in amplitude, and lengthening of the final syllable in the phrase sometimes accompanied by glottalization of that syllable and perhaps preceding syllables. In general, major phrase boundaries tend to be associated with longer pauses, greater tonal changes, and more final lengthening than minor boundaries.

Most text-to-speech systems that handle unrestricted text rely upon simple phrasing algorithms based upon orthographic indicators, keyword or part-of-speech spotting, and simple timing information to assign phrase boundaries (O'Shaughnessy 1989; Larreur, Emerard, and Marty 1989; Schnabel and Roth 1990). However, not every internal punctuation mark is properly mapped to a full intonational boundary. Consider, for example, ... *a prairie cornfield site near Broken Bow, Neb.* ..., *A November 27, 1991, memo* ..., and ... *his tankers usually use small, obscure border crossings*, from AP Newswire stories. Reading these phrases aloud with a full phrase boundary at each comma (the reader may try taking a breath at each comma to produce this effect) seems fairly odd. Nor is the absence of punctuation a reliable sign that a human reader would read a given sentence without internal major phrase boundaries as the reader may test simply by reading the current sentence aloud without any internal boundaries whatsoever.

More sophisticated rule-based systems have so far been implemented primarily for message-to-speech systems, where syntactic and semantic information is available during the generation process (Young and Fallside 1979; Danlos, LaPorte, and Emerard 1986). However, some general proposals have been made which assume the availability of more sophisticated syntactic and semantic information to use in boundary prediction (Altenberg 1987; Bachenko and Fitzpatrick 1990; Monaghan 1991; Quené and Kager 1992; Bruce, Granström, Gustafson, and House 1993), although no current proposal integrating such information into the phrase assignment process has been shown to work well, even from hand-corrected labeled input. The prediction of likely intonational phrase boundaries from text has a lengthy history in theoretical linguistics as well as in text-to-speech applications (Downing 1970; Bresnan 1971; Lea 1972; Lehiste 1973; O'Malley, Kloker, and Dara-Abrams 1973; Cooper and Sorenson 1977; Selkirk 1978; Streeter 1978; Wales and Toner 1979; Cooper and Paccia-Cooper 1980; Gee and Grosjean 1983; Bachenko and Fitzpatrick 1990; Schnabel and Roth 1990; Bruce, Granström, and House 1992). In recent years, attention has been given to deriving phrasing rules for text-to-speech systems from large labeled corpora (Altenberg 1987); most recently, attempts have been made to use self-organizing procedures to compute phrasing rules automatically from such corpora (Hirschberg 1991; Wang and Hirschberg 1991a, 1991b, 1992; Veilleux and Ostendorf 1992). Even if accurate syntactic and semantic information could be obtained automatically and in real time for text-to-speech, such hand-crafted rule systems are notoriously difficult to build and to maintain.

5.1. Phrasing Prediction from Labeled Speech Corpora

In Wang and Hirschberg (1991a, 1991b, 1992) and Hirschberg (1991), use of CART techniques to predict intonational phrasing from a large, hand-labeled corpus of spontaneous speech was described. Initially, 298 utterances from the ATIS corpus, described in Section 4.1, representing approximately 24 minutes of speech elicited from 26 speakers, were prosodically labeled by hand. This corpus was later expanded to include 773 spontaneous utterances, comprising approximately approximately 65 minutes of speech, from 31 speakers. In addition, a corpus of 478 read ATIS utterances, comprising approximately 31 minutes of speech from 20 speakers was analyzed separately. Prosodic features were labeled by hand for all utterances in these corpora as discussed in Section 2. The DARPA-supplied orthographic transcription was then examined to identify the set of syntactic, lexical, and orthographic features which best predicted the observed intonational boundaries, using features suggested in the literature as well as additional possibilities.

Given the variability in performance observed among speakers, an obvious feature to include in the analysis was speaker identity. While for applications like speaker-independent recognition, this variable would be uninstantiable, we nonetheless need to determine how important speaker idiosyncracy may be in boundary location. We have found no significant increase in predictive power when this variable is used; so, results presented below are speaker-independent.

One class of variable which is readily obtainable involves temporal information. Temporal variables include utterance and phrase duration, and distance of potential boundary from various strategic points in the utterance. Although it is tempting to assume that phrase boundaries represent a purely intonational phenomenon, we must allow for the possibility that processing constraints help govern their occurrence. That is, longer utterances may tend to be produced with more boundaries for mechanical production reasons. Accordingly, we measure the length of each utterance both in seconds and in words. The distance of the boundary site from the beginning and end of the utterance is another variable which appears likely to be correlated with boundary location. We speculate that the tendency to end a phrase might be affected by the position of the potential boundary site in the utterance. For example, it seems likely that positions very close to the beginning or end of an utterance might be less likely positions for intonational boundaries – again, perhaps, on grounds of production constraints. We measure this variable too, both in seconds and in words.

Gee and Grosjean (1983) and Bachenko and Fitzpatrick (1990) inter alia have noted the importance of phrase length in determining boundary location. Simply put, it seems possible that consecutive phrases often have roughly equal lengths. To capture this notion, we calculate the following fraction: the elapsed distance from the last boundary to the potential boundary site, divided by the length of the last phrase encountered. Unfortunately, to obtain this information from text analysis alone would require us to factor prior boundary predictions into subsequent predictions. As a first step, therefore, to test the utility of this information, we have used observed boundary locations in our current analysis. However, using prior boundary prediction would be feasible, if we discover that this information has predictive power. We employ both temporal and word count prediction for this variable.

Previous researchers have long considered syntactic information to be a good predictor of phrasing information (Gee and Grosjean 1983; Selkirk 1984; Steedman 1990). To investigate such possibilities, we consider first the sentence type to which the potential boundary site belongs. It is possible that certain classes of sentences, e.g. indirect questions, will contain more phrase boundaries than others, possibly because of some disambiguating function that phrasing might serve. We accommodate this possibility by dividing questions into three main categories – *wh*-questions, direct yes/no questions, and indirect yes/no questions. Certainly if question type turns out to be an important predictor for presence or absence of intonational boundary, results of our future examination of declaratives should prove even more interesting.

Part-of-speech information is another factor widely believed to predict boundary location, particularly in text-to-speech synthesis. For example, the belief that phrase boundaries rarely occur after function words forms the basis for most algorithms used to assign intonational phrasing for text-to-speech synthesis. Furthermore, we might expect that heads of phrases (e.g. prepositions and determiners) do not constitute the typical end to an intonational phrase. We explore these ideas by examining a window of four words surrounding each potential phrase break. We obtain part-of-speech information via Church's part-of-speech tagger (Church 1988), whose output has been modified slightly to predict preposition/particle distinctions.

In addition to part-of-speech information, we also investigate the importance of syntactic constituency to the prediction of boundary location. Intuitively, we want to test the notion that some constituents may be more or less likely than others to be internally separated by intonational boundaries. To test which constituents are *less* likely to be intonationally separated, we examine the class of the lowest node in the parse tree to dominate adjacent words w_i and w_j. To test which constituents are *more* likely to be separated, we determine the class of the highest node in the parse tree to dominate w_i, but not w_j, and the class of the highest node in the tree to dominate w_j but not w_i. In this way, we can determine whether constituency relations are important to intonational phrasing. We use the Fidditch parser (Hindle 1989) to provide information on syntactic constituency for this analysis.

Recall that each intermediate phrase is composed of one or more pitch accents plus a phrase accent, and each intonational phrase is composed of one or more intermediate phrases plus a boundary tone. Clearly, then, pitch accents form the main building blocks for phrases. In addition, informal observation suggests that phrase boundaries are more likely to occur in some accent contexts than in others. For example, phrase boundaries between words that are deaccented seem to occur much less frequently than boundaries between two accented words. To test this observation, we look at the pitch accent values of w_i and w_j for each $\langle w_i, w_j \rangle$. Since our prosodic labeling of the speech under analysis includes pitch accent location, we can use observed data values for the accent variables. However, since we would prefer to predict boundary location simply from information collected from text, we also substituted predicted pitch accent information for the observed data. We obtain accent predictions from text analysis procedures described in Hirschberg (1990). We use predicted accent as a binary feature (accented or not) or as a four-valued features (cliticized, deaccented, accented, or not available – denoted NA– for end of sentence), to see whether or not finer-grained information will prove useful in predicted boundary location. On the one hand, we want to discover from actual accent information how useful accent location can be in predicting boundary location. On the other hand, we want to determine whether information available from text analysis can help to automate our analysis.

The set of text-based phrasing predictors was further constrained to the set of features which could be identified from text automatically, which best predicted observed prosody. Data points included all potential boundary locations in an utterance, where a potential boundary location was defined as each pair of adjacent words in the utterance $\langle w_i, w_j \rangle$. The best automatically-obtainable predictors for intonational boundaries eventually included the following:

- part-of-speech (for a window of size four around the potential boundary, i.e. $\langle w_{i-1}, w_i, w_j, w_{j+1} \rangle$);
- constituency information, such as the identity of the smallest constituent dominating $\langle w_i, w_j \rangle$ and of the largest constituents dominating w_i but not w_j and dominating w_j but not w_i;
- predicted pitch accent for w_i and w_j (as determined via the hand-constructed rules described in Section 4.3);
- the mutual information between each word-pair in the potential boundary sequence $\langle w_{i-1}, w_i, w_j, w_{j+1} \rangle$;
- the total number of words in the utterance and the distance of $\langle w_i, w_j \rangle$ from both the beginning and the end of the utterance, and the same information calculated in syllables;
- the location of $\langle w_i, w_j \rangle$ with respect to surrounding noun phrases (NPs), including whether the potential boundary precedes or follows an NP, or whether it is within an NP and, if the latter, where it is located within that NP.

The best prediction trees generated for the spontaneous ATIS corpus predict with an estimated success rate of just over 90%, according to this calculation, which was independently verified on (held-out) ATIS test sets. The best trees obtained for prediction of boundary locations from automatically available information for the read ATIS speech predicted with 88.4% success when intonational and intermediate phrase boundaries were collapsed, and 90.6% success for intonational boundaries only. An example of such a tree is presented in Figure 11.3.

This tree is derived using predicted accent values (produced using the procedures described in Section 4 to estimate actual prominence. Using binary-valued accented predictions (are $\langle w_i, w_j \rangle$ accented or not?), we obtained a success rate for boundary prediction of 89%, and using a four-valued distinction for predicted accented (cliticized, deaccented, accented, not available – NA) we increased this slightly to 90.6%. The tree in Figure 11.3 presents results of the latter analysis. The variables used in this tree include:

`type`	type of question (WH, Yes/no, Indirect, Indirect and WH, WH and WH)
`tt`	total number of seconds in utterance
`tw`	total number of words in utterance
`st`	time (in seconds) between start and w_j
`et`	time (in seconds) between w_j and end
`sw`	distance (in words) between start and w_j
`ew`	distance (in words) between w_j and end
`la`	is w_i accented or not, cliticized, deaccented, accented, NA
`ra`	is w_j accented or not, cliticized, deaccented, accented, NA
`per`	ratio of [distance (in words) from last boundary] to [length (in words) of last phrase]
`tper`	ratio of [distance (in seconds) from last boundary] to [length (in seconds) of last phrase]
`j1, j2, j3, j4`	part-of-speech label for $w_{i-1}, w_i, w_j, w_{j+1}$, respectively v = verb b = *be* verb m = modifier f = function word n = noun p = preposition w = WH
`fs, fl, fr`	Fidditch's label for, respectively: s = smallest constituent that dominates w_i and w_j l = largest constituent that dominates w_i but not w_j r = largest constituent that dominates w_j but not w_i which can take values: m = modifier d = determiner v = verb p = preposition w = WH n = noun s = sentence f = function word

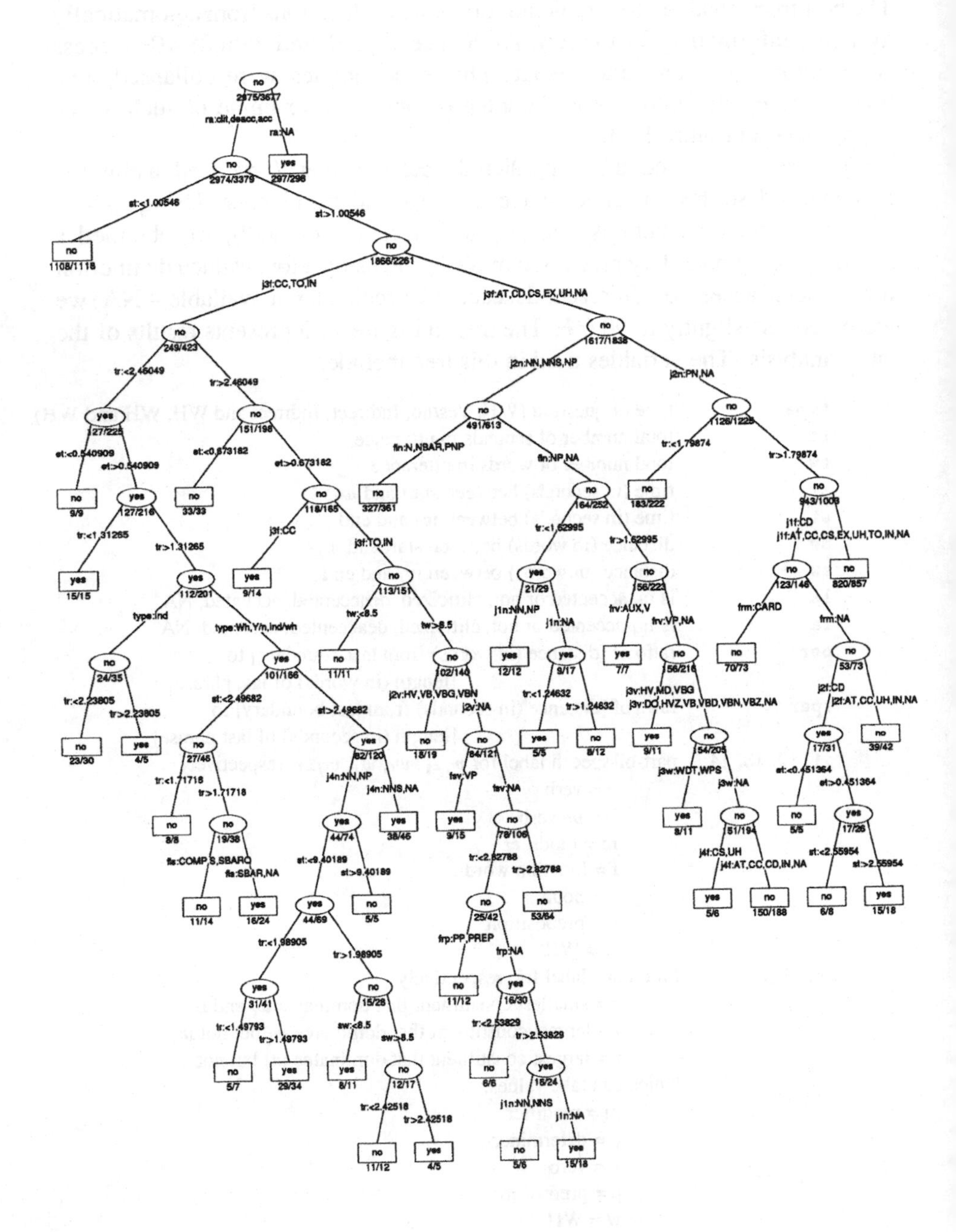

Figure 11.3. ATIS phrase-boundary prediction (90.6% correct).

Interestingly, predicted accent is used only trivially here, to indicate sentence-final boundaries (`ra` = 'NA'). The second split in the tree is explicitly based upon temporal distance of boundary site from beginning of utterance. Together these measurements correctly predict nearly 40% of the data (actually 38.2%). This tree next makes use of a variable which has not appeared in earlier trees – the part-of-speech of w_j. The tree shows that, where w_j is a function word other than *to*, *in* or a conjunction (true for about half of potential boundary sites), in the majority of cases (88%) a boundary does not occur. Part-of-speech of w_i and type of constituent dominating w_i but not w_j are further used to classify these items. This portion of the classification appears to support the notion of 'function word group' used commonly in assigning prosody in text-to-speech synthesis, in which phrases are defined, roughly, from one function word to the next. Overall rate of the utterance and type of utterance are made use of in the tree, in addition to part-of-speech and constituency information and distance of potential boundary site from beginning and end of utterance.

5.2. Phrasing Prediction Trained on Labeled Text

The procedure described above for speech data performed fairly well. However, the hand-labeling required for the training data (originally, labeling of phrase boundaries and pitch accents was employed, so both of these features had to be identified in the speech corpus) is enormously time-consuming and expensive, requiring well over one person-year to accomplish, for the phrasing procedure described in this section. But automatic labeling of prosodic features does not appear to be reliable enough yet to serve as a substitute, despite some progress made in this area in recent years (Ostendorf, Price, Bear, and Wightman 1990; Wightman and Ostendorf 1994).

While the phrasing decisions described in Section 5.1 modeled human decisions more accurately than TTS default phrasing rules for transcriptions of spontaneous speech, they were not always appropriate for other genres of text, such as news stories, or common TTS applications, such as name and address listing. Also, the process of acquiring orthographically transcribed speech corpora and of labeling prosodic features by hand is extremely time-consuming; so, the training corpus was fairly small. Modeling other genres or styles and even adding to the existing training data to correct errors would have involved an enormous investment of labor.

To address these problems, a simpler procedure for acquiring training data was employed for the generation of phrasing rules. Intonational boundary predictions were first obtained for new training text obtained from the AP newswire, using a prediction tree derived from the read ATIS corpus described in Section 4.1. The text was annotated with these predictions automatically. This annotated text was then corrected by hand, to eliminate what the human labeler deemed 'unacceptable' phrasings. These hand-corrected annotations were then used to provide the dependent variable for the production of a new prediction tree, with values for independent variables acquired from text analysis as in the studies described in Section 5.1. A series of trees produced from AP news text in this manner did not differ significantly in predictive power from CART prediction trees produced from prosodically labeled speech. The best tree predicts with 95.4% accuracy on an 89,103 word subset from the AP news.

To produce a phrasing procedure for a new application, domain, or language, then, on-line text from an appropriate domain in the language desired is first annotated by a native speaker of that language with plausible intonational boundaries, by identifying locations in the text where the annotator believes boundaries sound 'natural'. We have used newswire text from the English and Spanish AP for general TTS training purposes, but other text could be used for particular applications. The unannotated version of the text is itself analyzed to extract values for features at each potential boundary site (defined as each position between two orthographic words $\langle w_i, w_j \rangle$ in the input) which have been shown or appear likely to correlate with phrase boundary location – and which can be extracted automatically and in real time. Vectors of independent feature values plus the dependent (observed) value – the annotator's decision as to the plausibility of an intonational boundary between w_i and w_j – are then given as input to CART, as described above. For the phrasing module currently implemented in English TTS, a new matrix of feature vectors can be generated for new text simply by running TTS in training mode, so that it prints the inferred values for independent variables and the annotated values for the dependent variable. The resulting tree is then compiled automatically into C code, which can be used for simple prediction in a stand-alone procedure, or which can be substituted for an existing decision tree module in the TTS phrasing module.

By this method, the amount of time spent in processing new training corpora can be drastically reduced. Training data that would have taken months to record, transcribe, and label can now be produced from text in a few days at most. An experienced prosodic labeler takes about one minute to label one second of speech for phrasing and accent, and about half as long to label intonational boundaries only. However, additional time is spent performing error checks and corrections, since this process is prone to slips of the hand. While the labeling of the speech corpus used to construct the earlier phrasing module for TTS took a full person-year, a better comparison of the amount of time saved by the text-based procedure can be found by comparing more recent labeling of phrase boundaries only in spontaneous and read speech by a similarly trained labeler, using the ToBI system (Pitrelli, Beckman, and Hirschberg 1994). This labeler averaged 67 words per hour for a 3163-word corpus of recorded speech, compared to 4833 words per hour for the annotator assigning plausible boundaries to text. Note also that the speech labeler's labor does not include the time required to record and process the speech data, which is of course not necessary for the text-based approach. While some additional work was needed to select and pre-process the text for annotation, this was all done via some simple shell scripts, which are easily re-used. So, the work required to produce a new phrasing module for a TTS system was speeded up by a factor of 70 or better using the text-based approach. While this type of boundary labeling from text requires some initial instruction, this is much less than the training necessary for prosodic labeling, and the task is easily performed by subjects with no linguistic knowledge once they are given some pertinent examples. So, phrasing predictions can be trained on much larger corpora than have been possible in the past, and it is easy to produce separate phrasing modules for different applications and genres. The current English version of Bell Labs TTS contains a phrasing module which was produced automatically, using the procedure described above on a hand-annotated corpus of approximately 87,000 words of text taken from the AP newswire.

With the text-based method, it is thus possible to retrain the existing TTS phrasing procedure quickly, as deficiencies are uncovered, by the simple addition of exemplars of the (corrected) behavior to the training set. It is also possible to produce phrasing procedures easily for new domains or languages without recording or labeling a large corpus. The (Mexican) Spanish phrasing procedure developed (Hirschberg and Nakatani 1996) is a demonstration of this technique's versatility: a baseline version of this model which performed at about 90% correct was produced in only about a one and one-half person weeks.

6. Conclusion

In sum, it is possible to achieve fairly good performance in assigning intonational features such as phrase boundaries and pitch accent by training on large labeled speech corpora. The amount of time needed to construct such corpora has, however, been a limiting factor in the usefulness of this method. More recently, experiments attempting to emulate the value of labeled speech corpora by substituting text corpora labeled by native speakers with plausible intonational features have been performed, to create phrasing assignment modules for English and Spanish text-to-speech. Results of these experiments suggest that text corpora can indeed serve as successful substitutes for speech corpora, and cut the time needed for labeling the corpus quite dramatically.

Chapter 12

LEARNABLE PHONETIC REPRESENTATIONS IN A CONNECTIONIST TTS SYSTEM – II: PHONETICS TO SPEECH

Andrew D. Cohen
Independent Consultant

Abstract In an earlier chapter, we described the overall structure of the SOMtalk text-to-speech system and detailed results suggesting that non-symbolic ('phonetic') representations – based on trajectories through a 'phonetic' space derived from a self-organising map – may play a useful part in deriving pronunciations from text. A similar strategy suggests itself for the subsequent stage in which synthetic speech is produced from the 'phonetic' representation. This makes it possible to bypass a symbolic 'phonemic' stage in the overall, trained system. In this case, only a small database has been used for learning because of the high computational cost of training on spectral data, but some encouraging preliminary results have been obtained.

1. Introduction

In Chapter 8, we described the overall structure of the SOMtalk text-to-speech system. This consisted of two subsystems: a text-to-phonetics module (M_1) and a phonetics-to-speech module (M_2). Results were detailed from connectionist experiments on module M_1 in which non-symbolic ('phonetic') representations – based on trajectories through a 'phonetic' space derived from a self-organising map – were used in deriving pronunciations from text. In this chapter, we concentrate on the second (M_2) module which produces synthetic speech from the non-symbolic, phonetic representation.

The approach developed in Chapter 8 is in marked contrast to most text-to-speech (TTS) systems, which rely heavily on a symbolic representation of pronunciation in phonemic form. The main potential of the new approach is seen as the greater ease with which prior phonetic and phonological knowledge can be incorporated into the system. Further, techniques based on combining multiple

R. I. Damper (ed.), Data-Driven Techniques in Speech Synthesis, 275-282.

networks as in Chapter 8 are useful in providing ways to exploit these non-symbolic representations. Finding a good separation of modules and assigning an exact function to each of them is less important than the overall flow of data through the system, from levels which can be considered higher (phonetic) to those which can be termed lower (phonological structure of words, and finally the physical events). It is suggested that this direction of information flow (in contrast to the rule-based approaches in modular TTS systems which build a phoneme string and then consider what changes need to be made to account for coarticulation) can be better exploited in a connectionist than a rule-based system. At a sentence level, this phonological flow is even more important, as prosodic factors come into play.

SOMtalk is a concatenative synthesis system using diphones. This choice was motivated by the good tradeoff between quality and computational costs for concatenative synthesis, and the fact that diphones encode directly the perceptually important transitions between steady-state speech sounds. In the context of synthesis by diphone concatenation, we suppose that we already have values such as fundamental frequency contour `F0`, amplitude of voicing `AV` and linear prediction coding (LPC) coefficients (Makhoul 1975; Markel and Gray 1976) stored. So we wish only to learn in the neural networks the changes to be made when the phonetic and prosodic contexts alter, for as many cases as we have training data.

To deal with variability, we need to be able to train with large databases. Hence, model complexity (here the number of weights in our networks) becomes a concern. Undue complexity should not be necessary if our models and representations are properly matched, such that smoothing is not a major issue. In the phonetics-to-speech stage of TTS, the problem of training large databases is acute because of the enormous amount of data generated by training on LPC coefficients (or indeed any spectral representation). Two possible remedies explored here are:

1 The use of multiple networks: A single network can only be trained on a certain amount of data, up to an asymptotic point at which increasing data brings little or no improvement. Multiple networks give more information about a single input pattern than a single network and allow a greater degree of user control. We do not need to worry so much about individual networks overfitting, thus it is possible to use a validation set for early stopping of training, rather than the more expensive cross-validation (Weiss and Kulikowski 1991; Bishop 1995). Much of the overfitting due to variance can be averaged out in combining networks. In some ways, overfitting can be an advantage as the contribution made by bias becomes clearer. For more information on some advantages of overfitting, see Sollich and Krogh (1996).

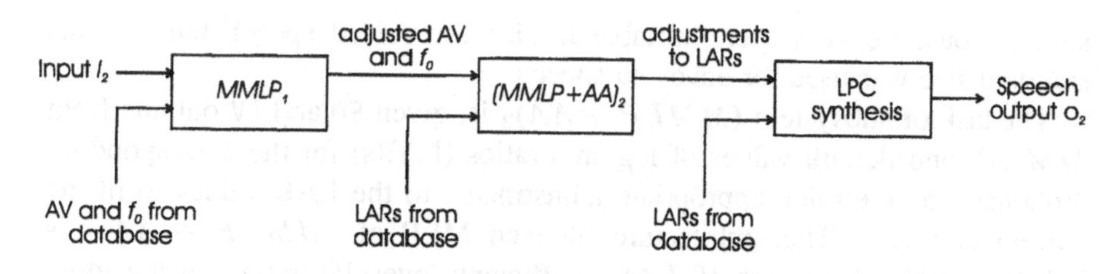

Figure 12.1. Architecture for phonetics-to-speech module M_2. See text for details.

2 The use of a storage-based technique such as diphone concatenation: With a storage cost of about 6.5 MB for 1150 diphones, it is attractive to use the networks to focus on learning the necessary changes for different phonetic contexts, rather than having to store data.

Separation of tasks into very separate modules (as in conventional TTS systems) leads to rigidity and excessive complexity. The multilayered approach has not worked well in phonology, probably for the same reasons. By shifting the emphasis from processes to representations, we gain many advantages in a connectionist (or other data-driven) approach. Using multiple networks and attractor basins is one way to exploit these advantages, as described in Chapter 8.

The remainder of this chapter is structured as follows. Immediately following, Section 2 outlines the architecture of the phonetics-to-text module which is the subject of this chapter. Section 3 details the training of the various subsystems of this module, and Section 4 presents results of some early experiments producing speech output. Finally, Section 5 concludes.

2. Architecture of Phonetics-to-Speech Module

Figure 12.1 shows the architecture of the phonetics-to-speech (M_2) module. The input is a sequence of diphone coordinates in D_dip form obtained following text-to-phonetics conversion (see Chapter 8). There are two multiple multilayer perceptron (MMLP) subsystems, the second of which also features an auto-associator (AA) corrector, used in a structure just like that of Fig. 8.2 in Chapter 12. They are identified $MMLP_1$ and $(MMLP + AA)_2$ in the figure. Multiple MLP networks were used as in Chapter 8 to make performance less sensitive to the partition of available data into test and train subsets.

The task of subsystem $MMLP_1$ is, given I_2 and default values of F0 and AV for the corresponding diphones taken from the speech database, to adjust these values to fit the current context. The architecture of each MLP is 230-62-2. There are 2 inputs for the F0 and AV values, plus 20 inputs for coordinates of the current diphone, plus 1 input representing the Euclidean distance from the previous diphone, over 10 frames – 230 inputs in all. Frames were pitch synchronous, i.e. they were variable in size for voiced speech but a 22 ms standard size was used for unvoiced speech.

The task of subsystem $(MMLP + AA)_2$ is, given F0 and AV outputs from $MMLP_1$ and default values of log area ratios (LARs) for the corresponding diphones, to determine appropriate adjustments to the LAR values to fit the current context. The architecture of each MLP of $(MMLP + AA)_2$ is is 110-55-100. There are 10 LAR coefficients over 10 frames in the input plus 1 (omitted from Fig. 12.1 for simplicity) to indicate if a frame is on a boundary (see below). This is because most adjustments will occur close to a boundary, so it is probably worth indicating this to the network. The outputs represent the vector of changes to be added to those 100 coefficients, scaled to lie between −0.5 and 0.5. The architecture of the (single) AA is 100-40-100.

Finally, the original LAR coefficients plus the adjustments determined by $(MMLP + AA)_2$ together with the F0 and AV values fed forward from MLP_1, are used to find LPC coefficients for actual synthesis.

3. Training and Alignment

Training was on natural, connected speech, in which diphone boundaries were unspecified. We therefore confront an alignment problem of the kind which arises frequently in speech technology. Alignment problems in speech synthesis have previously been addressed with hidden Markov models (HMMs) in conjunction with the Viterbi algorithm (Van Coile 1990), dynamic time warping (DTW) (Leung and Zue 1984; Hunt 1984) or learned with chunking procedures (Lucas and Damper 1992; Luk and Damper 1996). The use of DTW may lead to distortion and loss of perceptually important information, and HMMs have many well known difficulties. However, we wish to use a simpler procedure which does not require that all training data be presegmented and transcribed. Instead a small dataset is transcribed and labelled with segmentation points and the rest of the training data only broken up at a word level. We bootstrap from the set of 100 sentences of naturally-spoken speech, segmented by hand, and attempt to learn the differences between these and a purely linear segmentation based on making pro-rata changes to each diphone according to the overall length of the word. In a similar way, word-level segmentations are bootstrapped from 1000 hand-labelled naturally-spoken words. In the case of individually spoken words, the linear assumption works reasonably well – it appears that some speakers are more consistent than others in this regard. Only when prosodic factors are introduced do more radical, nonlinear changes become important. Therefore, the strategy is not to use a separate durations module for the sentence level, but to let this information flow through from the prosodic level. We only wish to have the networks learn the portion of the problem that is nonlinear, so that they are not also required to store the linear model.

Segmentation at a word level can be given with the natural training speech – this is much less labour-intensive than providing a fully-labelled and prosodically-marked database. We extract values for F0 and AV, and represent LPC coefficients in terms of the log area ratios (LARs) derived from 10th-order linear prediction. Although they may not have the best interpolation performance in terms of spectral distortion (Paliwal 1995), they are positive, have a uniform spectral sensitivity and a low computational cost, and will always provide stable LPC filters for resynthesis.

A novel attempt was made to avoid alignment of the natural speech and its diphone segmentation. Hence, a third MLP-based network – called $(MMLP+AA)_3$ – was introduced to predict diphone boundary information for training purposes. This information was not needed in on-line operation since the boundaries were implicit in the selection of diphones from the database.

The architecture of the MLPs in $(MMLP + AA)_3$ was 120-45-1. Inputs were F0, AV and 10 LAR coefficients over 10 frames, with one binary output unit to indicate whether the current frame is on a boundary. This predicted boundary information is needed for training $(MMLP+AA)_2$ on natural (hence unaligned) word-level data. Hence, $(MMLP + AA)_3$ is trained on word-level data only. As with $(MMLP + AA)_2$, there were 10 multiple MLP networks. Because this network has only a single output, the AA corrector operates on the hidden-layer activations of $(MMLP + AA)_2$.

After training $(MMLP+AA)_3$, $MMLP_1$ was next trained on sentence-level data. There were 100 training sentences in the database. Sentence-level training was necessary in order to capture prosodic variations in F0 and AV. Each of the multiple networks was trained on 20 sentences, so there were 5 multiple networks in total.

Since the task of $MMLP_1$ is to adjust F0 and AV values, it must be trained on targets which are representative of natural speech. These target data values are extracted from pitch-synchronous LPC. Input training data consist of a moving window of 10 frames generated from individual words, and the corresponding target data consist of the relevant values from the database of 100 sentences minus the input values, scaled to lie between −0.5 and 0.5. These data are hand-aligned, which is somewhat unsatisfactory as it is fairly labour-intensive, although there is no need to label phones. Where the synthetic speech is longer or shorter in duration than the natural speech, multiple artificial frames are mapped to a single natural frame or *vice versa*. Target values are generated by subtracting these values from the F0 contour of the artificial (concatenated) versions of these sentences, which were recorded onto tape before digitising, to match our method of diphone recording and to avoid the difficulties of handling a single very large file.

$(MMLP + AA)_2$ was subsequently trained on word-level data. This was done because there was too much sentence-level data to train the (much bigger) MLPs. Further, the kind of LAR adjustment intended is reasonably viewed as a local (word-level) issue. Each of the multiple MLP networks was trained on 100 words from a total of 1000 words in the training set, so there were 10 multiple networks in total. The same data were used for training the AAs.

When training is completed, all inputs in on-line operation will be in terms of concatenated speech which, of course, comes already segmented. Clearly, performance on unseen data is of great interest, so our test sentences have been drawn from unseen sentences, although some individual words may previously have been seen.

4. Phonetics-to-Speech Results

The system has not been formally evaluated by subjective listening tests, as off-training set performance has not yet developed to the point where such a test would be worthwhile. Also, a satisfactory test would be difficult to devise, and needs to be run on reasonable numbers of different subjects after each adjustment is made to the system to factor out subject adaptation. The widely-used modified rhyme test (House, Williams, Hecker, and Kryter 1965) only measures segmental quality – precisely what we do not need in a diphone system, where it is a given that segmental quality will be good. Naturalness is harder to quantify, and depends more on hearing longer stretches of connected speech, rather than isolated sentences with no semantic connection.

Overall, the speech quality is limited by the use of the simplest form of LPC analysis-resynthesis, in which the voicing source is synthesised by a single pulse. While this allows a low data rate, it would probably not be acceptable in most commercial systems. There are also limitations stemming from the use of the all-pole model in linear prediction, the fact that only 10th-order prediction is used, and shortcomings of the package used to do pitch extraction, which could all be addressed by more sophisticated techniques (e.g. Dutoit 1994a). The modest aim has been to show an improvement in performance over a simple diphone system, based on concatenation and simple smoothing, with the limited prosody derived from a network model. The real challenge is to integrate this form of prosody with a non-segmental system based on our non-symbolic phonetic representation. An improvement is seen on training set sentences, but because of the small database, results on unseen data are often excessively machine-like.

In the corrector networks, we have used the same methods for controlling the number of correction cycles as in Section 5.5 of Chapter8. Where stabilisation is used as a stopping criterion, this takes place at the level of output activations. Table 12.1 shows the best predicted performance on the unseen data for

Table 12.1. Predicted mean square errors for $(MMLP + AA)_2$. See Chapter 8 for definitions of error measures. All testing was at word-level, using 1000 test words.

	$\overline{E}$	$\overline{A}$	$\overline{E} - \overline{A}$
Maximise Ambiguity	32954.1	15822.2	17131.9
Maximise Ambiguity and Stabilise	33263.6	15917.5	17346.1
Minimise Ambiguity	21326.8	6978.7	14348.1

minimising ambiguity in $(MMLP + AA)_2$. In this table, the error measure is that of equation (2) of Chapter 8. As might be expected in this highly-complex task, the best results are obtained by distrusting (i.e. assigning the lowest weights to) networks with a high ambiguity – those that generate a highly irregular response. For comparison purposes, divine guidance gave a predicted mean square error of 11,663, although we have noticed that divine guidance does not always generate the best performance (see Chapter 8). In the cases of $(MMLP + AA)_3$ and $MMLP_1$, we found that maximising ambiguity gave the lowest predicted error.

The most obvious extension to our system would be reimplementation with multiband-excited pitch-synchronous overlap-add (PSOLA) resynthesis (Dutoit and Leich 1993; Dutoit 1997), or any of the higher-quality PSOLA-based approaches with resynthesis at constant pitch and constant initial phases to allow segment smoothing. TD-PSOLA is perhaps the easiest to implement, but lacks the capability for smoothing, and has a few other drawbacks including the fact that optimal pitch marking is not automatic. Theoretically, the line-spectral pair (LSP) representation offers the best interpolation properties (Paliwal 1995) but Dutoit (1994b) reports no detectable difference between conventional (partial correlation) and LSP implementations in his 18th-order LPC synthesiser.

5. Conclusions and Further Work

Much further work remains to be done, as any radically new approach based on non-standard representations clearly implies much necessary research before any improvement can be shown over state-of-the-art rule-based and concatenative systems. Because of the small training sets used, we cannot expect good performance on unseen data. However, performance has so far improved approximately linearly with increased training-set size in the text-to-phonetics experiments of Chapter 8. For prosody, the problem of small datasets, implying that many important phenomena are not represented, is especially acute. We have no general solution to this problem, except that, in this study collection of training data does not rely on a full transcription, but can be based on a simple word-level segmentation with no requirement to label phones. Using special-purpose hardware it would be possible to train much larger datasets, which would allow for a more thorough testing of many of our hypotheses, as well as probably improving performance. Given that there is no noise in this problem, it should be possible to achieve convergence on much larger datasets.

While there is rigidity in the proposed system in the direction of flow of information, there is considerable flexibility in the recurrent modules in the kind and amount of information flowing. We can alter the way of controlling the number of cycles through the corrector networks to obtain the optimal performance on different datasets. We can also adjust the method used to combine the network outputs as the amount of training data grows and so better represents the domain. The ultimate goal of a data-driven speech synthesis system must be to learn from unlabelled, natural speech. We believe the system outlined in this chapter and in Chapter 8 has, by incorporating significant prior speech knowledge in a modular fashion, taken useful steps towards this goal.

References

Abercrombie, D. (1981). Extending the Roman alphabet: Some orthographic experiments of the past four centuries. In R. E. Asher and E. Henderson (Eds.), *Towards a History of Phonetics*, pp. 207–224. Edinburgh, UK: Edinburgh University Press.

Adamson, M. J. and R. I. Damper (1996). A recurrent network that learns to pronounce English text. In *Proceedings of Fourth International Conference on Spoken Language Processing, ICSLP'96*, Volume 3, Philadelphia, PA, pp. 1704–1707.

Adamson, M. J. and R. I. Damper (1999). B-RAAM: A connectionist model which develops holistic internal representations of symbolic structures. *Connection Science 11*(1), 41–71.

Aha, D. W. (1997). Lazy learning. *Artificial Intelligence Review 11*(1–5), 7–10.

Aha, D. W., D. Kibler, and M. Albert (1991). Instance-based learning algorithms. *Machine Learning 6*(1), 37–66.

Aho, A. V. and M. J. Corasick (1975). Efficient string matching: An aid to bibliographic search. *Communications of the ACM 18*(6), 333–340.

Aho, A. V., R. Sethi, and J. D. Ullman (1986). *Compilers: Principles, Techniques and Tools*. Reading, MA: Addison-Wesley.

Ainsworth, W. A. (1973). A system for converting English text into speech. *IEEE Transactions on Audio and Electroacoustics AU-21*(3), 288–290.

Ainsworth, W. A. and B. Pell (1989). Connectionist architectures for a text-to-speech system. In *Proceedings of European Conference on Speech Communication and Technology, Eurospeech'89*, Volume 1, Paris, France, pp. 125–128.

Allen, J., M. S. Hunnicutt, and D. Klatt (1987). *From Text to Speech: The MITalk System*. Cambridge, UK: Cambridge University Press.

Altenberg, B. (1987). *Prosodic Patterns in Spoken English: Studies in the Correlation between Prosody and Grammar for Text-to-Speech*

Conversion, Volume 76 of *Lund Studies in English*. Lund, Sweden: Lund University Press.

Andersen, O. and P. Dalsgaard (1994). A self-learning approach to transcription of Danish proper names. In *Proceedings of International Conference on Spoken Language Processing, ICSLP'94*, Volume 3, Yokohama, Japan, pp. 1627–1630.

Anderson, J. A. (1995). *An Introduction to Neural Networks*. Cambridge, MA: MIT Press.

Anderson, M. D., J. B. Pierrehumbert, and M. Y. Liberman (1984). Synthesis by rule of English intonation patterns. In *Proceedings of IEEE International Conference on Acoustics, Speech, and Signal Processing, ICASSP'84*, San Diego, CA, pp. 2.8.1–2.8.4.

Aubergé, V. (1991). *La synthèse de la Parole: des Règles au Lexique*. PhD thesis, Université Stendhal, Grenoble, France.

Avesani, C., J. Hirschberg, and P. Prieto (1995). The intonational disambiguation of potentially ambiguous utterances in English, Italian, and Spanish. In *Proceedings of the XIIIth International Congress of Phonetic Sciences, ICPhS'95*, Volume 1, Stockholm, Sweden, pp. 174–177.

Bachenko, J. and E. Fitzpatrick (1990). A computational grammar of discourse-neutral prosodic phrasing in English. *Computational Linguistics 16*(3), 155–170.

Bagshaw, P. C. (1998). Phonemic transcription by analogy in text-to-speech synthesis: Novel word pronunciation and lexicon compression. *Computer Speech and Language 12*(2), 119–142.

Bailly, G. (1997). Introduction to Part III: Prosody in Speech Synthesis. In Y. Sagisaki, N. Campbell, and N. Higuchi (Eds.), *Computing Prosody: Computational Models for Processing Spontaneous Speech*, pp. 157–164. New York, NY: Springer.

Bailly, G., T. Barbe, and H. Wang (1992). Automatic labelling of large prosodic databases: Tools, methodology and links with a text-to-speech system. See Bailly, Benoît, and Sawallis (1992), pp. 323–333.

Bailly, G., C. Benoît, and T. R. Sawallis (Eds.) (1992). *Talking Machines: Theories, Models and Applications*. Amsterdam, The Netherlands: Elsevier (North-Holland).

Bakiri, G. (1991). Converting English text to speech: A machine learning approach. Technical Report 91-30-2, Department of Computer Science, Oregon State University, Corvallis, OR.

Beckman, M. E. and J. B. Pierrehumbert (1986). Intonational structure in Japanese and English. In C. Ewen and J. Anderson (Eds.), *Phonology Yearbook 3*, pp. 255–309. New York, NY: Cambridge University Press.

Bellman, R. (1957). *Dynamic Programming*. Princeton, NJ: Princeton University Press.

Belrhali, R. (1995). *Phonétisation Automatique d'une Lexique Général du Français: Systématique et Émergence Linguistic*. PhD thesis, Université Stendahl, Grenoble, France.

Benton, R. A., H. Tunoana, and A. Robb (1982). *Ko Ngā Kupu Pū Noa O Te Reo Māori/The First Basic Māori Word List*. Wellington, NZ: New Zealand Council for Education.

Bickley, C. A., K. N. Stevens, and D. R. Williams (1997). A framework for synthesis of segments based on pseudoarticulatory parameters. See van Santen, Sproat, Olive, and Hirschberg (1997), pp. 211–220.

Biggs, B. (1961). The structure of New Zealand Māori. *Anthropological Linguistics 3*(1), 1–54.

Bimbot, F. (1988). *Synthèse de la Parole – des Segments aux Règles avec Utilisation de la Décomposition Temporelle*. PhD thesis, ENST, Paris, France.

Bimbot, F., S. Deligne, and F. Yvon (1995). Unsupervised decomposition of phoneme strings into variable-length sequences, by multigrams. In *Proceedings of the XIIIth International Congress of Phonetic Sciences, ICPhS'95*, Volume 3, Stockholm, Sweden, pp. 270–273.

Bimbot, F., R. Pieraccini, E. Levin, and B. Atal (1994). Modèles de séquences à horizon variable: Multigrammes. In *Actes des XXèmes Journées d'Études sur la Parole*, Trégastel, France, pp. 467–472.

Bimbot, F., R. Pieraccini, E. Levin, and B. Atal (1995). Variable-length sequence modeling: Multigrams. *IEEE Signal Processing Letters 2*(6), 111–113.

Bishop, C. M. (1995). *Neural Networks for Pattern Recognition*. Oxford, UK: Clarendon Press.

Black, A. W. and A. J. Hunt (1996). Generating F_0 contours from ToBI labels using linear regression. In *Proceedings of Fourth International Conference on Spoken Language Processing, ICSLP'96*, Volume 3, Philadelphia, PA, pp. 1385–1388.

Black, A. W., K. Lenzo, and V. Pagel (1998). Issues in building general letter-to-sound rules. In *Proceedings of 3rd European Speech Communication Association (ESCA)/COCOSDA International Workshop on Speech Synthesis*, Jenolan Caves, Australia, pp. 77–80.

Black, A. W. and P. Taylor (1994). Assigning intonation elements and prosodic phrasing for English speech synthesis from high level linguistic input. In *Proceedings of International Conference on Spoken Language Processing, ICSLP'94*, Volume 2, Yokohama, Japan, pp. 715–718.

Blin, L. and L. Miclet (2000). Generating synthetic speech prosody with lazy learning in tree structures. In *Proceedings of Fourth Conference on Natural Language Learning and of Second Learning Language in Logic Workshop*, Lisbon, Portugal, pp. 87–90.

Bolinger, D. (1961). Contrastive accent and contrastive stress. *Language 37*(1), 83–96.

Bolinger, D. (1982). The network tone of voice. *Journal of Broadcasting 26*(3), 725–728.

Bolinger, D. (1986). *Intonation and its Parts*. Palo Alto, CA: Stanford University Press.

Bolinger, D. (1989). *Intonation and its Uses: Melody in Grammar and Discourse*. London, UK: Edward Arnold.

Bose, R. C. and D. K. Ray-Chaudhuri (1960). On a class of error-correcting binary group codes. *Information and Control 3*(1), 68–79.

Breiman, L. (1992). Stacked regressions. Technical Report 367, Department of Statistics, University of California, Berkeley, CA. (Revised June 1994).

Breiman, L. (1996a). Bagging predictors. *Machine Learning 24*(2), 123–140.

Breiman, L. (1996b). Bias, variance and arcing classifiers. Technical Report 460, Department of Statistics, University of California, Berkeley, CA.

Breiman, L., J. H. Friedman, R. A. Olshen, and C. J. Stone (1984). *Classification and Regression Trees*. Pacific Grove, CA: Wadsworth and Brooks.

Brennan, S. E., M. W. Friedman, and C. J. Pollard (1987). A centering approach to pronouns. In *Proceedings of the 25th Annual Meeting of the Association for Computational Linguistics*, Stanford University, Palo Alto, CA, pp. 155–162.

Bresnan, J. (1971). Sentence stress and syntactic transformations. *Language 47*(2), 257–281.

Brill, E. (1992). A simple rule-based part-of-speech tagger. In *Proceedings of the DARPA Speech and Natural Language Workshop*, Harriman, NY, pp. 112–116.

Brown, G. (1983). Prosodic structure and the given/new distinction. In D. R. Ladd and A. Cutler (Eds.), *Prosody: Models and Measurements*, pp. 67–78. Berlin, Germany: Springer Verlag.

Brown, P. and D. Besner (1987). The assembly of phonology in oral reading: A new model. In M. Coltheart (Ed.), *Attention and Performance XII: The Psychology of Reading*, pp. 471–489. Hillsdale, NJ: Lawrence Erlbaum Associates.

Bruce, G. (1977). *Swedish Word Accents in Sentence Perspective*. Lund, Sweden: Gleerup.

Bruce, G., B. Granström, K. Gustafson, and D. House (1993). Prosodic modelling of phrasing in Swedish. In *Working Papers: Proceedings of ESCA Workshop on Prosody*, Lund, Sweden, pp. 180–183. Lund University Department of Linguistics.

Bruce, G., B. Granström, and D. House (1992). Prosodic phrasing in Swedish speech synthesis. See Bailly, Benoît, and Sawallis (1992), pp. 113–125.

Burnage, G. (1990). *CELEX: A Guide for Users*. Centre for Lexical Information, Nijmegen, The Netherlands.

Byrd, R. J. and M. S. Chodorow (1985). Using an on-line dictionary to find rhyming words and pronunciations for unknown words. In *Proceedings of the 23rd Meeting of Association for Computational Linguistics*, Chicago, IL, pp. 277–283.

Campbell, N. (1998). Where is the information in speech? (and to what extent can it be modelled in synthesis?). In *Proceedings of 3rd European Speech Communication Association (ESCA)/COCOSDA International Workshop on Speech Synthesis*, Jenolan Caves, Australia, pp. 17–20.

Campbell, N. and A. W. Black (1997). Prosody and the selection of source units for concatenative synthesis. See van Santen, Sproat, Olive, and Hirschberg (1997), pp. 279–292.

Campbell, W. N. (1992). Syllable-based segmental duration. See Bailly, Benoît, and Sawallis (1992), pp. 211–227.

Carney, E. (1994). *A Survey of English Spelling*. London, UK: Routledge.

Carré, B. (1979). *Graphs and Networks*. Oxford, UK: Oxford University Press.

Catach, N. (1984). *La Phonétisation Automatique du Français*. Paris, France: Editions du CNRS.

Chafe, W. (1976). Givenness, contrastiveness, definiteness, subjects, topics, and point of view. In C. Li (Ed.), *Subject and Topic*, pp. 25–55. New York, NY: Academic Press.

Chalmers, D. J. (1990). Syntactic transformations on distributed representations. *Connection Science 2*(1–2), 53–62.

Charpentier, F. J. and M. G. Stella (1986). Diphone synthesis using an overlap-add technique for speech waveforms concatenation. In *Proceedings of IEEE International Conference on Acoustics, Speech, and Signal Processing, ICASSP'86*, Volume 3, Tokyo, Japan, pp. 2015–2018.

Cherkassky, V. and F. Mulier (1998). *Learning from Data*. New York, NY: John Wiley.

Chomsky, C. (1970). Reading, writing and phonology. *Harvard Educational Review 40*(2), 287–309.

Chomsky, N. (1969). Quines's empirical assumptions. In D. Davidson and J. Hintikka (Eds.), *Words and Objections: Essays on the Work of W. V. Quine*, pp. 53–68. Dordrecht, The Netherlands: D. Reidel.

Chomsky, N. and M. Halle (1968). *The Sound Pattern of English*. New York, NY: Harper and Row.

Church, K. W. (1988). A stochastic parts program and noun phrase parser for unrestricted text. In *Proceedings of the Second Conference on Applied Natural Language Processing*, Austin, TX, pp. 136–143. Association for Computational Linguistics.

Clark, H. H. and S. E. Haviland (1977). Comprehension and the given-new contract. In R. O. Freedle (Ed.), *Discourse Production and Comprehension*, pp. 1–40. Norwood, NJ: Ablex.

Cohen, A. D. (1995). Developing a non-symbolic phonetic notation for speech synthesis. *Computational Linguistics 21*(4), 567–575.

Cohen, A. D. (1997). *The Use of Learnable Phonetic Representations in a Connectionist Text-to-Speech System*. PhD thesis, Department of Cybernetics, University of Reading, Reading, UK.

Coker, C. H., K. W. Church, and M. Y. Liberman (1990). Morphology and rhyming: Two powerful alternatives to letter-to-sound rules for speech synthesis. In *Proceedings of European Speech Association (ESCA) Workshop on Speech Synthesis*, Autrans, France, pp. 83–86.

Coleman, J. (1992). 'Synthesis-by-rule' without segments or rewrite-rules. See Bailly, Benoît, and Sawallis (1992), pp. 43–60.

Coleman, J. and J. K. Local (1992). Monostratal phonology and speech synthesis. In P. Tench (Ed.), *Studies in Systemic Phonology*, pp. 183–193. London: Pinter Publishers.

Collier, R. and J. t'Hart (1983). The perceptual relevance of formant trajectories in diphthongs. In *Sound Structures: Studies for Antonie Cohen*. Dordrecht, The Netherlands: Foris Publications.

Coltheart, M. (1978). Lexical access in simple reading tasks. In G. Underwood (Ed.), *Strategies of Information Processing*, pp. 151–216. New York: Academic Press.

Coltheart, M. (1981). MRC Psycholinguistic Database. *Quarterly Journal of Experimental Psychology 33A*, 497–505. Available in electronic form from Oxford Text Archive at `http://ota.ahds.ac.uk/`. Compiled by Max Coltheart; deposited by Roger Mitton.

Coltheart, M. (1984). Writing systems and reading disorders. In L. Henderson (Ed.), *Orthographies and Reading*, pp. 67–79. London, UK: Lawrence Erlbaum Associates.

Coltheart, M., B. Curtis, P. Atkins, and M. Haller (1993). Models of reading aloud: Dual-route and parallel-distributed-processing approaches. *Psychological Review 100*(4), 589–608.

Conkie, A. D. and S. Isard (1997). Optimal coupling of diphones. See van Santen, Sproat, Olive, and Hirschberg (1997), pp. 293–304.

Cooper, W. and J. Paccia-Cooper (1980). *Syntax and Speech*. Cambridge, MA: Harvard University Press.

Cooper, W. E., S. J. Eady, and P. R. Mueller (1985). Acoustical aspects of contrastive stress in question-answer contexts. *Journal of the Acoustical Society of America 77*(6), 2142–2156.

Cooper, W. E. and J. M. Sorenson (1977). Fundamental frequency contours at syntactic boundaries. *Journal of the Acoustical Society of America 62*(3), 683–692.

Cost, S. and S. Salzberg (1993). A weighted nearest neighbor algorithm for learning with symbolic features. *Machine Learning 10*(1), 57–78.

Cover, T. M. and P. E. Hart (1967). Nearest neighbor pattern classification. *IEEE Transactions on Information Theory 13*(1), 21–27.

Cruttenden, A. (1986). *Intonation*. Cambridge, UK: Cambridge University Press.

Crystal, D. (1975). *The English Tone of Voice: Essays in Intonation, Prosody, and Paralanguage*. London, UK: Edward Arnold.

Daelemans, W. (1988). GRAFON: A grapheme-to-phoneme system for Dutch. In *Proceedings of Twelfth International Conference on Computational Linguistics (COLING-88)*, Budapest, Hungary, pp. 133–138.

Daelemans, W. (1996). Experience-driven language acquisition and processing. In M. van der Avoird and C. Corsius (Eds.), *Proceedings of the CLS Opening Academic Year 1996–1997*, pp. 83–95. Tilburg, The Netherlands: Center for Language Studies.

Daelemans, W., S. Gillis, and G. Durrieux (1994). The acquisition of stress: A data-oriented approach. *Computational Linguistics 20*(3), 421–451.

Daelemans, W. and A. van den Bosch (1992). Generalisation performance of backpropagation learning on a syllabification task. In M. F. J. Drossaers and A. Nijholt (Eds.), *TWLT3: Connectionism and Natural Language Processing*, Enschede, The Netherlands, pp. 27–37. Twente University.

Daelemans, W. and A. van den Bosch (1993). Tabtalk: Reusability in data-oriented grapheme-to-phoneme conversion. In *Proceedings of 3rd European Conference on Speech Communication and Technology, Eurospeech'93*, Volume 2, Berlin, Germany, pp. 1459–1466.

Daelemans, W. and A. van den Bosch (1997). Language-independent data-oriented grapheme-to-phoneme conversion. See van Santen, Sproat, Olive, and Hirschberg (1997), pp. 77–89.

Daelemans, W., A. van den Bosch, and T. Weijters (1997). IGTree: Using trees for compression and classification in lazy learning algorithms. *Artificial Intelligence Review 11*(1–5), 407–423.

Daelemans, W., A. van den Bosch, and J. Zavrel (1997). A feature-relevance heuristic for indexing and compressing large case bases. In M. van Someren and G. Widmer (Eds.), *Poster Papers of the Ninth European Conference on Machine Learning*, Prague, Czech Republic, pp. 29–38. University of Economics.

Daelemans, W., A. van den Bosch, and J. Zavrel (1999). Forgetting exceptions is harmful in language learning. *Machine Learning 34*(1–3), 11–43.

Damper, R. I. (1991). Statistical inferencing of text-phonemics correspondences. *Phonetics Experimental Research at the Institute of Linguistics, University of Stockholm, PERILUS XIII, Papers from the 5th National Phonetics Conference*, Stockholm, Sweden, pp. 109–112.

Damper, R. I. (1995). Self-learning and connectionist approaches to text-phoneme conversion. In J. Levy, D. Bairaktaris, J. Bullinaria, and P. Cairns (Eds.), *Connectionist Models of Memory and Language*, pp. 117–144. London: UCL Press.

Damper, R. I. and J. F. G. Eastmond (1996). Pronouncing text by analogy. In *Proceedings of 16th International Conference on Computational Linguistics*, Volume 2, Copenhagen, Denmark, pp. 268–273.

Damper, R. I. and J. F. G. Eastmond (1997). Pronunciation by analogy: Impact of implementational choices on performance. *Language and Speech 40*(1), 1–23.

Damper, R. I., Y. Marchand, M. J. Adamson, and K. Gustafson (1999). Evaluating the pronunciation component of text-to-speech systems for English: A performance comparison of different approaches. *Computer Speech and Language 13*(2), 155–176.

Danlos, L., E. LaPorte, and F. Emerard (1986). Synthesis of spoken messages from semantic representations. In *Proceedings of the 11th International Conference on Computational Linguistics*, Bonn, West Germany, pp. 599–604.

Dasarathy, B. V. (1980). Nosing around the neighborhood: A new system structure and classification rule for recognition in partially exposed environments. *IEEE Transactions on Pattern Analysis and Machine Intelligence 2*(1), 67–71.

de Mori, R., L. Lam, and M. Gilloux (1987). Learning and plan refinement in a knowledge-based system for automatic speech recognition. *IEEE Transactions on Pattern Analysis and Machine Intelligence 9*(2), 289–205.

Dedina, M. J. and H. C. Nusbaum (1986). PRONOUNCE: A program for pronunciation by analogy. Speech Research Laboratory Progress Report No. 12, Indiana University, Bloomington, IN.

Dedina, M. J. and H. C. Nusbaum (1991). PRONOUNCE: A program for pronunciation by analogy. *Computer Speech and Language 5*(1), 55–64.

Deligne, S. (1996). *Modèles de Séquences de Longueurs Variables: Application au Traitement du Langage Écrit et de la Parole*. PhD thesis, ENST, Paris, France.

Deligne, S. and F. Bimbot (1995). Language modeling by variable length sequences: Theoretical formulation and evaluation of multigrams. In *Proceedings of IEEE International Conference on Acoustics, Speech, and Signal Processing, ICASSP'95*, Volume 1, Detroit, MI, pp. 169–172.

Deligne, S. and F. Bimbot (1997). Inference of variable-length linguistic and acoustic units by multigrams. *Speech Communication 23*(3), 223–241.

Deligne, S., F. Bimbot, and F. Yvon (1995). Phonetic transcription by variable length sequences: Joint multigrams. In *Proceedings of 4th European Conference on Speech Communication and Technology, Eurospeech'95*, Volume 3, Madrid, Spain, pp. 2243–2246.

Deller, J. R., J. P. Proakis, and J. H. L. Hansen (1993). *Discrete-Time Processing of Speech Signals*. Englewood Cliffs, NJ: MacMillan.

Dempster, A. P., N. M. Laird, and D. B. Rubin (1977). Maximum-likelihood from incomplete data via the EM algorithm. *Journal of the Royal Statistics Society, Series B 39*(1), 1–38.

Deng, L. (1997). Speech recognition using autosegmental representation of phonological units with interface to the trended HMM. *Speech Communication 23*(3), 211–222.

Deng, L. and D. X. Sun (1994). A statistical approach to automatic speech recognition using the atomic speech units constructed from overlapping articulatory features. *Journal of the Acoustical Society of America 95*(5), 2702–2719.

Dietterich, T. G. (1997). Machine learning research: Four current directions. *AI Magazine 18*(4), 97–136.

Dietterich, T. G. and G. Bakiri (1991). Error-correcting output codes: A general method for improving multiclass inductive learning programs. In *Proceedings of the 9th National Conference on Artificial Intelligence*, Cambridge, MA, pp. 572–577. AAAI Press/MIT Press.

Dietterich, T. G. and G. Bakiri (1995). Solving multiclass learning problems via error-correcting output codes. *Journal of Artificial Intelligence Research* 2(1), 263–286.

Dietterich, T. G., H. Hild, and G. Bakiri (1990). A comparative study of ID3 and backpropagation for English text-to-speech mapping. In *Proceedings of the 7th International Conference on Machine Learning*, Austin, TX, pp. 24–31. Morgan Kaufmann.

Dietterich, T. G., H. Hild, and G. Bakiri (1995). A comparison of ID3 and backpropagation for English text-to-speech mapping. *Machine Learning 18*(1), 51–80.

Digital Equipment Corporation (1985). DECtalk DTC01 Owners Manual, Third Edition. Technical Report EK-DTC01-OM-003, Digital Equipment Corporation, Maynard, MA.

Dirksen, A. and J. S. Coleman (1997). All-prosodic speech synthesis. See van Santen, Sproat, Olive, and Hirschberg (1997), pp. 91–108.

Divay, M. and A. J. Vitale (1997). Algorithms for grapheme-phoneme translation for English and French: Applications for database searches and speech synthesis. *Computational Linguistics 23*(4), 495–523.

Djezzar, L. and J.-P. Haton (1995). Exploiting acoustic-phonetic knowledge and neural networks for stop recognition. In *Proceedings of 4th European Conference on Speech Communication and Technology, Eurospeech'95*, Volume 3, Madrid, Spain, pp. 2217–2220.

Donovan, R. E. (1996). *Trainable Speech Synthesis*. PhD thesis, Cambridge University Engineering Department, Cambridge, UK.

Donovan, R. E. and P. C. Woodland (1999). A hidden Markov-model-based trainable speech synthesizer. *Computer Speech and Language 13*(3), 223–241.

Dorffner, G., M. Kommenda, and G. Kubin (1985). GRAPHON – the Vienna speech synthesis system for arbitrary German text. In *Proceedings of IEEE International Conference on Acoustics, Speech, and Signal Processing, ICASSP'85*, Volume 2, Tampa, FL, pp. 774–747.

Downing, B. (1970). *Syntactic Structure and Phonological Phrasing in English*. PhD thesis, University of Texas, Austin, TX.

Duffy, S. A. and D. B. Pisoni (1992). Comprehension of synthetic speech produced by rule: A review and theoretical interpretation. *Language and Speech 35*(4), 351–389.

Dutoit, T. (1994a). On the ability of various speech models to smooth segment discontinuities in the context of text-to-speech synthesis by concatenation. In *Proceedings of EUSIPCO-94*, Edinburgh, UK, pp. 8–12.

Dutoit, T. (1994b). High quality text-to-speech synthesis – A comparison of four candidate algorithms. In *Proceedings of IEEE International Conference on Acoustics, Speech, and Signal Processing, ICASSP'94*, Volume 1, Adelaide, South Australia, pp. 565–568.

Dutoit, T. (1997). *Introduction to Text-to-Speech Synthesis*. Dordrecht, The Netherlands: Kluwer Academic Publishers.

Dutoit, T. and H. Leich (1993). MBR-PSOLA: Text-to-speech synthesis based on an MBE re-synthesis of the segments database. *Speech Communication 13*(3–4), 435–440.

Eefting, W. (1991). The effect of 'information value' and 'accentuation' on the duration of Dutch words, syllables, and segments. *Journal of the Acoustical Society of America 89*(1), 412–424.

Elman, J. L., E. A. Bates, M. H. Johnson, A. Karmiloff-Smith, D. Parisi, and K. Plunkett (1996). *Rethinking Innateness: A Connectionist Perspective on Development*. Cambridge, MA: MIT Press.

Elovitz, H. S., R. Johnson, A. McHugh, and J. E. Shore (1976). Letter-to-sound rules for automatic translation of English text to phonetics. *IEEE Transactions on Speech and Audio Processing ASSP-24*(6), 446–459.

Emerard, F., L. Mortamet, and A. Cozannet (1992). Prosodic processing in a text-to-speech synthesis system using a database and learning procedures. See Bailly, Benoît, and Sawallis (1992), pp. 225–254.

Fant, G. (1989). Speech research in perspective. In *Proceedings of European Conference on Speech Communication and Technology, Eurospeech'89*, Volume 1, Paris, France, pp. 3–4.

Federici, S., V. Pirrelli, and F. Yvon (1995). Advances in analogy-based learning: False friends and exceptional items in pronunciation by paradigm-driven analogy. In *Proceedings of International Joint Conference on Artificial Intelligence (IJCAI'95) Workshop on New Approaches to Learning for Natural Language Processing*, Montreal, Canada, pp. 158–163.

Fletcher, R. P., J. K. Local, and J. S. Coleman (1990). Speech synthesis – how to do it, and using graphics to get it right. In *Proceedings of the DECUS (UK, Ireland and Middle East) Conference*, Keele University, UK, pp. 101–111.

Forney, G. D. (1973). The Viterbi algorithm. *Proceedings of the IEEE 61*(3), 268–278.

Francis, W. N. and H. Kučera (1982). *Frequency Analysis of English Usage*. Boston, MA: Houghton Mifflin.

Fredkin, E. (1960). Trie memory. *Communications of the ACM 3*(9), 490–500.

Freund, Y. and R. E. Schapire (1996). Experiments with a new boosting algorithm. In L. Saitta (Ed.), *Proceedings of the 13th International Conference on Machine Learning*, San Francisco, CA, pp. 148–156. Morgan Kaufmann.

Fujisawa, K. and N. Campbell (1998). Prosody-based unit-selection for Japanese speech synthesis. In *Proceedings of 3rd European Speech Communication Association (ESCA)/COCOSDA International Workshop on Speech Synthesis*, Jenolan Caves, Australia, pp. 181–184.

Gates, G. W. (1972). The reduced nearest neighbor rule. *IEEE Transactions on Information Theory 18*(3), 431–433.

Gee, J. P. and F. Grosjean (1983). Performance structure: A psycholinguistic and linguistic apprasial. *Cognitive Psychology 15*(4), 411–458.

Gelb, I. J. (1952). *A Study of Writing*. Chicago, IL: University of Chicago Press.

Geman, S., E. Bienenstock, and R. Doursat (1992). Neural networks and the bias/variance dilemma. *Neural Computation 4*(1), 1–58.

Gildea, D. and D. Jurafsky (1996). Learning bias and phonological-rule induction. *Computational Linguistics* 22(4), 497–530.

Gimson, A. C. (1980). *An Introduction to the Pronunciation of English*. London, UK: Edward Arnold. Third Edition.

Glushko, R. J. (1979). The organization and activation of orthographic knowledge in reading aloud. *Journal of Experimental Psychology: Human Perception and Performance 5*(4), 674–691.

Glushko, R. J. (1981). Principles for pronouncing print: The psychology of phonography. See Lesgold and Perfetti (1981), pp. 61–84.

Gold, E. M. (1967). Language identification in the limit. *Information and Control 10*(5), 447–474.

Golding, A. R. (1991). *Pronouncing Names by a Combination of Case-Based and Rule-Based Reasoning*. PhD thesis, Stanford University, CA.

Goldsmith, J. (1976). An overview of autosegmental phonology. *Linguistic Analysis 2*(1), 23–68.

Gonzalez, R. C. and M. G. Thomason (1978). *Syntactic Pattern Recognition: An introduction*. Reading, MA: Addison-Wesley.

Granström, B. (1997). Applications of intonation – An overview. In *Proceedings of European Speech Communication Association (ESCA) Workshop on Intonation: Theory, Models and Applications*, Athens, Greece, pp. 21–24.

Grosz, B., A. K. Joshi, and S. Weinstein (1983). Providing a unified account of definite noun phrases in discourse. In *Proceedings of the 21st Annual Meeting of the Association for Computational Linguistics*, Cambridge, MA, pp. 44–50.

Grosz, B. J. and C. L. Sidner (1986). Attention, intentions, and the structure of discourse. *Computational Linguistics 12*(3), 175–204.

Halliday, M. A. K. and R. Hassan (1976). *Cohesion in English*. New York, NY: Longman.

Hamon, C., E. Moulines, and F. Charpentier (1989). A diphone system based on time-domain prosodic modifications of speech. In *Proceedings of IEEE International Conference on Acoustics, Speech, and Signal Processing, ICASSP'89*, Volume 1, Glasgow, Scotland, pp. 238–241.

Harlow, R. B. and A. H. F. Thornton (1986). *A Name and Word Index to Nga Moteatea*. Dunedin, NZ: Otago University Press.

Harris, C. M. (1953). A study of the building blocks in speech. *Journal of the Acoustical Society of America 25*(5), 962–969.

Hart, P. E. (1968). The condensed nearest neighbor rule. *IEEE Transactions on Information Theory 14*(3), 515–516.

Hassoun, M. H. (1995). *Fundamentals of Artificial Neural Networks*. Cambridge, MA: MIT Press.

Hawkins, S. and A. Slater (1994). Spread of CV and V-to-V coarticulation in British English: Implications for the intelligibility of synthetic speech. In *Proceedings of International Conference on Spoken Language Processing, ICSLP'94*, Volume 1, Yokohama, Japan, pp. 57–60.

Hemphill, C. T., J. J. Godfrey, and G. R. Doddington (1990). The ATIS spoken language systems pilot corpus. In *Proceedings of the DARPA Speech and Natural Language Workshop*, Hidden Valley, PA, pp. 96–101.

Henderson, L. (1985). On the use of the term 'grapheme'. *Language and Cognitive Processes 1*(2), 135–148.

Hindle, D. M. (1989). Acquiring disambiguation rules from text. In *Proceedings of the 27th Annual Meeting of the Association for Computational Linguistics*, Vancouver, Canada, pp. 118–125.

Hinton, G. E., J. L. McClelland, and D. E. Rumelhart (1986). Distributed representations. In D. E. Rumelhart and J. L. McClelland (Eds.), *Parallel Distributed Processing: Explorations in the Microstructure of Cognition, Volume 1 – Foundations*, pp. 77–109. Cambridge, MA: Bradford Books/MIT Press.

Hirschberg, J. (1990). Accent and discourse context: Assigning pitch accent in synthetic speech. In *Proceedings of the 8th National Conference of the American Association for Artificial Intelligence*, Boston, MA, pp. 952–957.

Hirschberg, J. (1991). Using text analysis to predict intonational boundaries. In *Proceedings of 2nd European Conference on Speech Communication and Technology, Eurospeech'91*, Volume 3, Genova, Italy, pp. 1275–1278.

Hirschberg, J. and D. Litman (1987). Now let's talk about *now*: Identifying cue phrases intonationally. In *Proceedings of the 25th Annual Meeting of the Association for Computational Linguistics*, Stanford University, Palo Alto, CA, pp. 163–171.

Hirschberg, J. and C. Nakatani (1996). A prosodic analysis of discourse segments in direction-giving monologues. In *Proceedings of the 34th Annual Meeting of the Association for Computational Linguistics*, Santa Cruz, CA, pp. 286–293.

Hirschberg, J. and P. Prieto (1994). Training intonational phrasing rules automatically for English and Spanish text-to-speech. In *Proceedings of Joint IEEE/European Speech Communication Association (ESCA) Workshop on Speech Synthesis*, Lake Mohonk, New Paltz, NY, pp. 159–162.

Hochberg, J., S. M. Mniszewski, T. Calleja, and G. J. Papçun (1991). A default hierarchy for pronouncing English. *IEEE Transactions on Pattern Analysis and Machine Intelligence 13*(9), 957–964.

Hocquenghem, A. (1959). Codes corecteurs d'erreurs. *Chiffres 2*, 147–156.

Holmes, J. N. (1973). The influence of glottal waveform on the naturalness of speech from a parallel formant synthesizer. *IEEE Transactions on Audio and Electroacoustics AU-21*(3), 298–305.

Holmes, J. N., I. Mattingly, and J. Shearme (1964). Speech synthesis by rule. *Language and Speech 7*(3), 127–143.

Hopcroft, J. E. and J. D. Ullman (1979). *Introduction to Automata Theory, Languages and Computation*. Reading, MA: Addison-Wesley.

Hornby, A. S. (1974). *Oxford Advanced Learner's Dictionary of Current English* (Third ed.). Oxford, UK: Oxford University Press. Available in electronic form from Oxford Text Archive at `http://ota.ahds.ac.uk/`. Compiled by Roger Mitton.

Horne, M. (1987). Towards a discourse-based model of English sentence intonation. Technical Report 32, Department of Linguistics, Lund University, Sweden.

House, A. S., C. E. Williams, M. H. L. Hecker, and K. D. Kryter (1965). Articulation-testing methods: Consonantal differentiation with a closed-response set. *Journal of the Acoustical Society of America 37*(1), 158–166.

Hu, J., W. Turin, and M. K. Brown (1997). Language modeling using stochastic automata with variable-length contexts. *Computer Speech and Language 11*(1), 1–16.

Huang, X., A. Acero, J. Adcock, H.-W. Hon, J. Goldsmith, J. Liu, and M. Plumbe (1996). Whistler: A trainable text-to-speech system. In *Proceedings of Fourth International Conference on Spoken Language Processing, ICSLP'96*, Volume 4, Philadelphia, PA, pp. 2387–2390.

Humphreys, G. W. and L. J. Evett (1985). Are there independent lexical and non-lexical routes in word processing? An evaluation of the dual route theory of reading. *Behavioral and Brain Sciences 8*(4), 689–739.

Hunnicutt, S. (1976). Phonological rules for a text-to-speech system. *American Journal of Computational Linguistics Microfiche 57*, 1–72.

Hunnicutt, S. (1980). Grapheme-to-phoneme rules: A review. *Speech Transmission Laboratory Quarterly Progress and Status Report, Royal Institute of Technology (KTH), Stockholm STL-QPSR 2-3/1980*, 38–60.

Hunt, A. J. and A. W. Black (1996). Unit selection in a concatenative speech synthesis using a large speech database. In *Proceedings of IEEE*

International Conference on Acoustics, Speech, and Signal Processing, ICASSP'96, Volume 1, Atlanta, GA, pp. 373–376.

Hunt, M. (1984). Time alignment of natural to synthetic speech. In *Proceedings of IEEE International Conference on Acoustics, Speech, and Signal Processing, ICASSP'84*, San Diego, CA, pp. 2.5.1–2.5.4.

Hyman, L. (1978). Tone and/or Accent. In D. J. Napoli (Ed.), *Elements of Tone, Stress, and Intonation*, Chapter 1, pp. 1–20. Washington DC: Georgetown University Press.

Iles, J. and N. Ing-Simons (1993). Klatt cascade-parallel formant synthesizer v.3.0. Available as `klatt.3.04.tar.gz` via anonymous ftp from `ftp://svr-ftp.eng.cam.ac.uk/pub/comp.speech/synthesis/`.

International Phonetic Association (1999). *Handbook of the International Phonetic Association: A Guide to the Use of the International Phonetic Alphabet*. Cambridge, UK: Cambridge University Press.

Jackendoff, R. S. (1972). *Semantic Interpretation in Generative Grammar*. Cambridge, MA: MIT Press.

Jacobs, R. A. (1988). Increased rates of convergence through learning rate adaptation. *Neural Networks 1*(4), 295–307.

Jacobs, R. A., M. I. Jordan, S. J. Nowlan, and G. E. Hinton (1991). Adaptive mixtures of local experts. *Neural Computation 3*(1), 79–87.

Jelinek, F. (1990). Self-organized language modeling for speech recognition. In A. Waibel and K.-F. Lee (Eds.), *Readings in Speech Recognition*, pp. 450–506. San Mateo, CA: Morgan Kaufmann.

Jelinek, F., L. R. Bahl, and R. L. Mercer (1975). Design of a linguistic statistical decoder for the recognition of continuous speech. *IEEE Transactions on Information Theory IT-21*(3), 250–256.

Jensen, U., R. K. Moore, P. Dalsgaard, and B. Lindberg (1994). Modeling intonation contours at the phrase level using continuous density HMMs. *Computer Speech and Language 8*(3), 247–260.

Johnson, C. D. (1972). *Formal Aspects of Phonological Description*. The Hague, Netherlands: Mouton.

Jones, D. (1996). *Analogical Natural Language Processing*. London, UK: UCL Press.

Jordan, M. A. and R. A. Jacobs (1994). Hierarchical mixtures of experts and the EM algorithm. *Neural Computation 6*(2), 181–214.

Kaplan, R. M. and M. Kay (1994). Regular models of phonological rule systems. *Computational Linguistics 20*(3), 331–378.

Karaali, O., G. Corrigan, and I. Gerson (1996). Speech synthesis with neural networks. In *Proceedings of the World Congress on Neural Networks*, San Diego, CA, pp. 45–50.

Katz, L. and L. B. Feldman (1981). Linguistic coding in word recognition: comparisons between a deep and a shallow orthography. See Lesgold and Perfetti (1981), pp. 85–106.

Kewley-Port, D. (1983). Time-varying features as correlates of place of articulation in stop consonants. *Journal of the Acoustical Society of America 73*(1), 322–335.

Kingston, J. and M. E. Beckman (Eds.) (1990). *Papers in Laboratory Phonology I: Between the Grammar and Physics of Speech*. Cambridge, UK: Cambridge University Press.

Kirsch, A. (1996). *An Introduction to the Mathematical Theory of Inverse Problems*. New York, NY: Springer-Verlag.

Klatt, D. H. (1980). Software for a cascade/parallel formant synthesizer. *Journal of the Acoustical Society of America 67*(3), 971–995.

Klatt, D. H. (1987). Review of text-to-speech conversion for English. *Journal of the Acoustical Society of America 82*(3), 737–793.

Klatt, D. H. and L. C. Klatt (1990). Analysis, synthesis, and perception of voice quality variations among female and male talkers. *Journal of the Acoustical Society of America 87*(2), 820–857.

Klatt, D. H. and D. W. Shipman (1982). Letter-to-phoneme rules: A semi-automatic discovery procedure. *Journal of the Acoustical Society of America 72*(Suppl. 1), S48.

Knill, K. and S. Young (1997). Hidden Markov models in speech and language processing. See Young and Bloothooft (1997), pp. 27–68.

Kohler, K. J. (1990). Macro and micro F0 in the synthesis of intonation. See Kingston and Beckman (1990), pp. 115–138.

Kohonen, T. (1982). Self-organized formation of topologically correct feature maps. *Biological Cybernetics 43*(1), 59–69.

Kohonen, T. (1986). Dynamically expanding context, with application to the correction of symbol strings in the recognition of continuous speech. In *Proceedings of the 8th International Conference on Pattern Recognition*, Paris, France, pp. 27–31.

Kohonen, T. (1990). The self-organising map. *Proceedings of the IEEE 78*(9), 1464–1480.

Koskenniemi, K. (1984). A general computational model for wordform recognition and production. In *Proceedings of the Tenth International*

Conference on Computational Linguistics/22nd Annual Conference of the ACL, Stanford University, Palo Alto, CA, pp. 178–181.

Kraft, V. and J. R. Andrews (1992). Design, evaluation and acquisition of a speech database for German synthesis-by-concatenation. In *Proceedings of Fourth Australian International Conference on Speech Science and Technology, SST-92*, Brisbane, Australia, pp. 724–729.

Krogh, A. and J. Vedelsby (1995). Neural network ensembles, cross-validation and active learning. In G. Tesauro, D. S. Touretzky, and T. K. Leen (Eds.), *Advances in Neural Information Processing Systems 7, (Denver, 1994)*, Cambridge, MA, pp. 231–238. MIT Press.

Kučera, H. and W. N. Francis (1967). *Computational Analysis of Present-Day American English*. Providence, RI: Brown University Press.

Ladd, D. R. (1996). *Intonational Phonology*. Cambridge, UK: Cambridge University Press.

Lagana, A., F. Lavagetto, and A. Storace (1996). Visual synthesis of source acoustic speech through Kohonen neural networks. In *Proceedings of Fourth International Conference on Spoken Language Processing, ICSLP'96*, Volume 4, Philadelphia, PA, pp. 2183–2186.

Lakoff, G. (1971). Presupposition and relative well-formedness. In *Semantics: An Interdisciplinary Reader in Philosophy, Linguistics, and Psychology*, pp. 329–340. Cambridge, UK: Cambridge University Press.

Lancaster-Oslo/Bergen Corpus (1978). Department of English, University of Oslo, Norway.

Larreur, D., F. Emerard, and F. Marty (1989). Linguistic and prosodic processing for a text-to-speech synthesis system. In *Proceedings of European Conference on Speech Communication and Technology, Eurospeech'89*, Volume 1, Paris, France, pp. 510–513.

Lawrence, S. G. C. and G. Kaye (1986). Alignment of phonemes with their corresponding orthography. *Computer Speech and Language 1*(2), 153–165.

Lea, W. A. (1972). *Intonational Cues to the Constituent Structure and Phonemics of Spoken English*. PhD thesis, Indiana University, Bloomington, IN.

Lehiste, I. (1973). Phonetic disambiguation of syntactic ambiguity. *Glossa 7*(2), 107–121.

Lehnert, W. G. (1987). Case-based problem solving with a large knowledge base of learned cases. In *Proceedings of 6th National Conference of American Association for Artificial Intelligence (AAAI-87)*, Seattle, WA, pp. 301–306.

Lesgold, A. M. and C. A. Perfetti (Eds.) (1981). *Interactive Processes in Reading*. Hillsdale, NJ: Lawrence Erlbaum Associates.

Leung, H. and V. Zue (1984). A procedure for automatic alignment of phonetic transcriptions with continuous speech. In *Proceedings of IEEE International Conference on Acoustics, Speech, and Signal Processing, ICASSP'84*, San Diego, CA, pp. 2.7.1–2.7.4.

Levenshtein, V. I. (1966). Binary codes capable of correcting deletions, insertions and reversals. *Soviet Physics Doklady 10*(8), 707–710.

Liberman, I., A. Liberman, I. Mattingly, and D. Shankweiler (1980). Orthography and the beginning reader. In J. Kavanagh and R. Venezky (Eds.), *Orthography, Reading and Dyslexia*, pp. 137–153. Baltimore, OH: University Park Press.

Liberman, M. and A. Prince (1977). On stress and linguistic rhythm. *Linguistic Inquiry 8*(2), 249–336.

Liberman, M. and R. Sproat (1992). The stress and structure of modified noun phrases in English. In I. Sag (Ed.), *Lexical Matters*, pp. 131–182. Chicago, IL: University of Chicago Press.

Lin, S. and D. J. Costello (1983). *Error Control Coding: Fundamentals and Applications*. Englewood Cliffs, NJ: Prentice-Hall.

Litman, D. and J. Hirschberg (1990). Disambiguating cue phrases in text and speech. In *Proceedings of 13th International Conference on Computational Linguistics*, Helsinki, Finland, pp. 251–256.

Lloyd, D. (1989). *Simple Minds*. Cambridge, MA: Bradford Books/MIT Press.

Local, J. K. (1992). Modelling assimilation in a non-segmental rule-free phonology. In G. J. Docherty and D. R. Ladd (Eds.), *Papers in Laboratory Phonology II*, pp. 190–223. Cambridge, UK: Cambridge University Press.

Local, J. K. (1994). Phonological structure, parametric phonetic interpretation and natural-sounding synthesis. In E. Keller (Ed.), *Fundamentals of Speech Synthesis and Speech Recognition*, pp. 253–269. Chichester, UK: John Wiley.

Logan, J. S., B. G. Greene, and D. B. Pisoni (1989). Segmental intelligibility of synthetic speech produced by rule. *Journal of the Acoustical Society of America 86*(2), 566–581.

Lucas, S. M. and R. I. Damper (1992). Syntactic neural networks for bi-directional text-phonetics translation. See Bailly, Benoît, and Sawallis (1992), pp. 127–141.

Lucassen, J. M. (1983). Discovering phonemic baseforms automatically: An information theoretic approach. Technical Report RC-9833, IBM T. J. Watson Research Center, Yorktown Heights, NY.

Lucassen, J. M. and R. L. Mercer (1984). An information theoretic approach to the automatic determination of phonemic baseforms. In *Proceedings of IEEE International Conference on Acoustics, Speech, and Signal Processing, ICASSP'84*, San Diego, CA, pp. 42.5.1–42.5.4.

Luk, R. W. P. and R. I. Damper (1991). Stochastic transduction for English text-to-phoneme conversion. In *Proceedings of 2nd European Conference on Speech Communication and Technology, Eurospeech'91*, Volume 2, Genova, Italy, pp. 779–782.

Luk, R. W. P. and R. I. Damper (1992). Inference of letter-phoneme correspondences by delimiting and dynamic time warping techniques. In *Proceedings of IEEE International Conference on Acoustics, Speech, and Signal Processing, ICASSP'92*, Volume 2, San Francisco, CA, pp. II.61–II.64.

Luk, R. W. P. and R. I. Damper (1993a). Inference of letter-phoneme correspondences with pre-defined consonant and vowel patterns. In *Proceedings of IEEE International Conference on Acoustics, Speech, and Signal Processing, ICASSP'93*, Volume 2, Minneapolis, MN, pp. II.203–II.206.

Luk, R. W. P. and R. I. Damper (1993b). Inference of letter-phoneme correspondences using generalized stochastic transducers. In *Proceedings of the IEEE International Conference on Signal Processing '93*, Volume 1, Beijing, China, pp. 650–653.

Luk, R. W. P. and R. I. Damper (1993c). Experiments with silent-*e* and affix correspondences in stochastic phonographic transduction. In *Proceedings of 3rd European Conference on Speech Communication and Technology, Eurospeech'93*, Volume 2, Berlin, Germany, pp. 917–920.

Luk, R. W. P. and R. I. Damper (1994). A review of stochastic transduction. In *Proceedings of Joint IEEE/European Speech Communication Association (ESCA) Workshop on Speech Synthesis*, Lake Mohonk, New Paltz, NY, pp. 248–251.

Luk, R. W. P. and R. I. Damper (1996). Stochastic phonographic transduction for English. *Computer Speech and Language 10*(2), 133–153.

Luk, R. W. P. and R. I. Damper (1998). Computational complexity of a fast Viterbi decoding algorithm for stochastic letter-phoneme transduction. *IEEE Transactions on Speech and Audio Processing 6*(3), 217–225.

Macchi, M., M. J. Altom, D. Kahn, S. Singhal, and M. Spiegel (1993). Intelligibility as a function of speech coding method for template-based speech synthesis. In *Proceedings of 3rd European Conference on Speech Communication and Technology, Eurospeech'93*, Volume 2, Berlin, Germany, pp. 893–896.

Makhoul, J. (1975). Linear prediction: A tutorial review. *Proceedings of the IEEE 63*(4), 561–580.

Malécot, A. (1972). New procedures for descriptive phonetics. In A. Valdemann (Ed.), *Papers in Linguistics and Phonetics to the Memory of P. Delattre*, pp. 344–355. The Hague, The Netherlands: Mouton.

Malfrère, F., T. Dutoit, and P. Mertens (1998). Automatic prosody generation using suprasegmental unit selection. In *Proceedings of 3rd European Speech Communication Association (ESCA)/COCOSDA International Workshop on Speech Synthesis*, Jenolan Caves, Australia, pp. 323–328.

Marchand, Y. and R. I. Damper (2000). A multi-strategy approach to improving pronunciation by analogy. *Computational Linguistics 26*(2), 195–219.

Markel, J. D. and A. H. Gray (1976). *Linear Prediction of Speech*. Berlin, Germany: Springer-Verlag.

McCulloch, N., M. Bedworth, and J. Bridle (1987). NETspeak – a re-implementation of NETtalk. *Computer Speech and Language* 2(3/4), 289–301.

McIlroy, M. D. (1973). *Synthetic English Speech by Rule*. Computer Science Technical Report No. 14, Bell Telephone Laboratories, Murray Hill, NJ.

McKeown, K. R. and S. M. Pan (2000). Prosody modelling in concept-to-speech generation. *Philosophical Transactions of the Royal Society of London: Series A – Mathematical, Physical and Engineering Sciences 358*(1769), 1419–1430.

Meier, H. (1967). *Deutsche Sprachstatistik*. Hildesheim, Germany: Georg Olms.

Meng, H. M. (1995). *Phonological Parsing for Bi-directional Letter-to-Sound/Sound-to-Letter Generation*. PhD thesis, Massachusetts Institute of Technology, Cambridge, MA.

Meng, H. M., S. Seneff, and V. W. Zue (1994). Phonological parsing for reversible letter-to-sound/sound-to-letter generation. In *Proceedings of IEEE International Conference on Acoustics, Speech, and Signal Processing, ICASSP'94*, Volume 2, Adelaide, South Australia, pp. 1–4.

Mercier, G., D. Bigorgne, L. Miclet, L. le Guennec, and M. Querre (1989). Recognition of speaker-dependent continuous speech with KEAL. *IEE Proceedings–I Communications, Speech and Vision 136*(2), 145–154.

Miller, G. A. (1981). *Language and Speech*. San Francisco, CA: W. H. Freeman.

Mohanan, K. P. (1986). *The Theory of Lexical Phonology*. Dordrecht, The Netherlands: D. Reidel.

Mohri, M. (1997). Finite-state transducers in language and speech processing. *Computational Linguistics 23*(2), 269–311.

Monaghan, A. (1991). *Intonation in a Text-to-Speech Conversion System*. PhD thesis, University of Edinburgh, Edinburgh, UK.

Moulines, E. and F. Charpentier (1990). Pitch-synchronous waveform processing techniques for text-to-speech synthesis using diphones. *Speech Communication 9*(5/6), 453–324.

Nakajima, S. (1994). Automatic synthesis unit generation for English speech synthesis based on multi-layered context-dependent clustering. *Speech Communication 14*(4), 313–324.

Nakajima, S. and H. Hamada (1988). Automatic generation of synthesis units based on context-dependent clustering. In *Proceedings of IEEE International Conference on Acoustics, Speech, and Signal Processing, ICASSP'88*, New York, NY, pp. 659–662.

Nilsson, N. J. (1982). *Principles of Artificial Intelligence*. Berlin, Germany: Springer-Verlag.

Oakey, S. and R. C. Cawthorne (1981). Inductive learning of pronunciation rules by hypothesis testing and correction. In *Proceedings of International Joint Conference on Artificial Intelligence, IJCAI-81*, Vancouver, Canada, pp. 109–114.

Ogden, C. K. (1937). *Basic English*. London, UK: Kegan Paul.

Olive, J. P. (1990). A new algorithm for a concatenative speech synthesis system using an augmented acoustic inventory of speech sounds. In *Proceedings of European Speech Association (ESCA) Workshop on Speech Synthesis*, Autrans, France, pp. 25–29.

Olive, J. P. (1997). Section introduction. Concatenative synthesis. See van Santen, Sproat, Olive, and Hirschberg (1997), pp. 261–262.

Olive, J. P., A. Greenwood, and J. Coleman (1993). *Acoustics of American English Speech: A Dynamic Approach*. New York, NY: Springer-Verlag.

O'Malley, M. M., D. Kloker, and B. Dara-Abrams (1973). Recovering parentheses from spoken algebraic expressions. *IEEE Transactions on Audio and Electroacoustics AU-21*(3), 217–220.

O'Shaughnessy, D. (1989). Parsing with a small dictionary for applications such as text to speech. *Computational Linguistics 15*(2), 97–108.

Ostendorf, M., P. Price, J. Bear, and C. W. Wightman (1990). The use of relative duration in syntactic disambiguation. In *Proceedings of the DARPA Speech and Natural Language Workshop*, Hidden Valley, PA, pp. 26–31. Morgan Kaufmann.

Ostendorf, M., P. Price, and S. Shattuck-Hufnagel (1995). The Boston University Radio News Corpus. Technical Report ECS-95-001, Department of Electrical, Computer and Systems Engineering, Boston University, Boston, MA.

Paliwal, K. K. (1995). Interpolation properties of linear prediction parametric representations. In *Proceedings of 4th European Conference on Speech Communication and Technology, Eurospeech'95*, Volume 2, Madrid, Spain, pp. 1029–1032.

Pallet, D., J. Fiscus, W. Fisher, J. Garofolo, B. Lund, and M. Przbock (1994). 1993 Benchmark Tests for the ARPA Spoken Language Program. In *Proceedings of the ARPA Workshop on Human Language Technology*, Plainsboro, NJ, pp. 49–54.

Parfitt, S. H. and R. A. Sharman (1991). A bidirectional model of English pronunciation. In *Proceedings of 2nd European Conference on Speech Communication and Technology, Eurospeech'91*, Volume 2, Genova, Italy, pp. 800–804.

Paz, A. (1971). *Introduction to Probabilistic Automata*. New York, NY: Academic Press.

Peeters, W. J. M. (1987). Acoustic structure and perceptual relevance of 'steady states' and 'glides' within formant trajectories of diphthongs, complex vowels, and vowel clusters. In *European Conference on Speech Technology*, Volume 1, Edinburgh, UK, pp. 42–45.

Pérennou, G., M. de Calmès, I. Ferrane, and J.-M. Pécatte (1992). Le projet BDLEX de base de données lexicales du Francais Écrit et parlé. Actes du Séminaire Lexique, Toulouse.

Perrone, M. P. and L. N. Cooper (1994). When networks disagree: Ensemble methods for hybrid neural networks. In R. J. Mammone (Ed.), *Artificial Neural Networks for Speech and Vision*, pp. 126–142. London: Chapman and Hall.

Pierrehumbert, J. and M. E. Beckman (1988). *Japanese Tone Structure*. Cambridge, MA: MIT Press.

Pierrehumbert, J. and J. Hirschberg (1990). The meaning of intonation contours in the interpretation of discourse. In P. Cohen, J. Morgan, and M. Pollack (Eds.), *Intentions in Communication*, pp. 271–311. MIT Press.

Pierrehumbert, J. B. (1980). *The Phonology and Phonetics of English Intonation*. PhD thesis, Massachusetts Institute of Technology, Cambridge, MA. (Distributed by the Indiana University Linguistics Club).

Pierrehumbert, J. B. (1981). Synthesizing intonation. *Journal of the Acoustical Society of America 70*(4), 985–995.

Pinker, S. (1999). *Words and Rules: The Ingredients of Language*. London, UK: Weidenfeld and Nicolson.

Pirrelli, V. and S. Federici (1994). On the pronunciation of unknown words by analogy in text-to-speech systems. In *Proceedings of the Second Onomastica Research Colloquium*, London, UK, pp. 43–50.

Pirrelli, V. and S. Federici (1995). You'd better say nothing than something wrong: Analogy, accuracy and text-to-speech applications. In *Proceedings of 4th European Conference on Speech Communication and Technology, Eurospeech'95*, Volume 1, Madrid, Spain, pp. 855–858.

Pirrelli, V. and F. Yvon (1999). The hidden dimension: A paradigmatic view of data-driven NLP. *Journal of Experimental and Theoretical Artificial Intelligence 11*(3), 391–408.

Pisoni, D. B., H. C. Nusbaum, and B. G. Greene (1985). Perception of synthetic speech produced by rule. *Proceedings of the IEEE 73*(11), 1665–1676.

Pitrelli, J., M. Beckman, and J. Hirschberg (1994). Evaluation of prosodic transcription labeling reliability in the ToBI framework. In *Proceedings of International Conference on Spoken Language Processing, ICSLP'94*, Volume 2, Yokohama, Japan, pp. 123–126.

Plumbe, M. and S. Meredith (1998). Which is more important in a concatenative text-to-speech system–pitch, duration or spectral discontinuity. In *Proceedings of 3rd European Speech Communication Association (ESCA)/COCOSDA International Workshop on Speech Synthesis*, Jenolan Caves, Australia, pp. 231–235.

Portele, T., F. Höfer, and W. J. Hess (1997). A mixed inventory structure for German concatenative synthesis. See van Santen, Sproat, Olive, and Hirschberg (1997), pp. 263–277.

Post, E. (1943). Formal reductions of the general combinatorial problem. *American Journal of Mathematics 65*(2), 197–215.

Prieto, P., J. van Santen, and J. Hirschberg (1995). Tonal alignment patterns in Spanish. *Journal of Phonetics 23*(4), 429–451.

Prince, A. S. and P. Smolensky (1993). Optimality theory: Constraint interaction in generative grammar. Technical Report RuCCs 2, Rutgers University Center for Cognitive Science, Piscataway, NJ.

Prince, E. F. (1981). Toward a taxonomy of given-new information. In P. Cole (Ed.), *Radical Pragmatics*, pp. 223–255. New York, NY: Academic Press.

Quené, H. and R. Kager (1992). The derivation of prosody for text-to-speech from prosodic sentence structure. *Computer Speech and Language 6*(1), 77–98.

Quinlan, J. R. (1979). Discovering rules by induction from large collections of examples. In D. Michie (Ed.), *Expert Systems in the Micro Electronic Age*, pp. 168–201. Edinburgh, UK: Edinburgh University Press.

Quinlan, J. R. (1982). Semi-autonomous acquisition of pattern-based knowledge. In D. Michie, J. E. Michie, and Y.-H. Pao (Eds.), *Machine Intelligence, Volume 10*, pp. 159–172. Chichester, UK: Ellis Horwood.

Quinlan, J. R. (1983). Learning efficient classification procedures and their application to chess endgames. In R. S. Michalski, J. Carbonell, and T. M. Mitchell (Eds.), *Machine Learning: An Artificial Intelligence Approach*, Volume 1, pp. 463–482. Palo Alto, CA: Tioga Press.

Quinlan, J. R. (1986). Induction of decision trees. *Machine Learning 1*(1), 81–106.

Quinlan, J. R. (1993). C4.5*: Programs for Machine Learning*. San Francisco, CA: Morgan Kaufmann.

Quinlan, J. R. (1996). Bagging, boosting, and C4.5. In *Proceedings of the Thirteenth National Conference on Artificial Intelligence*, Cambridge, MA, pp. 725–730. AAAI Press/MIT Press.

Quirk, R., J. Svartvik, A. P. Duckworth, J. P. L. Rusiecki, and A. J. T. Colin (1964). Studies in the correspondence of prosodic to grammatical features in English. In H. Lunt (Ed.), *Proceedings of the 9th International Congress of Linguists*, The Hague, The Netherlands, pp. 679–691. Mouton.

Rabiner, L. R. (1989). A tutorial on hidden Markov models and selected applications in speech recognition. *Proceedings of the IEEE 77*(2), 257–285.

Reidi, M. (1995). A neural-network-based model of segmental duration for speech synthesis. In *Proceedings of 4th European Conference on Speech Communication and Technology, Eurospeech'95*, Volume 1, Madrid, Spain, pp. 599–602.

Riccardi, G., R. Pieraccini, and E. Bocchieri (1996). Stochastic automata for language modeling. *Computer Speech and Language 10*(4), 265–293.

Riley, M. D. (1989). Some applications of tree-based modelling to speech and language. In *Proceedings of the DARPA Speech and Natural Language Workshop*, Cape Cod, MA, pp. 339–348. Morgan Kaufmann.

Rooth, M. (1985). *Association with Focus*. PhD thesis, University of Massachusetts, Amherst, MA.

Rosenfeld, R. (2000). Incorporating linguistic structure into statistical language models. *Philosophical Transactions of the Royal Society of London: Series A – Mathematical, Physical and Engineering Sciences 358*(1769), 1311–1324.

Ross, K. (1995). *Modeling of Intonation for Speech Synthesis*. PhD thesis, College of Engineering, Boston University, Boston, MA.

Ross, K. and M. Ostendorf (1994). A dynamical system model for generating F_0 for synthesis. In *Proceedings of Joint IEEE/European Speech Communication Association (ESCA) Workshop on Speech Synthesis*, Lake Mohonk, New Paltz, NY, pp. 131–134.

Rosson, M. (1985). The interaction of pronunciation rules and lexical representation in reading aloud. *Memory and Cognition 13*(1), 90–99.

Rumelhart, D. E., G. E. Hinton, and R. Williams (1986). Learning representations by back-propagating errors. *Nature 323*(9), 533–536.

Sagisaka, Y. and N. Iwahashi (1995). Objective optimization in algorithms for text-to-speech synthesis. In W. B. Kleijn and K. K. Paliwal (Eds.), *Speech Coding and Synthesis*, pp. 686–706. Amsterdam, The Netherlands: Elsevier Science.

Sagisaki, Y. (1988). Speech synthesis by rule using an optimal selection of non-uniform synthesis units. In *Proceedings of IEEE International Conference on Acoustics, Speech, and Signal Processing, ICASSP'88*, New York, NY, pp. 679–682.

Sakoe, H. and S. Chiba (1978). Dynamic programming algorithm optimization for spoken word recognition. *IEEE Transactions on Acoustics, Speech and Signal Processing ASSP-26*(1), 43–49.

Sampson, G. (1985). *Writing Systems*. London, UK: Hutchinson.

Sanford, A. J. (1985). Aspects of pronoun interpretation: Evaluation of search formulations of inference. In G. Rickheit and H. Strohner (Eds.), *Inferences in Text Processing*, pp. 183–204. Amsterdam, The Netherlands: North-Holland.

Schmerling, S. F. (1976). *Aspects of English Sentence Stress*. Austin, TX: University of Texas Press. (Revised 1973 thesis, University of Illinois at Urbana).

Schnabel, B. and H. Roth (1990). Automatic linguistic processing in a German text-to-speech synthesis system. In *Proceedings of European Speech Association (ESCA) Workshop on Speech Synthesis*, pp. 121–124.

Scragg, D. G. (1975). *A History of English Spelling*. Manchester, UK: Manchester University Press.

Segre, A. M., B. A. Sherwood, and W. B. Dickerson (1983). An expert system for the production of phoneme strings from unmarked English text using machine-induced rules. In *Proceedings of the 1st Conference of the European Chapter of the Association for Computational Linguistics*, Pisa, Italy, pp. 35–42.

Seidenberg, M. S. (1985). The time course of phonological code activation in two writing systems. *Cognition 19*(1), 1–30.

Seidenberg, M. S. and J. L. Elman (1999). Networks are not 'hidden rules'. *Trends in Cognitive Science 3*(8), 288–289.

Sejnowski, T. J. and C. R. Rosenberg (1987). Parallel networks that learn to pronounce English text. *Complex Systems 1*(1), 145–168.

Selkirk, E. O. (1978). On prosodic structure and its relation to syntactic structure. In T. Fretheim (Ed.), *Nordic Prosody II*, pp. 111–140. Trondheim, Norway: TAPIR.

Selkirk, E. O. (1984). *Phonology and Syntax: The Relation Between Sound and Structure*. Cambridge, MA: MIT Press.

Seneff, S. (1992). TINA: A natural language system for spoken language applications. *Computational Linguistics 18*(1), 61–86.

Seneff, S., H. Meng, and V. W. Zue (1992). Language modelling for recognition and understanding using layered bigrams. In *Proceedings of International Conference on Spoken Language Processing, ICSLP'92*, Volume 1, Banff, Canada, pp. 317–320.

Shadle, C. H. and R. I. Damper (2001). Prospects for articulatory synthesis: A position paper. In *Proceedings of 4th International Speech Communication Association (ISCA) Workshop on Speech Synthesis*, Blair Atholl, Scotland. (In press).

Shipman, D. W. and V. W. Zue (1982). Properties of large lexicons: Implications for advanced isolated word recognition systems. In *Proceedings of IEEE International Conference on Acoustics, Speech, and Signal Processing, ICASSP'82*, Paris, France, pp. 546–549.

Sidner, C. L. (1983). Focusing in the comprehension of definite anaphora. In M. Brady and R. Berwick (Eds.), *Computational Models of Discourse*, pp. 267–330. Cambridge, MA: MIT Press.

Silverman, K. (1987). *The Structure and Processing of Fundamental Frequency Contours*. PhD thesis, Cambridge University, Cambridge, UK.

Silverman, K., M. Beckman, J. Pitrelli, M. Ostendorf, C. Wightman, P. Price, J. Pierrehumbert, and J. Hirschberg (1992). ToBI: A standard for labeling English prosody. In *Proceedings of International Conference on Spoken Language Processing, ICSLP'92*, Volume 2, Banff, Canada, pp. 867–870.

Silverman, K. and J. Pierrehumbert (1990). The timing of pre-nuclear high accents in English. See Kingston and Beckman (1990), pp. 72–106.

Simpson, A. M. (1995). A tool for the complete production of copy synthesis from natural tokens. In *Proceedings of the XIIIth International Congress of Phonetic Sciences, ICPhS'95*, Volume 2, Stockholm, Sweden, pp. 350–352.

Skousen, R. (1989). *Analogical Modeling of Language*. Dordrecht, The Netherlands: Kluwer Academic Publishers.

Slater, A. and J. Coleman (1996). Non-segmental analysis and synthesis based on a speech database. In *Proceedings of Fourth International Conference on Spoken Language Processing, ICSLP'96*, Volume 4, Philadelphia, PA, pp. 2379–2382.

Sollich, P. and A. Krogh (1996). Learning with ensembles: How over-fitting can be useful. In D. S. Touretzky, M. C. Mozer, and M. E. Hasselmo (Eds.), *Advances in Neural Information Processing Systems 8, (Denver, 1995)*, pp. 190–196. Cambridge, MA: MIT Press.

Spärk Jones, K. I. B., G. J. M. Gazdar, and R. M. Needham (2000). Introduction: Combining formal theories and statistical data in natural language processing. *Philosophical Transactions of the Royal Society of London: Series A – Mathematical, Physical and Engineering Sciences 358*(1769), 1227–1237.

Spiegel, M. F., M. J. Macchi, and K. D. Gollhardt (1989). Synthesis of names by a demisyllable-based speech synthesizer (SPOKSMAN). In *Proceedings of European Conference on Speech Communication and Technology, Eurospeech'89*, Volume 1, Paris, France, pp. 117–120.

Sproat, R. (1990). Stress assignment in complex nominals for English text-to-speech. In *Proceedings of European Speech Association (ESCA) Workshop on Speech Synthesis*, Autrans, France, pp. 129–132.

Sproat, R. (Ed.) (1998). *Multilingual Text-to-Speech Synthesis: The Bell Labs Approach*. Dordrecht, The Netherlands: Kluwer Academic Publishers.

Sproat, R., J. Hirschberg, and D. Yarowsky (1992). A corpus-based synthesizer. In *Proceedings of International Conference on Spoken Language Processing, ICSLP'92*, Volume 1, Banff, Canada, pp. 563–566.

Sproat, R., B. Möbius, K. Maeda, and E. Tzoukermann (1998). Multilingual text analysis. See Sproat (1998), pp. 31–87.

Sproat, R., J. van Santen, and J. P. Olive (1998). Further issues. See Sproat (1998), pp. 245–254.

Sproat, R. W. and J. P. Olive (1997). A modular architecture for multilingual text-to-speech. See van Santen, Sproat, Olive, and Hirschberg (1997), pp. 565–573.

Stanfill, C. and D. Waltz (1986). Toward memory-based reasoning. *Communications of the ACM 29*(12), 1213–1228.

Stanfill, C. W. (1987). Memory-based reasoning applied to English pronunciation. In *Proceedings of 6th National Conference on Artificial Intelligence, AAAI-87*, Seattle, WA, pp. 577–581.

Stanfill, C. W. (1988). Learning to read: A memory-based model. In *Proceedings of Case-Based Reasoning Workshop*, Clearwater Beach, FL, pp. 406–413.

Steedman, M. (1990). Structure and intonation in spoken language understanding. In *Proceedings of the 28th Annual Meeting of the Association for Computational Linguistics*, Pittsburgh, PA, pp. 9–16.

Stevens, K. N. (1972). The quantal nature of speech: Evidence from articulatory-acoustic data. In E. E. David and P. B. Denes (Eds.), *Human Communication: A Unified View*, pp. 51–66. New York, NY: McGraw-Hill.

Stevens, K. N. and C. A. Bickley (1991). Constraints among parameters simplify control of Klatt formant synthesizer. *Journal of Phonetics 19*(1), 161–174.

Stevens, K. N. and S. E. Blumstein (1978). Invariant cues for place of articulation in stop consonants. *Journal of the Acoustical Society of America 64*(5), 1358–1368.

Streeter, L. (1978). Acoustic determinants of phrase boundary perception. *Journal of the Acoustical Society of America 63*(6), 1582–1592.

Sullivan, K. P. H. (1992). *Synthesis-by Analogy: A Psychologically-Motivated Approach to Text-to-Speech Conversion*. PhD thesis, Department of Electronics and Computer Science, University of Southampton, UK.

Sullivan, K. P. H. (1995). Text-to-speech conversion for Māori: A choice of methods? *PHONUM: Reports from the Department of Phonetics, Umeå University 3*, 111–118.

Sullivan, K. P. H. and R. I. Damper (1989). The relevance of psychological models of oral reading to computer text-to-speech conversion. Technical Report VSSP89/1, Department of Electronics and Computer Science, University of Southampton, UK.

Sullivan, K. P. H. and R. I. Damper (1990). A psychologically-governed approach to novel-word pronunciation within a text-to-speech system. In *Proceedings of IEEE International Conference on Acoustics, Speech, and Signal Processing, ICASSP'90*, Volume 1, Albuquerque, NM, pp. 341–344.

Sullivan, K. P. H. and R. I. Damper (1992). Novel-word pronunciation with a text-to-speech system. See Bailly, Benoît, and Sawallis (1992), pp. 183–195.

Sullivan, K. P. H. and R. I. Damper (1993). Novel-word pronunciation: A cross-language study. *Speech Communication 13*(3–4), 441–452.

Takeda, K., K. Abe, Y. Sagisaki, and H. Kuwabara (1989). Adaptive manipulation of non-uniform synthesis units. In *Proceedings of European Conference on Speech Communication and Technology, Eurospeech'89*, Volume 2, Paris, France, pp. 195–198.

Talkin, D. (1989). Looking at speech. *Speech Technology 4*(Apr-May), 74–77.

Tatham, M., K. Morton, and E. Lewis (2000). Modelling speech prosodics for synthesis – Perspectives and trials. In *IEE Seminar on State of the Art in Speech Synthesis*, Number 058 in Series 2000, pp. 1/1–1/4. Institution of Electrical Engineers, London.

Taylor, P. (1994). The Rise/Fall/Connection model of intonation. *Speech Communication 15*(1/2), 169–186.

Taylor, P. (2000). Analysis and synthesis of intonation using the Tilt model. *Journal of the Acoustical Society of America 107*(3), 1697–1714.

Taylor, P. and A. W. Black (1994). Synthesizing conversational intonation from a linguistically rich input. In *Proceedings of Joint IEEE/European Speech Communication Association (ESCA) Workshop on Speech Synthesis*, Lake Mohonk, New Paltz, NY, pp. 175–178.

Taylor, P., S. King, S. Isard, H. Wright, and J. Kowtko (1997). Using intonation to constrain language models in speech recognition. In *Proceedings of 5th European Conference on Speech Communication and Technology, Eurospeech'97*, Volume 5, Rhodes, Greece, pp. 2763–2766.

Terken, J. and S. G. Nooteboom (1987). Opposite effects of accentuation and deaccentuation on verification latencies for given and new information. *Language and Cognitive Processes* 2(3/4), 145–163.

Tesar, B. (1995). *Computational Optimality Theory*. PhD thesis, Department of Computer Science, University of Colorado, Boulder, CO.

Torkolla, K. (1993). An efficient way to learn English grapheme-to-phoneme rules automatically. In *Proceedings of IEEE International Conference on Acoustics, Speech, and Signal Processing, ICASSP'93*, Volume 2, Minneapolis, MN, pp. II.199–II.202.

Traber, C. (1992). F0 generation with a database of F0 patterns and a neural network. See Bailly, Benoît, and Sawallis (1992), pp. 287–304.

Tubach, J.-P. and L.-J. Boë (1985). Un corpus de transcriptions phonétiques : Constitution et exploitation statistique. Technical Report 85D001, ENST, Paris.

Tzoukermann, E. (1993). Progress in French text-to-speech synthesis. Work-Project 311402-2228/File Case 60011, AT&T Bell Laboratories, Murray Hill, NJ.

Van Coile, B. (1990). Inductive learning of grapheme-to-phoneme rules. In *Proceedings of International Conference on Spoken Language Processing, ICSLP'90*, Volume 2, Kobe, Japan, pp. 765–768.

Van Coile, B., S. Lyes, and L. Mortier (1992). On the development of a name pronunciation system. In *Proceedings of International Conference on Spoken Language Processing, ICSLP'92*, Volume 1, Banff, Canada, pp. 487–490.

van den Bosch, A. (1997). *Learning to Pronounce Written Words: A Study in Inductive Language Learning*. PhD thesis, University of Maastricht, The Netherlands.

van den Bosch, A. and W. Daelemans (1993). Data-oriented methods for grapheme-to-phoneme conversion. In *Proceedings of 6th Conference of the European Chapter of the Association for Computational Linguistics*, Utrecht, The Netherlands, pp. 45–53.

van den Bosch, A., W. Daelemans, and A. Weijters (1996). Morphological analysis as classification: An inductive-learning approach. In *Proceedings of Conference on New Methods in Natural Language Processing (NeMLaP-2'96)*, Ankara, Turkey, pp. 79–89.

van den Bosch, A., A. Weijters, H. J. van den Herik, and W. Daelemans (1997). When small disjuncts abound, try lazy learning. In *Proceedings of Proceedings of the 7th Belgian-Dutch Conference on Machine Learning, BENELEARN-97*, Tilburg, The Netherlands, pp. 109–118.

van der Wouden, T. (1990). Celex: Building a multifunctional polytheoretical lexical data base. In T. Magay (Ed.), *Proceedings of BudaLex'88*, Budapest, Hungary, pp. 363–373. Akadémiai Kladó.

van Santen, J. and R. Sproat (1998). Introduction. See Sproat (1998), pp. 1–6.

van Santen, J. P. H. (1992). Contextual effects on vowel duration. *Speech Communication 11*(6), 513–546.

van Santen, J. P. H. (1993). Perceptual experiments for diagnostic testing of text-to-speech systems. *Computer Speech and Language 7*(1), 49–100.

van Santen, J. P. H. and J. Hirschberg (1994). Segmental effects on timing and height of pitch contours. In *Proceedings of International Conference on Spoken Language Processing, ICSLP'94*, Volume 2, Yokohama, Japan, pp. 719–722.

van Santen, J. P. H., R. W. Sproat, J. P. Olive, and J. Hirschberg (Eds.) (1997). *Progress in Speech Synthesis*. New York, NY: Springer-Verlag.

Varga, A. and F. Fallside (1987). A technique for using multipulse linear predictive speech synthesis in text-to-speech type systems. *IEEE Transactions on Acoustics, Speech and Signal Processing 35*(4), 586–587.

Veilleux, N. M. and M. Ostendorf (1992). Probability parse scoring based on prosodic phrasing. In *Proceedings of the DARPA Speech and Natural Language Workshop*, Harriman, NY, pp. 429–434.

Venezky, R. L. (1965). *A Study of English Spelling-to-Sound Correspondences on Historical Principles*. Ann Arbor, MI: Ann Arbor Press.

Venezky, R. L. (1970). *The Structure of English Orthography*. The Hague, The Netherlands: Mouton.

Vidal, E. (1994). Grammatical inference: An introductory survey. In *Second International Colloquium on Grammatical Inference and its Applications, ICGI-94, Alicante, Spain*, Berlin, Germany, pp. 1–4. Springer-Verlag.

Viterbi, A. J. (1967). Error bounds for convolutional codes and an asymptotically optimum decoding algorithm. *IEEE Transactions on Information Theory IT-13*(2), 260–269.

Wales, R. and H. Toner (1979). Intonation and ambiguity. In W. E. Cooper and E. C. Walker (Eds.), *Sentence Processing: Psycholinguistic Studies Presented to Merrill Garrett*, pp. 135–158. New York, NY: Halsted Press.

Wang, M. Q. and J. Hirschberg (1991a). Predicting intonational boundaries automatically from text: The ATIS domain. In *Proceedings of the DARPA Speech and Natural Language Workshop*, Pacific Grove, CA, pp. 378–383. Morgan Kaufmann.

Wang, M. Q. and J. Hirschberg (1991b). Predicting intonational phrasing from text. In *Proceedings of the 29th Annual Meeting of the Association for Computational Linguistics*, Berkeley, CA, pp. 285–292.

Wang, M. Q. and J. Hirschberg (1992). Automatic classification of intonational phrase boundaries. *Computer Speech and Language* 6(2), 175–196.

Wang, W. J., W. N. Campbell, N. Iwahashi, and Y. Sagisaki (1993). Tree-based unit selection for English speech synthesis. In *Proceedings of IEEE International Conference on Acoustics, Speech, and Signal Processing, ICASSP'93*, Volume 2, Minneapolis, MN, pp. II.191–II.194.

Ward, G. (1985). *The Semantics and Pragmatics of Preposing*. PhD thesis, University of Pennsylvania, Philadelphia, PA.

Weijters, A. (1991). A simple look-up procedure superior to NETtalk? In *Proceedings of International Conference on Artificial Neural Networks (ICANN-91)*, Volume 2, Espoo, Finland, pp. 1645–1648.

Weiss, S. and C. Kulikowski (1991). *Computer Systems that Learn*. San Mateo, CA: Morgan Kaufmann.

Wettschereck, D. and T. G. Dietterich (1992). Improving the performance of radial basis function networks by learning center locations. In J. E. Moody, S. J. Hanson, and R. P. Lippmann (Eds.), *Advances in Neural Information Processing Systems 4, (Denver, 1991)*, San Francisco, CA, pp. 1133–1140. Morgan Kaufmann.

Wightman, C. and N. Campbell (1994). Automatic labeling of prosodic structure. Technical Report TR-IT-0061, ATR Interpreting Telecommunications Laboratories, Kyoto, Japan.

Wightman, C. W. and M. Ostendorf (1994). Automatic labeling of prosodic patterns. *IEEE Transactions on Speech and Audio Processing* 2(4), 469–481.

Wolpert, D. H. (1990). Constructing a generalizer superior to NETtalk via a mathematical theory of generalization. *Neural Networks* 3(4), 445–452.

Wolpert, D. H. (1992). Stacked generalization. *Neural Networks* 5(2), 241–259.

Wolpert, D. H. (Ed.) (1995). *The Mathematics of Generalization*. Reading, MA: Addison-Wesley.

Wolpert, D. H. and W. G. Macready (1995). No free lunch theorems for search. Technical Report SFI-TR-95-02-010, Santa Fe Institute, Santa Fe, NM.

Wolpert, D. H. and W. G. Macready (1997). No free lunch theorems for optimization. *IEEE Transactions on Evolutionary Computation 1*(1), 67–82.

Yoshida, Y., S. Nakajima, K. Hakoda, and T. Hirokawa (1996). A new method of generating speech synthesis units based on phonological knowledge and clustering technique. In *Proceedings of Fourth International Conference on Spoken Language Processing, ICSLP'96*, Volume 3, Philadelphia, PA, pp. 1712–1715.

Young, S. and G. Bloothooft (Eds.) (1997). *Corpus-Based Methods in Language and Speech Processing*. Dordrecht, The Netherlands: Kluwer Academic Publishers.

Young, S. J. and F. Fallside (1979). Speech synthesis from concept: A method for speech output from information systems. *Journal of the Acoustical Society of America 66*(3), 685–695.

Yvon, F. (1996a). *Prononcer par Analogie: Motivations, Formalisations et Évaluations*. PhD thesis, ENST, Paris, France.

Yvon, F. (1996b). Grapheme-to-phoneme conversion using multiple unbounded overlapping chunks. In *Proceedings of Conference on New Methods in Natural Language Processing (NeMLaP-2'96)*, Ankara, Turkey, pp. 218–228.

Yvon, F., P. Boula de Mareüil, C. d'Alessandro, V. Aubergé, M. Bagein, G. Bailly, F. Béchet, S. Foukia, J.-F. Golman, E. Keller, D. O'Shaugnessy, V. Pagel, F. Sannier, J. Véronis, and B. Zellner (1998). Objective evaluation of grapheme-to-phoneme conversion for text-to-speech synthesis in French. *Computer Speech and Language 12*(4), 393–410.